T0139919

Challenges and Advances in Computational Chemistry and Physics

Volume 31

Series Editor

Jerzy Leszczynski
Department of Chemistry and Biochemistry
Jackson State University, Jackson, MS, USA

This book series provides reviews on the most recent developments in computational chemistry and physics. It covers both the method developments and their applications. Each volume consists of chapters devoted to the one research area. The series highlights the most notable advances in applications of the computational methods. The volumes include nanotechnology, material sciences, molecular biology, structures and bonding in molecular complexes, and atmospheric chemistry. The authors are recruited from among the most prominent researchers in their research areas. As computational chemistry and physics is one of the most rapidly advancing scientific areas such timely overviews are desired by chemists, physicists, molecular biologists and material scientists. The books are intended for graduate students and researchers.

All contributions to edited volumes should undergo standard peer review to ensure high scientific quality, while monographs should be reviewed by at least two experts in the field. Submitted manuscripts will be reviewed and decided by the series editor, Prof. Jerzy Leszczynski.

More information about this series at http://www.springer.com/series/6918

Tadeusz Andruniów · Massimo Olivucci
Editors

QM/MM Studies of Light-responsive Biological Systems

Springer

Editors
Tadeusz Andruniów
Faculty of Chemistry
Wrocław University of Science
and Technology
Wrocław, Poland

Massimo Olivucci
Department of Biotech, Chemistry
and Pharma
University of Siena
Siena, Italy

ISSN 2542-4491   ISSN 2542-4483   (electronic)
Challenges and Advances in Computational Chemistry and Physics
ISBN 978-3-030-57723-0   ISBN 978-3-030-57721-6   (eBook)
https://doi.org/10.1007/978-3-030-57721-6

This Springer imprint is published by the registered company Springer Nature Switzerland AG
The registered company address is: Gewerbestrasse 11, 6330 Cham, Switzerland

Preface

The understanding of the mechanism and dynamics of chemical processes requires the use of quantum mechanical calculations. In fact, quantum mechanics (QM) is necessary for the description of basic events such as the covalent bond-breaking and bond-making in thermal chemistry or the production of electronically excited states in photochemical, as well as in thermally induced chemiluminescent reactions [1]. In contrast with bonding-conserving conformational changes, these processes are characterized by large electronic structure variations that accompany the nuclei motion effectively driving the chemical transformation of the system. These electronic and nuclear changes are only observed indirectly and in too long timescale through experimental investigations. Accordingly, QM calculations (also known as quantum chemical or electronic structure calculations) provide, even when applied within the Born-Oppenheimer approximation, information still not accessible at the experimental level.

In spite of their high value, the application of QM calculations to chemical reactions occurring in biological macromolecules is troublesome. This is indeed the case for proteins and DNA strands that contain a too large number of electrons to be treatable with the past, and regrettably, present QM technologies. However, more than 30 years ago, a key technical advancement made possible, for the first time, the modeling of chemical events in large molecules (e.g., proteins) and molecular assemblies (e.g., molecules in solution and supramolecular materials). Such key advancement is called the quantum mechanics/molecular mechanics (QM/MM) hybrid method and it was first reported by Arieh Warshel and Michael Levitt in 1976 [2], ultimately yielding their 2013, Nobel prize for chemistry. The story of the discovery made at the Medical Research Council (MRC) in the UK, is recounted impressively well in a paragraph of the biography of Arieh Warshel (see https://www.nobelprize.org/prizes/chemistry/2013/warshel/biographical/):

"...I arrived in the fall of 1974 at the MRC, with Tami, Merav and our second daughter, Yael, and started to focus on my efforts on modeling enzymatic reactions. My trial and error attempts led to the realization that the only way to progress is to introduce the explicit effect of the charges and dipoles of the environment into the quantum mechanical Hamiltonian. This led to the breakthrough development of the

QM/MM approach, where the QM and MM where consistently coupled in contrast to the previous QM+MM attempts. Our advances also included the development of the first consistent models for electrostatic effects in proteins. This model, that was later called the protein-dipole Langevin-dipoles model, represented explicitly (although in a simplified way) all the electrostatic elements of the protein plus the surrounding water system and thus evaded all the traps that eluded the subsequent macroscopic electrostatic models..."

Warshel and Levitt [2] proposed the simple, but at that time, original and ingenious idea of dividing the model of a biological macromolecule into two interacting regions: (i) a subsystem where the electronically driven chemical process takes place which is treated at the QM level of theory, and (ii) the subsystem formed by the surroundings described by a classical molecular mechanics force field where the electrons are not explicitly treated and the only transformation processes correspond to conformational changes or are driven by weak interactions. The first version of such QM/MM technology was applied to the study of the thermal reaction mechanism characterizing the lysozyme catalytic action. Almost at the same time, its effectiveness was demonstrated for conjugated (i.e., light-absorbing) molecules such as the retinal chromophore of visual photoreceptors, for which the chromophore photochemically relevant π-electrons were treated at the QM level of theory, whereas the σ-bonds at the MM level [3]. Since then both the development and the application of QM/MM technologies have become common research activities when studying the reactivity of biological macromolecules, with special emphasis on enzymatic reactions.

Nowadays the original QM/MM method is seen as part of a larger array of technologies called multi-scale methods [4–5]. Multi-scale methods can include more than two subsystems which can be treated at different MM and QM levels. For instance, it is possible to have two QM subsystems, one treated at the *ab initio* level and the other at the faster semiempirical or density functional theory (DFT) level. It is also possible to have, again in the same multi-scale molecular model, two MM subsystems with one treated using *atomistic* MM force fields and the other using *coarse-grain* force fields. The term multi-scale refers to the size of these subsystems, and therefore, to their spatial scale. For instance, typically, the subsystem with the smaller scale is the one treated with the most accurate but also most computationally "expensive" QM level. Then comes the larger subsystem treated at the *semiempirical* level, which is much smaller than the one treatable at the *atomistic* force field level to get finally to the largest portion of the molecule treated at the *coarse-grain* level. Notice that such division in different subsystem (also called "layers") also allows, within certain limits, for running molecular dynamics (MD) simulations (i.e., trajectory calculations) at desired timescales. The smaller is the expensive QM level calculation, the longer is the simulation that the researcher can afford. The spatial multi-scale technology does, therefore, allow multi-timescale studies. In other words, one can perform shorter time scale MD simulations using multi-scale models with large QM subsystems and these calculations may then be followed by longer timescale MD simulations calculations using QM/MM models with smaller QM subsystems. In the limit, the longest MD simulations can be performed with atomistic and coarse-grain MM force fields and no QM subsystem.

As Arieh Warshel stresses in his biography (see above), even at the basic two-layers model, the essence of the QM/MM technology, as well as the quality of the simulation, depends not only on the selected QM and MM methods, but most relevantly, on the accuracy of the treatment of the interaction between the QM and MM subsystems. A brute-force attempt for overcoming the problem of correctly describing such interaction is that of extending as much as possible the size of the QM subsystem also including part of the environment surrounding the reactive part. In this way the most important pair interactions (i.e., the interaction between pairs of "reactive or electronically active" and "spectator" atoms) are treated quantum mechanically, and substantially, at full accuracy. In the case of a protein hosting a reactive (e.g., a catalytic or light-absorbing) site, this is equivalent to including clusters of amino acids surrounding the activated-complex or chromophore in the QM subsystem. Of course, this treatment could be extended up to the point of treating the entire macromolecule at the QM level. While the latter strategy would make the application of QM/MM modeling of lesser importance, it would require an exceptional computing power not currently provided by conventional computing machines. However, one must reckon that hardware developments have recently made such type of calculations more accessible when using workstations equipped with graphical processing units (GPUs) [6]. On the other hand, such a technology is still impractical, if not impossible, when highly accurate multi-configurational QM methods needed for studying, for instance, photochemical reactions have to be used. In these cases, one has to rely on small QM subsystems and the rest of the macromolecule has to be treated at the MM level. In these cases the accuracy can only be increased by introducing less approximate ways of treating the interaction between the QM and MM subsystems.

As already stressed above, the interaction between the QM and MM subsystems represents, arguably, the most important technical challenge that the QM/MM computational investigation of biological macromolecules faces. Primarily, this calls for the development of more realistic MM force fields to describe the protein and its interaction with the QM subsystem. The importance of polarizable embedding has been pointed out in several pilot studies applied to the calculation of the excitation energies in photoreceptor proteins. However, while several polarizable embedding implementations exist and have been used to study spectroscopic properties, there is little experience in mechanistic or dynamics studies using polarizable embedding yet. The same can be said for DFT embedding schemes, which are also powerful and accurate but need to be further developed before they can be used in mechanistic/dynamic studies of, for instance, biological or synthetic photoreceptors. Finally, another promising approach for an unconventional QM/MM description is the Effective Fragment Potential (EFP) method [7]. In this approach, a potential for solvent molecules is generated by calculating several parameters from *ab initio* QM calculations.

Sophisticated and accurate QM/MM technologies have been applied to the investigation of light-driven processes in biological photoreceptors (i.e., specialized proteins) and nucleic acids. One example, is the description of their spectroscopy and light-induced ultrafast dynamics in condensed phase. This requires, in principle, the accurate computation of both the ground and electronically excited Born-Oppenheimer potential energy surfaces and of their interaction. A case-study by Filippi and Rothlisberger groups [8], illustrates this type of QM/MM treatment when looking at the absorption spectroscopy of the dim-light visual pigment rhodopsin. The study reveals the ingredients affecting the accuracy of the simulation of the absorption band: (i) size of the QM subsystem, including not just the light-absorbing retinal chromophore but also the nearby residues with a total of over 250 atoms, (ii) extensive thermal sampling, and (iii) mutual polarization of the QM chromophore and the MM protein environment hosting it by using polarizable MM force fields [9]. Some authors [10], even postulate that multipoles up to quadrupoles and anisotropic polarizabilities are required in the MM force field to obtain accurate values for excitation energies at the QM/MM level. Unfortunately, such computationally demanding protocols preclude their systematic usage especially when highly-correlated post-Hartree-Fock (post-HF) ab initio electronic structure methods are employed. In order, to make the study more practical, time-dependent density functional theory (TDDFT) QM methods are used in practice. In fact, highly-correlated post-HF multireference configuration interaction and multireference second order perturbation theory methods, are often used in QM/MM technique but at the price of a severely reduced size of the QM subsystem and limiting the description of the QM and MM subsystem interaction to electrostatic embedding based on point charges only, and thus, not accounting for the environment polarizability.

The aim of the present book *QM/MM Studies of Light-responsive Biological Systems* is to review, on the basis of diverse methodological and applicative case studies, in which way QM/MM models of biological molecules are being currently employed in mechanistic and dynamical studies involving electronic excitations. Firstly, the different book Chapters will expose the characteristics of the models selected as the best compromise between accuracy and computational cost; a "selection" process depending on the specific information sought by the researcher. For instance, if the researcher is looking for the simulation or prediction of properties trends rather than property values, the QM/MM model employed may have different features. Secondly, as the book title suggests, another feature of the revised models is that these must be suitable for dealing with electronic excitations. As anticipated above, this imposes strict constraints to the type of QM method used for describing the QM subsystem. Thirdly, the applications will not only regard the calculation of static features like excitation energies, computation of the structure of elusive reactive intermediates and fluorescent species but also the calculations of excited state and nonadiabatic trajectories describing the transition from excited to ground state. In conclusion, the reader will be guided through a wealth of diverse examples of the design, setup, and application of QM/MM models to the mechanistic and dynamics study of processes involving electronically excited states.

The three initial Chapters deal, at a certain extent, with QM/MM method and protocol development. The chapter "On the Automatic Construction of QM/MM Models for Biological Photoreceptors: Rhodopsins as Model Systems" by Laura Pedraza-González, Maria del Carmen Marín, Luca de Vico, Xuchun Yang, and Massimo Olivucci, discusses the possibility to automate the parallel construction of QM/MM models of entire sets of rhodopsin photoreceptors and of their mutants mainly for predicting the trends of their spectroscopic properties. The models generated for a set of wild-type and rhodopsin mutants, reveal that the corresponding automated protocol, based on an *ab initio* multi-configurational QM method, is capable of reproducing the trends in the wavelength of the absorption maximum. The proposed protocol exemplifies an effort toward the standardization, reproducibility, and transferability of rhodopsin QM/MM models with the target of reducing the consequences of human-errors during the QM/MM model building. The authors also show the applicability of their technique in studies relevant to optogenetics such as the engineering of microbal rhodopsins with enhanced fluorescence. Throughout the chapter the QM/MM methodology used is the Complete Active Space Self-Consistent Field (CASSCF) and Complete Active Space Second Order Perturbation Theory (CASPT2) QM methods coupled with the Amber MM force field and interacting via an electrostatic embedding scheme.

In the second chapter "Photo-Active Biological Molecular Materials: From Photoinduced Dynamics to Transient Electronic Spectroscopies", Irene Conti, Matteo Bonfanti, Artur Nenov, Ivan Rivalta, and Marco Garavelli review a computational strategy for characterizing transient spectra constructed on the basis of static and dynamic exploration of the potential energy surfaces. It is shown that in all discussed systems, the applied QM/MM model can successfully reproduce and study the spectral signatures measured by (two-dimensional) steady state and transient optical spectroscopy in rhodopsin pigments and DNA-related systems in their native environments. The authors also mention the possibility to alleviate the QM size problem present in multi-chromophoric systems by introducing an exciton model effectively integrated into static and dynamic QM/MM engines. Throughout the chapter the QM/MM methodology used is DFT, TDDFT, CASSCF, CASPT2, Restricted Active Space Self-Consistent Field (RASSCF), and Restricted Active Space Second Order Perturbation Theory (RASPT2) QM methods coupled with the Amber MM force field and interacting via an electrostatic embedding scheme.

Finally, in the chapter "Polarizable Embedding as a Tool to Address Light-Responsive Biological Systems" Peter Hartmann, Peter Reinholdt, and Jacob Kongsted discuss the theoretical background and derivation of a polarizable embedding scheme and how the latter can be formulated within QM response theory to account for environmental effects using both discrete and polarizable models. The authors illustrate their methodology on a Nile Red chromophore model incorporating local fields effects and structural dynamics (sampling). They show that these effects are both crucial when the target is to simulate/predict in an unbiased fashion one- and two-photon absorption spectra of the chromophore in condensed phase.

The next four chapters are generally devoted to spectroscopic characterization of various light-responsive biological molecules such as photoreceptors, fluorescent and bioluminescent proteins, as well as DNA-related systems. The section opens with the chapter "Computational Studies of Photochemistry in Phytochrome Proteins" by Jonathan R. Church, Aditya G. Rao, Avishai Barnoi, Christian Wiebeler, and Igor Schapiro in which the authors discuss the essential ingredients of a QM/MM based computational protocol relevant for the accurate prediction of absorption spectra of stable and meta-stable states of phytochromes. It is shown how QM/MM calculations may provide insight into chromophore conformation. Throughout the chapter the QM/MM methodology used is the semiempirical, DFT, TDDFT, Resolution of Identity Approximate Coupled Cluster Second Order (RI-CC2), and Resolution of Identity Algebraic Diagrammatic Construction to Second Order (RI-ADC(2)) QM methods coupled with the Amber MM force field and interacting via an electrostatic embedding scheme.

The Chapter "QM/MM Study of Bioluminescent Systems" by Isabelle Navizet gives a critical summary of the entire process of constructing reliable QM/MM models for investigating the chemistry behind bioluminescent reactions. This contribution discusses aspects of the model building complementary to ones stressed in the preceding chapters. In fact, it deals with the key steps in the procedure related to the choice of: (i) the size of the QM subsystem, (ii) the QM method employed to treat the QM subsystem and the force field employed to describe the MM subsystem, (iii) the interaction scheme between the QM and MM subsystems, (iv) the QM/MM frontier atom(s), (v) the solvation model, and finally, (vi) the representative snapshots when a conformational sampling is necessary. Examples of such computational procedure that use either *ab initio* (wavefunction based) multi-configurational QM methods or TDDFT methods, in conjunction with suitable MM force fields to account for environmental effects, to calculate vertical excitation energies, are shown in this contribution. In particular, an impact of the surrounding on color modulation in different firefly species is revised. In summary, throughout the chapter the QM/MM methodology used is the DFT, CASSCF, and CASPT2 QM methods coupled with the Amber MM force field and interacting via an electrostatic scheme.

The following Chapter "QM/MM Approaches Shed Light on GFP Puzzles" by Alexander V. Nemukhin and Bella L. Grigorenko discusses how flexible the EFP method is when this is employed to comprehensively dissect the mechanism of chemical events occurring in the green fluorescent protein (GFP). In particular, it is shown that when using EFP, QM/MM modeling can correctly predict: the chemical reactions implicated in the GFP chromophore maturation, the reaction route linking the neutral and anionic forms of the wild-type protein, and the irreversible photochemical reactions, such as decarboxylation and photobleaching, leading to GFP malfunctioning. In summary, throughout the chapter the QM/MM methodology used is the DFT, CASSCF, and Extended Multi-Configuration Quasi- Degenerate Perturbation Theory (XMCQDPT2) QM methods coupled with the Amber MM force field and interacting via flexible effective fragment approach.

The book's last chapter "DNA Photodamage and Repair: Computational Photobiology in Action" is from Antonio Frances-Monerris, Natacha Gillet, Elise Dumont, and Antonio Monari. In the chapter, the authors revise results that show how properly selected multi-scale protocols including QM/MM based protocols, provide insight into the mechanism driving the formation of photoinduced DNA lesions and the subsequent DNA recognition and repair mechanism by dedicated enzymes. In fact, the authors stress that, in spite of considerable progress achieved in modeling DNA systems, a number of challenges still need to be addressed. Accordingly, it is discussed how these challenges including those related to the reduction of high computational costs and investigation of the role of noncanonical environment (e.g., inhomogenous or highly charged nucleosomal environment) on the production of DNA lesions, are tackled. In summary, throughout the chapter the QM/MM methodology used is the DFT, and TDDFT QM methods coupled with the Amber MM force field and interacting via an electrostatic embedding scheme.

In conclusion, the purpose of *QM/MM Studies of Light-responsive Biological Systems* is to revise recent achievements in the modeling of excited states in proteins and nucleic acids. Thus, we believe that the book is of interest to both students and experienced researchers involved in fundamental and application-oriented research.

Wrocław, Poland   Tadeusz Andruniów
Siena, Italy   Massimo Olivucci

References

1. Gozem S, Luk HL, Schapiro I, Olivucci M (2017) Theory and simulation of the ultrafast double-bond isomerization of biological chromophores. Chem Rev 117:13502–13565
2. Warshel A, Levitt M (1976) Theoretical studies of enzymic reactions: dielectric, electrostatic and steric stabilization of the carbonium ion in the reaction of lysozyme. J Mol Biol 103:227–249
3. Warshel A (1976) Bicycle-pedal model for the first step in the vision process. Nat. (London) 260:679–683
4. Senn HM, Thiel W (2009) QM/MM methods for biomolecular systems. Angew Chem Int Ed 48:1198–1229
5. Kamerlin SCL, Warshel A. (2011) Multiscale modeling of biological functions. Phys Chem Chem Phys 13:10401–10411
6. Snyder JW Jr, Curchod BFE, Martínez TJ (2016) GPU-accelerated state-averaged complete active space self-consistent field interfaced with ab initio multiple spawning unravels the photodynamics of provitamin D3. J Phys Chem Lett 7:2444–2449
7. Gurunathan PK, Acharya A, Ghosh D, Kosenkov D, Kaliman I, Shao Y, Krylov AI, Slipchenko LV (2016) Extension of the effective fragment potential method to macromolecules. J Phys Chem B 120:6562–6574
8. Valsson O, Campomanes P, Tavernelli I, Rothlisberger U, Filippi C (2013) Rhodopsin absorption from first principles: bypassing common pitfalls. J Chem Theory Comput 9:2441–2454

9. Loco D, Lagardère L, Caprasecca S, Lipparini F, Mennucci B, Piquemal J-P (2017) Hybrid QM/MM molecular dynamics with amoeba polarizable embedding. J Chem Theory Comput 13:4025–4033
10. Söderhjelm P, Husberg C, Strambi A, Olivucci M, Ryde U (2009) Protein Influence on Electronic Spectra Modeled by Multipoles and Polarizabilities. J Chem Theory Comput 5:649–658

Contents

Contributors

Avishai Barnoy Fritz Haber Center for Molecular Dynamics Research Institute of Chemistry, The Hebrew University of Jerusalem, Jerusalem, Israel

Matteo Bonfanti Dipartimento di Chimica Industriale Toso Montanari, Universita di Bologna, Bologna, Italy

Jonathan R. Church Fritz Haber Center for Molecular Dynamics Research Institute of Chemistry, The Hebrew University of Jerusalem, Jerusalem, Israel

Irene Conti Dipartimento di Chimica Industriale Toso Montanari, Universita di Bologna, Bologna, Italy

Luca De Vico Biotechnology, Chemistry and Pharmacy Department, University of Siena, Siena, Italy

Elise Dumont Univ Lyon, ENS de Lyon, CNRS UMR 5182, Université Claude Bernard Lyon 1, Laboratoire de Chimie, Lyon, France;
Institut Universitaire de France, Paris, France

Antonio Francés-Monerris Université de Lorraine and CNRS, LPCT UMR 7019, Nancy, France;
Departamento de Química Física, Universitat de València, Burjassot, Spain

Marco Garavelli Dipartimento di Chimica Industriale Toso Montanari, Universita di Bologna, Bologna, Italy

Natacha Gillet Univ Lyon, ENS de Lyon, CNRS UMR 5182, Université Claude Bernard Lyon 1, Laboratoire de Chimie, Lyon, France

Bella L. Grigorenko Department of Chemistry, Lomonosov Moscow State University, Moscow, Russia;
Emanuel Institute of Biochemical Physics, Russian Academy of Sciences, Moscow, Russia

Peter Hartmann Department of Physics, Chemistry and Pharmacy, University of Southern Denmark, Odense, Denmark

Jacob Kongsted Department of Physics, Chemistry and Pharmacy, University of Southern Denmark, Odense, Denmark

María del Carmen Marín Biotechnology, Chemistry and Pharmacy Department, University of Siena, Siena, Italy;
Chemistry Department, Bowling Green State University, Ohio, USA

Antonio Monari Université de Lorraine and CNRS, LPCT UMR 7019, Nancy, France

Isabelle Navizet Laboratoire Modélisation et Simulation Multiéchelle, MSME UMR 8208, Université Gustave Eiffel, UPEC, CNRS, Marne-la-Vallée, France

Alexander V. Nemukhin Department of Chemistry, Lomonosov Moscow State University, Moscow, Russia;
Emanuel Institute of Biochemical Physics, Russian Academy of Sciences, Moscow, Russia

Artur Nenov Dipartimento di Chimica Industriale Toso Montanari, Universita di Bologna, Bologna, Italy

Massimo Olivucci Biotechnology, Chemistry and Pharmacy Department, University of Siena, Siena, Italy;
Chemistry Department, Bowling Green State University, Ohio, USA;
Universitè de Strasbourg, USIAS Institut d'Études Avanceés, Strasbourg, France

Laura Pedraza-González Biotechnology, Chemistry and Pharmacy Department, University of Siena, Siena, Italy

Aditya G. Rao Fritz Haber Center for Molecular Dynamics Research Institute of Chemistry, The Hebrew University of Jerusalem, Jerusalem, Israel

Peter Reinholdt Department of Physics, Chemistry and Pharmacy, University of Southern Denmark, Odense, Denmark

Ivan Rivalta Dipartimento di Chimica Industriale Toso Montanari, Universita di Bologna, Bologna, Italy;
Univ Lyon, Ens de Lyon, CNRS, Université Lyon 1, Laboratoire de Chimie, Lyon, France

Igor Schapiro Fritz Haber Center for Molecular Dynamics Research Institute of Chemistry, The Hebrew University of Jerusalem, Jerusalem, Israel

Christian Wiebeler Fritz Haber Center for Molecular Dynamics Research Institute of Chemistry, The Hebrew University of Jerusalem, Jerusalem, Israel

Xuchun Yang Chemistry Department, Bowling Green State University, Ohio, USA

On the Automatic Construction of QM/MM Models for Biological Photoreceptors: Rhodopsins as Model Systems

Laura Pedraza-González, María del Carmen Marín, Luca De Vico, Xuchun Yang, and Massimo Olivucci

Abstract The automatic building of quantum mechanical/molecular mechanical models (QM/MM) of rhodopsins has been recently proposed. This is a prototype of an approach that will be expanded to make possible the systematic computational investigation of biological photoreceptors. QM/MM models represent useful tools for biophysical studies and for protein engineering, but have the disadvantage of being time-consuming to construct, error prone and, also, of not being consistently constructed by researchers operating in different laboratories. These basic issues impair the possibility to comparatively study hundreds of photoreceptors of the same family, as typically required in biological or biotechnological studies. Thus, in order to carry out systematic studies of photoreceptors or, more generally, light-responsive proteins, some of the authors have recently developed the Automatic Rhodopsin Modeling (ARM) protocol for the fast generation of combined QM/MM models of photoreceptors formed by a single protein incorporating in its cavity a single chromophore. In this chapter, we review the results of such research effort by revising the building protocol and benchmark studies and by discussing selected applications.

Keywords QM/MM · Rhodopsin · Photochemistry · Photobiology · Optogenetics

Authors Laura Pedraza-González, María del Carmen Marín have equally contributed to the present chapter.

L. Pedraza-González · M. d. C. Marín · L. De Vico · M. Olivucci
Biotechnology, Chemistry and Pharmacy Department, University of Siena, Siena, Italy

M. d. C. Marín · X. Yang · M. Olivucci
Chemistry Department, Bowling Green State University, Ohio, USA

M. Olivucci (✉)
Universitè de Strasbourg, USIAS Institut d'Études Avanceés, Strasbourg, France
e-mail: molivuc@bgsu.edu

T. Andruniów and M. Olivucci (eds.), *QM/MM Studies of Light-responsive Biological Systems*, Challenges and Advances in Computational Chemistry and Physics 31,
https://doi.org/10.1007/978-3-030-57721-6_1

1

1 Introduction: Automatic QM/MM Model Generation for Biological Photoreceptors is Needed in the Genomics age

Presently, genomics is increasingly applied in fields as diverse as drug design [114, 123], agricultural and food production [16], as well as in environmental and biotechnological research [13, 49]. These applications all involve accessing functional information for proteins such as receptors or enzymes, starting from the corresponding amino acid sequences. For this reason, faster and cheaper molecular biology tools for gene sequencing, protein expression and analysis have been continuously developed and applied to systematic studies [24, 69]. As a consequence, a growing number of protein sequences are becoming available that, however, usually requires high-throughput functional characterization to be converted into useful information. Computational and theoretical studies provide, in principle, a highly effective strategy for accomplishing such a task. Typically, such studies start with the building of the three-dimensional protein structure associated with the retrieved sequence. This is usually carried out via comparative modeling methods [74] where unknown structures are built starting from a template corresponding to a set of atomic coordinates obtained experimentally via X-ray crystallographic techniques.

In order to get information on the protein dynamics and functions, such as conformational flexibility and chemical reactivity, the output of comparative modeling is often used as a guess to construct more complex models capable of providing information on properties going, for instance, from chemical catalysis to electronic spectroscopy to photochemistry. More specifically, when information on properties involving an electronic reorganization must be simulated or predicted (e.g., a bond breaking or an electronic transition), the model shall deal with both nuclear and electronic motions. In this case, and when not considering the GPU-based full quantum mechanical models of biological molecules still under development (see, for instance, Ref. [129]), quantum mechanics (QM) is used to treat the reactive part of the protein while its remaining part is treated classically using a molecular mechanics (MM) force field. This leads to the so-called hybrid quantum mechanics / molecular mechanics (QM/MM) models that are the subject of the present book.

In order to exploit the abundance of genomics data (i.e., protein sequences), a researcher must carry out comparative studies on large sets of homologous proteins. It is apparent that the rapid conversion of a set of homologous sequences into three-dimensional structures would lead to new knowledge on the functionally important protein parts that must be present (i.e., conserved) in all members of the set. The construction of the corresponding QM/MM models would, additionally, provides information on the conservation and variability of functional parts of the protein, depending on both their electronic and geometrical structures and also, within a certain time-scale, on their dynamical behavior. However, in order to perform a meaningful comparative investigation at the QM/MM level, it is necessary: (i) to build congruous QM/MM models according to a well- defined protocol with a known error bar and (ii) to employ a building procedure allowing the construction of hundreds if not thousands QM/MM models. It is apparent that, in order to satisfy points i–ii, the construction of QM/MM models needs to be automated.

In the present book chapter, we focus on the construction and application of a type of QM/MM models that, due to their basic architecture, can be automatically built and used for relatively fast screening studies. In particular, we will discuss how such an automation has been designed, implemented, and benchmarked for a specific but ecologically widespread and functionally diverse group of natural light-responsive proteins: rhodopsins [28]. We will see how such models have not been designed to be the most accurate model possible (see, for instance, the models of references [14, 48, 137, 138] that are designed for spectroscopic studies), but as partially constrained, gas-phase models aimed at the prediction of the relationship between variations in amino acid sequence and variations in spectroscopic and photochemical properties that, in principle, do not require the modification of the protein backbone (e.g., light absorption or ultrafast photochemical transformations). We will also see that they are specialized QM/MM models, in the sense that the building protocol is specifically designed and parametrized for the study of rhodopsin proteins, focused on the description of their prosthetic group (i.e., the chromophore), the protonated Schiff base of retinal (rPSB), and its interactions with the surrounding amino acids forming the chromophore cavity. For this reason, the building protocol has been called Automatic Rhodopsin Modeling (ARM) protocol.

While focusing on QM/MM models for a single protein family may appear to be of limited interest, rhodopsins form a vast family of biological photoreceptors capable of carrying out different biological functions. In fact, members of the rhodopsin family are found in many diverse organisms and, thus, constitute widespread light-responsive proteins driving a variety of fundamental functions in vertebrates and invertebrates, as well as in microorganisms [28]. The recent discovery of a new family of light-sensing microbial rhodopsins called heliorhodopsins [112] indicates that we do not still fully comprehend the vast ecological distribution and functional diversity of these light exploiting chemical systems, which appear to harness, globally, an amount of sun-light energy as large as that harnessed by photosynthesis [42]. In spite of their functional diversity, rhodopsins display a remarkably constant architecture featuring, as detailed below, a cavity hosting the rPSB chromophore [28]. The different protein functions are invariably initiated by a highly stereoselective photoisomerization of the chromophore triggered by the absorption of light of specific wavelengths [44, 46, 62, 80, 128].

There is an additional and more practical reason for constructing rhodopsin QM/MM models automatically: the search for novel and more accurate optogenetics tools [20, 61]. Optogenetics uses, mainly, genetically encoded and often specifically engineered rhodopsins to control physiological processes. Nowadays, optogenetics tools are mainly used in the investigation of the nervous system at the cellular and tissue level. They are employed as light-driven actuators (i.e., action potential triggers), light-driven silencers (i.e., action potential quenchers), and as fluorescent reporters (i.e., action potential probes) [61]. Actuators are capable of starting, at a chosen point in time and space, an action potential. Silencers can stop a moving action potential and reporters are instead able to signal the passage of the action potential in a specific cell body. One target in the development of optogenetic tools is that of increasing the variety of absorption wavelengths thus enabling simultaneous optical control by different colors of light. As such, various rhodopsin genes have been screened in order to find additional colors [9, 70]. In particular, while many blue-absorbing rhodopsin at $\lambda_{max}^a < 500$ nm have been reported [8] and even applied to optogenetics [70], the longer absorption maxima are limited in $\lambda_{max}^a < 600$ nm. Random point mutations identify the types of amino acid mutation that are effective for color tuning [27, 68]. Although numerous mutations causing bathochromic shift without disrupting protein function were identified in this way, the degree of redshifts achieved so far have been insufficient for application, and comprehensive experimental screening is difficult because of the large number of possible mutations. Thus, rhodopsin models capable to simulate light absorption and emission such as the proposed QM/MM models are expected to help to rapidly estimate the absorption energy [90], mainly allowing the screening of large sets of mutants.

Finally, there is a further motivation for focusing on the research endeavor outlined above. The general structure of rhodopsins, which is substantially made by a protein hosting a single chromophore molecule, is representative of both natural and artificial photoresponsive proteins. These would comprise, for instance, Xanthopsin (PYP), light-oxygen-voltage-sensing domains, cryptochromes, and phytochromes. An example of artificial systems is given by rhodopsin mimics constituted by soluble proteins hosting covalently bounded natural or synthetic retinal-like chromophores [18, 104]. It is thus clear that the automated construction of QM/MM models of rhodopsins and the kind of applications revised in the following sections could, in the future, be extended to other natural or artificial protein families.

This chapter is organized as follows. In Sect. 2, essentially a technical section, we introduce the concept of QM/MM models of photoresponsive proteins with emphasis on rhodopsin structure and on the possibility of automating the building of rhodopsin models. Accordingly, in Sect. 2.1 we revise the general structure of a QM/MM model (see Fig. 1a) and briefly describe the set of choices that stands at the basis of the different type of models found in the literature, while in Sect. 2.2 we motivate the currently proposed protocol for constructing rhodopsin models and for approaching the problem of automating such construction. In Sect. 3, we revise/discuss the development and features of an ARM model, focusing on the current and latest workflow for its automated construction called "a-ARM rhodopsin model building" and "original ARM rhodopsin model building" protocols (see Fig. 1b). In Sect. 3.1 we describe the a-ARM input file generator that automatically produces the input necessary for the QM/MM model generator performed by the ARM protocol. This section explains how, within a-ARM, it is possible to construct a less biased input with respect to the one provided by a user. In Sect. 3.2, we, instead, revise the workflow for the automated assembly of the QM/MM model. This process is performed automatically by the ARM protocol. In Sect. 3.3, we focus on the benchmarking of different ARM models, focusing on the comparison with experimental λ_{max}^a values. Accordingly, we discuss the different performance of ARM models generated using the ARM protocol and a manual input (ARM$_{manual}$) and by the more automated a-ARM default and customized models (a-ARM$_{default}$ and a-ARM$_{customized}$, respectively). Finally,

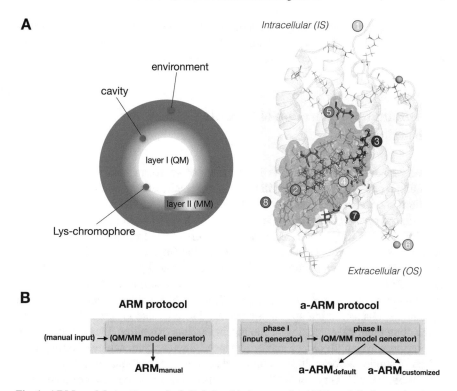

Fig. 1 ARM model structure. a Left. Relationship between the ARM model's three subsystems and two multi-scale layers. Right. Detailed description of the components of the three subsystems (for a microbial rhodopsin from the DA state of *bacteriorhodopsin*). (1) main protein chain (cyan cartoon), (2) rPSB chromophore (green ball-and-sticks), (3) Lysine side-chain covalently linked to the retinal chromophore (blue ball-and-sticks), (4) main chromophore counterion (cyan tubes), (5) residues with non-standard protonation states (violet tubes), (6) IS and OS external Cl⁻ (green balls) counterions, (7) crystallographic water molecules (red/white tubes), and the (8) amino acid residues forming the chromophore cavity subsystem (red frames and surface). The external OS and IS charged residues are shown in frame representation. The environment subsystem is formed by parts 1 and 6. The Lys-chromophore subsystem, which includes the H-link-atom located along the only bond connecting blue and green atoms, is formed by parts 2 and 3. The cavity subsystem is composed of parts 4 and 8. Part 7 corresponding to water molecules may be part of either the environment or cavity subsystems. **b** Relationship between the currently developed ARM protocols and ARM models. Adapted with permission from [106]. Copyright 2019 American Chemical Society

in Sect. 3.4 we summarize the present assets and limitations of the ARM models and in Sect. 3.5 we discuss the possibility of building such models via a web-based interface called Web-ARM. In Sect. 4, we focus on the applications of ARM models to the study of different aspects of rhodopsin functions, emphasizing their possible use for the development of optogenetics tools. Thus, in Sect. 4.1 we deal with the engineering of fluorescent mutants, in Sect. 4.2 with the engineering of microbial rhodopsins absorbing in the red, in Sect. 4.3 with infrared light absorbing rhodopsins via two-photon absorption, in Sect. 4.4 with the use of ARM models in modeling some aspects of rhodopsin photoisomerization and photocycle in Sect. 4.5 we focus on the sensitivity to the pH. Finally, in Sect. 5 we summarize the chapter materials and provide a perspective on the development of the field.

2 Automatic Generation of QM/MM Models for Rhodopsins

Above, we have explained that the development of a fast and standardized QM/MM modeling technology would be critical for a systematic understanding of rhodopsin functions, as well as for the rational design of rhodopsin variants useful in basic research and optogenetics [28, 145]. Such models have to be accurate enough to provide information in line with experimental observations. In fact, the ARM models revised in this chapter have been developed with the target of reproducing/predicting trends in spectroscopic, photophysical and photochemical properties and, therefore, to provide an atomic-level interpretation (i.e., mechanism) for phenomena such as spectral tuning and fluorescence, [41, 59, 85, 91, 106] as well as to design rhodopsin variants to enhance/inhibit specific properties. Indeed, there is presently an interest in the computational screening of rhodopsin variants exhibiting a longer maximum absorption wavelength (λ^a_{max}) and/or enhanced fluorescence, achieved through the effects of single and/or multiple amino acid mutations of a template wild-type structure [1, 59, 63, 85]. This would allow to select candidates for posterior experimental expression and characterization. Actually, the computation of accurate trends in spectral properties as derived from the screening of arrays of homologue wild-type (WT) and/or mutant rhodopsins, is more valuable than the computation of absolute values when the target is the rational design of rhodopsin variants [59, 85, 91, 106]. Recent studies pursuing these targets will be reported in Sect. 4.

The above foreseen computational research and applications require a standardized methodology, with a known error bar, allowing the study of hundreds, if not thousands, of wild-type and/or mutant rhodopsins in a comparative fashion. In order to design an effective screening methodology, it is fundamental to identify: (i) the main parts of the systems to be modeled (i.e., of the rhodopsin protein family) and (ii) the spectroscopic and/or photochemical properties whose trends have to be computed (i.e., light emission and photoisomerization of the retinal chromophore).

With respect to point (i), rhodopsins share a common architecture constituted by seven transmembrane helices, which form a cavity hosting the rPSB chromophore. rPSB is covalently linked via a specific lysine residue (located in the middle of helix VII and helix G for animal and microbial rhodopsins respectively), via an imine (–C=N–) linkage. However, in spite of such similarity, there are presently three known rhodopsin families with large differences in sequence (i.e. they are not considered homologues) as well as in certain structural features: animal (type II) rhodopsins, microbial (type I) rhodopsins [28], and heliorhodopsins [112]. For instance, the difference between animal rhodopsins and the other families, is the fact that the photoisomerization of the retinal chromophore involves the transformation of the 11-*cis* rPSB chromophore dominating their equilibrium dark adapted (DA) state to its all-*trans* configuration, while in microbial and heliorhodopsin families the DA state is dominated by an all-*trans* rPSB chromophore, which is usually transformed into the 13-*cis* configuration upon light absorption. On the other hand, the animal and microbial rhodopsin families differ from the heliorhodopsin family in terms of the orientation of the N-terminal and C-terminal residues with respect to the orientation of the biological membrane. In the first two families, the N-terminal amino acid is exposed to the inner (intracellular) part of the cell membrane, while in heliorhodopsins the same residue is exposed to the extracellular matrix. With respect to point (ii), it is well-known that the process driving the diverse functions carried out by rhodopsins is the photoisomerization of the rPSB chromophore occurring immediately after the absorption of a photon of appropriate wavelength. It is also known that the, usually weak, fluorescence is exploited in optogenetics. Therefore, important properties that the chosen QM/MM model shall reproduce/predict are the rhodopsin color (or absorption wavelength), fluorescence and photochemical reactivity. This last property includes the simulation of the excited state dynamics, decay to the ground state, and primary photoproduct formation on a total time-scale of few picoseconds [28, 44, 62]. This last process is ultimately accompanied by geometrical perturbation of the residues (i.e., amino acids, water molecules) belonging to the chromophore cavity, and mediated by the electrostatic/steric interactions of the rPSB chromophore with the residue side-chains. However, the residues located far away from the chromophore cavity are, in general, less prone to influence or being influenced by the chromophore isomerization.

When taking into account the structural features revised in point (i) and limiting our interest on the properties mentioned in point (ii), it is possible to subdivide the rhodopsin structure in three parts or subsystems, requiring an increasing level of accuracy for their description. These subsystems are: the (protein) environment, the chromophore cavity, and the Lys-chromophore (see Fig. 1a). These subsystems also define a multi-scale computational approach, which uses different levels of theory for treating different layers. As we will discuss below, the proposed ARM model features three subsystems and two layers. In such a basic QM/MM model, an MM method is used for treating the environment, cavity, and the Lys part of the Lys-chromophore subsystem. As it will be detailed in the following, the rPSB chromophore is instead treated at the multi-configurational QM level [2, 3, 91, 103, 106, 145].

In the present section, we introduce a protocol for the fast and congruous production of ARM models for wild-type and mutant rhodopsin variants. The model general structure and choice of specific QM and MM method is described in Sect. 2.1. The following Sect. 2.1 describes how to employ such QM/MM tools for the effective description of rhodopsins, and how to assemble them into an efficient algorithm.

2.1 QM/MM Partitioning

The total potential energy of a QM/MM model is calculated by (see Eq. 1)

$$E_{Total} = E_{QM} + E_{MM} + E_{QM/MM} \tag{1}$$

where E_{QM} is the energy of the QM subsystem, E_{MM} is the energy of the MM subsystem, and $E_{QM/MM}$ is the energy contribution coming from the interaction between them.

A general, first level, classification of QM/MM methods is based on the way the electrostatic interactions between the QM and MM subsystems is treated:

- *Mechanical Embedding*: the QM calculation is performed without the MM surrounding charges and the interaction between the QM and MM subsystems is treated at the MM level for both the electrostatics and the van der Waals interactions.
- *Electrostatic Embedding*: the QM calculation is performed in the presence of MM surrounding charges and the electrostatic interactions between the subsystems enter the QM Hamiltonian, while bonded and van der Waals interactions are treated at the MM level. This is the most common level of QM/MM theory used.
- *Polarizable Embedding*: the QM calculation is performed in the presence of surrounding polarizable MM charges and the electrostatic interactions between the subsystems enter, again, the QM Hamiltonian, while bonded and van der Waals interactions are treated at the MM level. This is a technology still under benchmarking even if a number of polarizable force fields and applications are available [60].

The selection of the best QM/MM approach relies on the characteristics of the problem under study and/or the success achieved by a method for the investigation of similar problems. In the QM/MM approach described in Sect. 3.2, the QM region is treated at the well-studied CASPT2//CASSCF level of theory [4, 31] (i.e., energy//geometry), with the 6-31G(d) basis set and an active space including the full π-system (12 electrons in 12 orbitals) of the retinal chromophore bound to the lysine via protonated Schiff base linkage. By default, in ARM, the excitation energies are computed using three-root state-averaged singlet wave functions thus providing the energies of the first three singlet electronic states (S_0, S_1, and S_2). For the MM region, the Amber94 force field [17] is used. The interaction between the QM and MM regions is based on an electrostatic embedding scheme involving an unconventional treatment called Electrostatic Potential Fitted (ESPF) [29]. In ESPF, the electrostatic interaction energy between a QM derived charge distribution and an arbitrary external potential is approximated using the expectation value of the distributed multipole operator fitted to the electrostatic potential grid derived from the electronic wave function computed in the presence of the MM charges. This operator, by construction, yields the best possible representation of the interaction energy in the framework of a limited multi-centered multipole expansion of the interacting charge distributions. The use of this operator is advantageous in QM/MM calculations, because it ensures a smooth transition between QM and MM modeling of the electrostatic interaction between the corresponding multi-scale layers.

2.1.1 Selection of QM Method

The selection of a suitable QM method for a QM/MM approach follows the same criteria as in pure QM models (accuracy *versus* computational cost). With this in mind, semi-empirical QM methods have been one of the most popular choices (see for instance Refs. [139] and [14]), and remain important for the study of large systems (i.e., proteins) where the computational costs is high. In this regard, *Density Functional Theory* (DFT) is the workhorse in many contemporary QM/MM studies, and correlated *ab initio* DFT methods are increasingly used in electronically demanding cases or in the quest for high accuracy. Here we mention the State-interaction State-average Spin-restricted Ensemble Kohn-Sham (SI-SA-REKS) [33, 34] method and the Multi-reference Spin-flip Time-dependent DFT (MR-SP-TDDFT) method [51] that have been tested for minimal rPSB chromophores in the gas-phase [55].

A multi-configurational quantum chemical method called *Multi-Configurational Self-Consistent Field* (MCSCF) is useful for systems where more than one electron configuration has to be taken into account to describe electron unpairing (i.e., ground and excited states, bond breaking, open-shell species). One of the most popular and benchmarked MCSCF methods is the *Complete Active Space Self-Consistent Field* (CASSCF) [118]. CASSCF is equivalent to the *full configuration Interaction* (FCI) method applied to a selected subset of electrons and orbitals called the *Active Space*, while the remaining electrons and orbitals are treated at the *Hartree-Fock* (HF) level, although they are optimized together with the active orbitals. CASSCF has a reasonable accuracy in the description of the excited states of any kind, performing qualitatively well even in the presence of charge transfer (CT) states (it is better for diradical (DIR) states) and it was used successfully for an extensive number of mechanistic photochemical studies. CASSCF has a major drawback: dynamic electron correlation, which is most important for CT states, is not properly described, leading to poor agreement with experimental measurements such as those from absorption and emission spectroscopies. Applying a multi-reference perturbative correction to the energy can fix this problem.

Indeed, the CASPT2 method (*Complete Active Space Second-order Perturbation Theory*) [4] is commonly used in order to include dynamic electron correlation effects on top of CASSCF. This is usually done according to the widely tested CASPT2//CASSCF protocol [45] where the geometrical structure of the model is determined at the CASSCF level while the potential energy of the electronic states of interest is determined at the CASPT2 level. CASPT2 can be viewed as a conventional perturbative method where the *zeroth*-order electronic wave function is a CASSCF function designed to include the electronic states of interest. Computing CASPT2 energies evaluated on CASSCF optimized geometries has shown to provide an acceptable agreement with experimental measurable quantities such as excitation energies and reaction barriers and, thus, provide a description of the spectroscopy and reactivity of the system. It has also been shown that such an agreement (usually within few kcal mol^{-1} or tenths of eV in the blue-shifted direction) is due, when using a limited basis set such as a double-zeta Gaussian basis with polarization functions (most frequently 6-31G(d)), to a cancelation of errors [32]. Here we shall mention that MS-CASPT2 [35], XMS-CASPT2 [7] and the substantially equivalent XMCQDPT2 [47] are significant improvements of CASPT2 as they allow for state interaction (not exclusively energy corrections) providing a smoother and more correct PES mapping. Furthermore, XMS-CASPT2 gradients [105] have been recently made available allowing for geometry optimization at that level of theory. CASPT2//CASSCF as well as XMS-CASPT2//CASSCF protocols are expensive from the computational point of view and cannot be applied to large chromophores. Fortunately, the rPSB chromophore of rhodopsins is small enough and can be described at these levels of theory (still using CASSCF gradients and therefore CASSCF optimized geometries). XMS-CASPT2 gradients are still far too expensive for systematic studies of systems featuring the entire rPSB chromophore. Some evidence on the suitability of CASSCF gradients for dynamics studies have been reported [83] with an acceptable computational costs on, for instance, a modern computer cluster. On the other hand, more complex chromophores such as, for instance, the tetrapyrrole chromophore of phytochromes or the carotenoid of the orange carotenoid protein can only be treated with active spaces that are reduced with respect to the ones comprising the full π-system [42]. We shall mention that attempts to treat a larger portion of a rhodopsin at the QM level have been reported using technologies exploiting GPUs rather than CPUs [150]. However, these have not used CASSCF. In fact, a special version of density functional theory such as the SI-SA-REKS method [33, 34] have been employed. In this case [150] for instance for the case of bacteriorhodopsin, a QM region consisting in the full rPSB chromophore, the CεH2 unit from the Lys216 side-chain, and the counterion cluster to the Schiff base, which includes three water molecules (Wat401, Wat402, and Wat406) and the side-chains of Asp-85 and Asp-212 (74 atoms in total) have been used. The calculations comprised two active electrons and two active orbitals and were run on GPUs. We shall also mention that methods suitable for treating larger active spaces and automatically selecting them have been recently reported [130]. These methods can describe π-systems larger than that required by the only retinal chromophore and can be employed to study if and how aromatic residues could significantly interact with the chromophore at the electronic wave function level [96, 136].

2.1.2 Selection of MM Method

Established MM force fields are available for biomolecular applications (e.g., CHARMM, AMBER, GROMOS, and OPLS) and for explicit solvent studies (e.g., TIP3P or SPC for water). However, it is often necessary to derive some additional force field parameters whenever the QM/MM calculations target situations in the active-site region that are not covered by the standard force field parameters (e.g., the frontier between covalently bounded QM and MM layers) which are, in the case of proteins, with the protein backbone and natural amino acid side-chains. The classical biomolecular force fields contain bonded terms as well as non-bonded electrostatic and van der Waals interactions. Electrostatics is normally treated using fixed point charges at the MM atoms. The charge distribution in the MM region is thus unpolarizable, which may limit the accuracy of the QM/MM results. The logical next step toward enhanced accuracy should thus be the use of polarizable force fields that are currently developed by several groups using various classical models (e.g., induced dipoles, fluctuating charges, or charge-on-spring models). The QM/MM formalism has been adapted to handle polarizable force fields [40, 125], but one may expect corresponding large-scale QM/MM applications only after these new force fields are firmly established. In the meantime, essential polarization effects in the active-site environment may be taken into account in QM/MM studies by a suitable extension of the QM region (at increased computational cost, of course). This has not extensively explored/done for rhodopsins due to the significant increase of the QM subsystem and, consequently, computational cost. When the additional cavity residues also contain electrons that need to be correlated with the one of the chromophore then an expansion of the active space is requested, which will make the cost even higher [136].

2.1.3 Boundary Treatment

The treatment of the frontier between the covalently bounded QM and MM subsystems provides a second level of classification of the QM/MM methodology. In certain QM/MM applications where there is no covalent bonding between the two subsystems, no boundary system is necessary as the embedding schemes revised above will take care of it. However, in many QM/MM studies, it is unavoidable that the QM/MM boundary cuts through a covalent bond. The frontier of the QM/MM system is set across the lysine $C\varepsilon$-$C\delta$ bond (see Fig. 2). The resulting dangling bond must be capped to satisfy the valency of the QM atom at the frontier, and in the case of electrostatic or polarized embedding, one must prevent over-polarization of the QM density by the MM charges close to the cut [30]. Link-atom (LA) schemes (see Fig. 2) introduce an additional atomic center (usually a hydrogen atom) that is not part of the real system and is covalently bonded to the QM frontier atom. Each link-atom generates three artificial nuclear degrees of freedom that are handled differently by different authors. The most common procedure is to fix the position of the link-atom such that it lies in the bond being cut, at some well-defined distance from the QM frontier atom, and to redistribute the forces acting on it to the two atoms of the bond being cut. This effectively removes the artificial degrees of freedom since the link-atom coordinates are fully determined by the positioning rule rather than being propagated according to the forces acting on them. Concerning the possible over-polarization in link-atom schemes, several protocols have been proposed to mitigate this effect which involves, for example, deleting or redistributing or smearing certain MM charges in the link region. The bonded interactions along the QM/MM

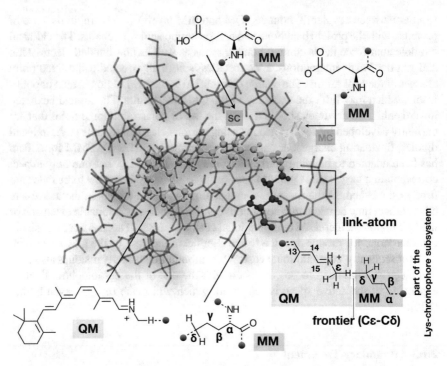

Fig. 2 Structure of the cavity and Lys-chromophore subsystems of a bovine rhodopsin QM/MM model constructed with *a*-ARM. The cavity subsystem, which contains the negatively charged counterion (GLU113) of the cationic rPSB moiety, and the LYS296 side-chain up to the C_δ atom, belong to the MM layer and are treated using the AMBER force field. The rPSB chromophore moiety and the remaining part of the of the LYS296 side-chain comprising the $C_\varepsilon H_2$ group and imine N atoms is treated at the QM level of theory. The dashed lines in the Lewis substructures represent dangling bonds connected to QM or MM atoms indicated by dark spheres. MC and SC correspond to the assigned primary and secondary chromophore (i.e., internal) counterions, respectively

boundary are treated with the corresponding standard MM potential, except for the re-parametrized C_{15}-N-Cε-Cδ torsional parameter [32]. The non-bonded interactions use the standard or re-parametrized van der Waals parameters to obtain the QM/MM van der Waals energy. The specific choices taken by the ARM model will be summarized in Sect. 3.2.

2.2   On the Development of a Standardized QM/MM Protocol

QM/MM models of rhodopsins have been widely employed to tackle problems in computational photochemistry and photobiology [145]. A well-documented case is that of bovine rhodopsin (Rh), which is considered a reference system for assessing the quality of QM/MM models, since it was the first wild-type animal rhodopsin (Rh-

WT) for which an X-ray crystallographic structure became available. Presently, Rh-WT represents an experimentally well-documented system in terms of spectroscopic and photochemical properties and their variation in different photocycle intermediates and mutants. Accordingly, different construction protocols for QM/MM models of Rh-WT aimed at reproducing its observed λ_{max}^a value (498 nm) have been reported. The result of these efforts illustrate how the differences in the adopted construction protocol affect the properties of QM/MM models and, in turn, the calculated λ_{max}^a which result in variations up to a 36 nm [2, 3, 5, 12, 39, 133, 137]. These results demonstrate the importance of a standardized protocol for QM/MM model generation. Such protocol should not strictly aim at the prediction of the absolute values of observable properties such as the λ_{max}^a and excited state lifetime, but to the description of their changes along sets of different rhodopsins and rhodopsin mutants.

Table 1 reports the λ_{max}^a obtained from nine different QM/MM models. The models listed in the table differ from each other for: the level of theory used for the QM calculation of the QM layer including the type of atomic basis set employed, the force field used to compute the MM layer and the approach used to describe the frontier connecting the QM and MM layers and, finally, the type of interaction between the QM and MM layers. These characteristics of the QM/MM model are not the only factors influencing the λ_{max}^a value. Indeed, the specific model construction protocol (i.e., including the different steps necessary to build the input for the final QM/MM calculation), is a factor impacting the model quality. Such choices are, for instance, the extension of the described protein environment (i.e., inside an explicit membrane or in the gas-phase (i.e., in isolated conditions) with electrostatics simulated with external counterions), the assignment of the protonation state for the ionizable protein residues (i.e., Asp → Ash, Glu → Glh, His → Hid or Hie, and Lys → Lyd), the selection of the side-chain conformation of specific residues (i.e., amino acids with different rotamers), the choice of the residues forming the chromophore cavity (i.e., the definition of such cavity), the constraints imposed on the initial structure (i.e., relaxed or fixed backbone) and the treatment of the internal water molecules (i.e., retention of the only crystallographic water molecules or the inclusion of possible missing ones).

From the description above, it is evident that a standardized protocol for the fast and automated production of congruous QM/MM models which can thus be replicated in any laboratory, must be based on two well-defined phases that here are called "generators": (i) input file generator and (ii) the QM/MM model generator. In Sect. 3, we report the development, implementation, and benchmarking of a construction protocol designed to deal automatically with phases (i) and (ii).

Table 1 A non-exhaustive list of QM/MM model setups reported in the literature for bovine rhodopsin (Rh). Calculated maximum absorption wavelength ($\lambda_{max}^{a,calc}$)

Reference	Template	QM/MM	Force field	QM/Basis set	$\lambda_{max}^{a,calc}$ (nm)	$\Delta^{Exp}\lambda_{max}^{a}$ (nm)[g]
Andruniow et al.[a] [5]	1HZX [132]	Gaussian 98/	AMBER94	CASPT2//CASSCF/	479	19
		TINKER		6-31G(d)		
Altun et al.[b] [2, 3]	1U19 [101]	Gaussian 03/	AMBER94	TD-B3LYP//B3LYP/	503	-5
		AMBER		6-31G(d)		
Tomasello et al.[c] [133]	1U19 [101]	Gaussian 03/	AMBER8	CASPT2//CASSCF/	496	2
		AMBER		6-31G(d)		
Sekharan et al.[d] [124]	1U19 [101]	Gaussian 03	CHARMM	CASPT2//ANO/	502	-4
				SCC-DFTB		
Bravaya et al.[e] [12]	1HZX [132]	GAMESS/	AMBER99	aug-MCQDPT2//PBE0/	515	-17
		TINKER		cc-pVDZ		
Fujimoto et al.[e] [39]	1L9H [100]	Gaussian 03,	AMBER99	SAC-CI//B3LYP/	506	-8
		GAMESS		D95		
Valsson et al.[f] [137]	1U19 [101]	MOLCAS	AMBER96	CASPT2//BLYP/	482	16
				ANO-L-VDZP		
Fujimoto et al.[e] [38]	1U19 [101]	Gaussian 03,	AMBER94	SAC-CI//HF/D95(d)	486	12
Melaccio et al.[g] [91]	1U19 [101]	MOLCAS/	AMBER94	CASPT2//CASSCF/	474	24
		TINKER		6-31G(d)		
Walczak et al.[f] [144]	1U19 [101]	MOLCAS/	AMBER94	CASPT2//B3LYP/	513	24
		TINKER		6-31G(d)		
Pedraza-González et al.[a] [106]	1U19 [101]	MOLCAS/	AMBER94	CASPT2//CASSCF/	496	2
		TINKER		6-31G(d)		

[a] Chain A. Ash (83), Glh (122,181), Hid (211)
[b] Chain A. Ash (83), Glh (122,181,249), Hid (152,195,278), Hie (211)
[c] Chain A. Ash (83), Glh (122,181), Hid (152,195,211)
[d] Chain B. No details on protonation states.
[e] Chain A. No details on protonation states.
[f] Chains A and B. Ash (83), Glu (122), Hid (100,211), Hie (65,152,195,278).
[f] Methodology selected from benchmark of different QM/MM model setups.
[g] Chain A. Ash (83), Glh (122), Hid (211)
[g] $\Delta^{Exp}\lambda_{max}^{a} = \lambda_{max}^{a,expc} - \lambda_{max}^{a,calc}$; ($\lambda_{max}^{a,expc} = 498$ nm)

3 Automatic Rhodopsin Modeling: ARM and *a*-ARM Protocols

As mentioned above the original ARM rhodopsin model building protocol has recently been introduced as a specialized tool for the automated generation of basic but congruous QM/MM models (ARM models) of wild-type and mutant rhodopsins. In ARM, rhodopsins are modeled as monomeric (single-chain), gas-phase, and globally uncharged proteins binding the rPSB chromophore. ARM employs a two-layer QM/MM approach based on electrostatic embedding and the H-link-atom frontier connecting the QM and MM layers [91, 106] (see Figs. 1 and 2). The term "gas phase" refers to the fact that while rhodopsins are membrane proteins exposed to the the transmembrane electrostatic field generated by an asymmetric distribution of the surface ions, the membrane and surrounding solvent are not incorporated in the model and such interaction is, therefore, not explicitly described. In fact, ARM models mimic the environment electrostatics by adding Cl^- and/or Na^+ as external counterions in the vicinity of the positively and/or negatively charged surface amino acids (i.e., ionizable residues Asp, Glu, Arg, Lys, His), in both the intracellular (IS) and extracellular (OS) protein surfaces (see Fig. 1a).

As stated in the caption of Fig. 1a, ARM models are composed of three subsystems: environment, cavity, and Lys-chromophore. The environment subsystem is a monomeric chain with the amino acid residues and few molecules of water, fixed at the crystallographic or comparative model structure, and includes external Cl^- and/or Na^+ counterions also fixed at pre-optimized positions. The cavity subsystem contains amino acid residues which exhibit fixed backbone and relaxed side-chains, and specific water molecules also free to relax. The Lys-chromophore subsystem contains the atoms of the lysine side-chain covalently linked to the chromophore in contact (through $C\delta$) with the QM/MM frontier and the entire QM subsystem (i.e., the N-methylated rPSB). All the Lys-chromophore atoms are free to relax during the QM/MM calculation. (see Fig. 2).

From the above description, it is apparent that ARM models are "conservative" models, since the environment subsystem preserve the structural information obtained from the X-ray crystallographic or comparative modeling. However, their basic definition makes the models more exposed to potential pitfalls with respect to more complex QM/MM models, as will be discussed in Sect. 3.4. On the other hand, ARM is not designed to generate the most accurate QM/MM models possible (see, for instance, [137]). Instead, its objective is to efficiently generate full arrays of QM/MM models suitable for comparative studies and for the predictions of trends of spectroscopic and photochemical properties.

ARM models feature: *automation*, so as to reduce building errors and avoid biased QM/MM modeling; *speed and parallelization*, so as to deal with large sets of rhodopsins wild-type and mutants simultaneously; *transferability*, so as to treat rhodopsins with large differences in sequence (i.e., rhodopsins belonging to both eubacterial, archaea, and eukarya domains); and *documented accuracy and reproducibility*, so as to be able to translate results into hypotheses that can be experimen-

A   Phase I: *Input File Generator*

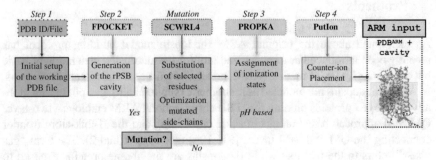

B   Phase II: *QM/MM Model Generator*

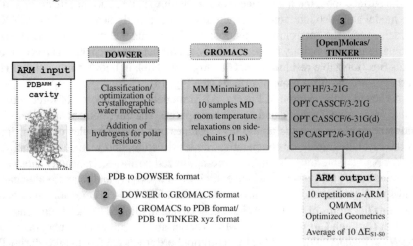

Fig. 3 General workflow of the *a*-ARM protocol. The *a*-ARM protocol comprehends two phases corresponding to **a** the use of the input file generator (Sect. 3.1) and **b** the QM/MM model generator (Sect. 3.2)

tally assessed. The features (see Fig. 1b) above have been introduced in the most automated protocol for building ARM models: a-ARM rhodopsin model building protocol. *a*-ARM yield the *a*-ARM$_{default}$ and *a*-ARM$_{customized}$ models mentioned in Sect. 1, through the development of, so-called, generators. As illustrated in Figs. 1b and 3, the *a*-ARM protocol requires two sequential phases (Phase I and Phase II) each one corresponding to the use of one generator. These are the *input file generator* (Fig. 3a) and the *QM/MM model generator* (Fig. 3b), respectively. The latest generator was introduced in 2016 and, when used with a manual input, corresponds to the ARM protocol, [91]. Thus, when using the ARM protocol the input file requires user decisions/operations (here referred to as "manual") and, thus, is not based on standardized and/or automated procedure. As a result, the input file for the same rhodopsin could display non negigible differences when prepared by different users. The reproducibility in the final QM/MM models (called ARM_{manual}) is, therefore, not guaranteed.

In the following, we will explain how, when using the original ARM rhodopsin model building protocol (Sect. 3.1.1), the input is manually generated and how substantially full automation is achieved with the *a*-ARM protocol which features the *input file generator* (Sect. 3.1.2). In Sect. 3.2, we instead describe the *QM/MM model generator* that is employed by both protocols.

3.1 Input File Generator

(Most of the content of Sects. 3.1.1 and 3.1.2 is reproduced/adapted with permission from *J. Chem. Theory Comput.* 2019, 15, 3134-3152. Copyright 2019 American Chemical Society). The general ARM input is composed of two files: a file in PDB format (PDB^{ARM}) containing the 3D structure of the rhodopsin under investigation, and a file containing the list of amino acid residues forming its chromophore cavity (see red residues in Fig. 2). More specifically, the PDB^{ARM} file contains information on the selected monomeric chain structure including the chromophore and excluding membrane lipids and ions; the protonation states for all the internal and surface ionizable amino acid residues; suitable external counterions (Cl^-/Na^+) needed to neutralize both IS and OS protein surfaces; and crystallographic/comparative waters (see also Fig. 1).

3.1.1 ARM: Manual Input File Preparation

In the original ARM protocol [91], the input file preparation required a few hours user's manipulation of the template protein structure (i.e., a skilled user completes the preparation of an ARM input for a new rhodopsin in not less than 3 hours and after taking a series of decisions based on his/her chemical/physical knowledge and intuition). In fact, the procedure employed to generate an ARM input comprises four steps, each containing user decisions and/or manipulations:

- *Step 1: Initial setup of the working PDB file. Selection of the protein chain and residue rotamers.*
  The initial step for the manual generation of a ARM input (PDB^{ARM}) consists in the modification of the PDB file that contains the original rhodopsin crystallographic/comparative structure, in order to delete irrelevant information concerning to unwanted protein chains, membrane lipids, and nonfunctional ions. Moreover, the identification of multiple side-chain rotamers for a single amino acid residue, and their subsequent selection, is based on a qualitative user decision without following a rigorous procedure.

Furthermore, a related chemical manipulation that also requires the user decision is the identification/selection of the residue number corresponding to the retinal chromophore for the posterior definition of the Lys-QM subsystem illustrated in Fig. 2.

Finally, the coordinates of the chosen protein chain (chain A is used by default), crystallographic/comparative waters, selected side-chain rotamers and retinal chromophore are written in the new PDB^{ARM} file, with the proper manual adjustment of atoms and residues numbering.

- *Step 2: Definition of the retinal chromophore cavity.*

  Although this step does not require user decisions, since the assignment of the amino acid residues belonging to the retinal chromophore cavity is fully carried out via the Web-based tool CASTp [25] the cavity file that contains the residue number of these amino acids is manually created. In cases when the main (MC) and second (SC) chromophore counterions (the SC is a second negatively charged residue located in the chromophore cavity at a larger distance from the protonated C=N group with respect to MC that may exist in certain rhodopsins) are not identified by CASTp as part of the cavity, these are manually included.

- *Step 3: Assignment of the protonation states of the ionizable residues.*

  The assignment of standard/nonstandard protonation states for the ionizable amino acid residues (*i.e.*, Asp, Glu, Lys, His, Arg) in the selected protein chain, is performed by using the PROPKA3.1 [102] software. First, the PDB^{ARM} file is processed by manually executing the PROPKA3.1 command-line. Then, the output file that contains information on the calculated (pK_a^{Calc}) and model (pK_a^{Model}) pK_a values and burying percentage for each amino acid residue is manually analyzed. For instance, the user has to manually check if the difference between the calculated and model pK_a values of each residue is higher than a shift value established in the range 1.5–2.0. Moreover, the burying percentage should be higher than 55%. Remarkably, there is no general consensus for the shift value and burying percentage.

  Then, the identified residues with nonstandard protonation states are manually labeled in the PDB^{ARM} file as follows: Asp → ASH, Glu → GLH, His → {HID, HIE, HIP}, Lys→ LYD

- *Step 4: Neutralization of the protein with external counterions. Selection of the initial location of the protein external counterions.*

  This step is the most time-demanding and it is subject to personal user choices, since it requires visual inspection and manual manipulation of the protein structure. Again, the final type and coordinates of the selected counterions are manually added to the PDB^{ARM} file.

A detailed list of the tasks involved in each of the above described steps is provided in Fig. 4. In this figure, the {M} mark symbolizes manual decision with manual action and the N.A. mark represent action not applicable in ARM protocol since this action was only introduced with the *a*-ARM protocol.

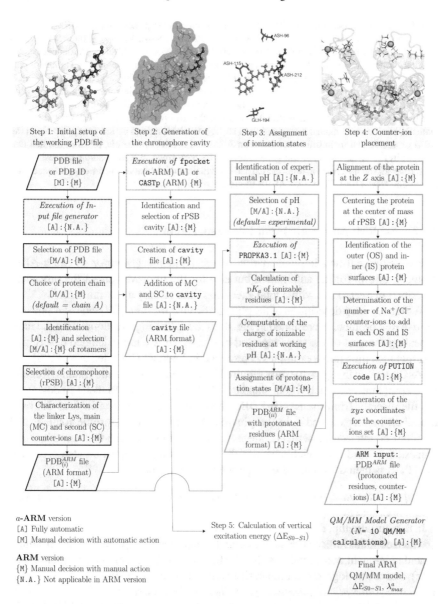

Fig. 4 (Continued)

◄ **Fig. 4 Input file generator workflow**. After the selection of the protein chain, a-ARM generates either automatically (a-ARM$_{default}$) or semiautomatically (a-ARM$_{customized}$) the ARM input files with complete information on the side-chain rotamers, water molecules, protonation states, external counterion placement, and chromophore cavity. The parallelograms represent input or output data, the continuous line squares refer to processes or actions, and the dashed line squares mean software executions. For the a-ARM [106] protocol, the [A] mark symbolizes fully automation, the [M] mark represent manual decision with automatic action and the [M/A] mark indicates situation that may be either manual or automated. For the ARM [91] protocol, the {M} mark symbolizes manual decision with manual action and the {N.A.} mark represents action not applicable in the ARM protocol. The main difference between ARM and a-ARM protocols, relies in the fact that whereas in the latest the input file generation can be performed in a fully automatic fashion, in the former all the described actions are carried out via manual decisions/operations. Adapted with permission from [106]. Copyright 2019 American Chemical Society

3.1.2 a-ARM: Automatic Input File Preparation

The development and implementation of the *input file generator* in the a-ARM protocol, allows for the automation but also for the improvement of the procedure described in Sect. 3.1.1 with a total execution time of only a few minutes.

A Step-by-Step description of the pipeline driving the *input file generator* phase in the a-ARM protocol is presented in Fig. 4. In the following, we summarize the most appealing features of the steps reported in Fig. 4, while a more detailed description that includes illustrative examples is provided in Pedraza-González et al. [106]

As illustrated in Fig. 4, the only required input is either a crystallographic structure (or a comparative model) file in standard PDB format, or the PDB ID to download the file directly from the RCSB PDB. Therefore, user decisions such as the ones described in Steps 1 and 4 in the previous subsection (e.g., residue rotamer and in initial counterion placement), are not required here. In case that mutations are requested, a third file which specifies the list of mutations must be also provided. This PDB is then manipulated through the following steps:

- *Step 1: Automatic Identification of Protein Chain, rPSB, Chromophore Bounded Lys, MC, and SC.*
  In this initial setup, the program generates a cleaned file in PDB format (PDB$_i^{ARM}$), which contains information on the structure of the selected protein chain (a), amino acid side-chain conformations (b), rPSB, MC and SC chromophore counterions (c), and crystallographic/comparative water molecules (see Fig. 5).
- *Step 2: Automatic Generation of the Chromophore Cavity.*
  One crucial step for the systematic and congruous construction of suitable QM/MM models for rhodopsins, is the automatic identification of the amino acid residues forming the chromophore cavity. In a-ARM, the command-line Fpocket [76] software recognizes the residues of the chromophore cavity based on the well-known theory of tessellation and alpha spheres, that is built on top of the package Qhull [76]. To this aim, a-ARM automatically executes the Fpocket software, over the previously generated PDB$_{(i)}^{ARM}$, using the default options/parameters [77].

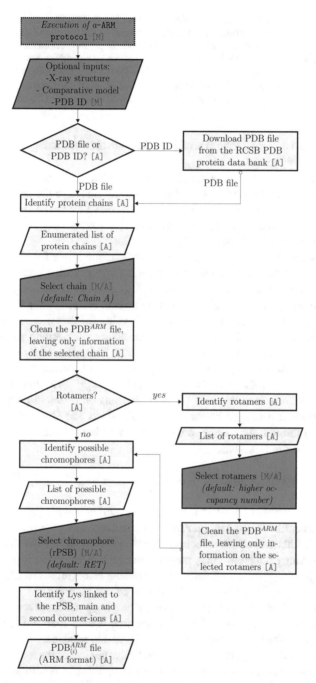

Fig. 5 (Continued)

◄ **Fig. 5 Step 1. Initial preparation of the input file.** The process initiates with an input 3D rhodopsin structure in PDB format and ends with a cleaned PDB structure in specific PDB ARM format ($PDB_{(i)}^{ARM}$). The parallelograms represent input or output data, the continuous line squares refer to processes or actions, and the dashed lines mean software executions. For the *a*-ARM [106] protocol, the [A] mark symbolizes fully automation, the [M] mark represent manual decision with automatic action. The red filled boxes represent tasks in which the user may interact with the program. Adapted with permission from [106]. Copyright 2019 American Chemical Society

As illustrated in Fig. 6, several protein pockets are obtained along with their scores sorted from highest to lowest.

Then, the pocket selected as the chromophore cavity is identified based on two criteria: the pocket i) contains the Lys covalently linked to the rPSB and ii) has assigned the highest score. Subsequently, a new `cavity` file [A] is automatically generated, containing the identifiers of those residues forming the cavity.

- *Step 3: Automatic Assignment of Ionization States.*

The assignment of standard/nonstandard protonation states for all the ionizable amino acid residues in their particular protein environment, is performed through the calculation of the pKa value (see Fig. 7) at a given pH (e.g., crystallographic pH). In such a basic approach, for each amino acid residue with a titratable group (*i.e.*, Asp, Glu, Lys, Arg, His) a model pK_a value (pK_a^{Model}) is computed, [147] and interpreted as the pK_a exhibited when the other protein amino acid side-chains are in their neutral protonation state. This pK_a^{Model} is affected by the interaction between the residue and its actual environment, causing a variation from the model value to the real pK_a value (see Eq. 2) called pK_a^{Calc}. The magnitude of this variation, known as shift value (ΔpK_a), depends on different parameters (i.e., the presence of hydrogen bonds, desolvation effects and Coulomb interactions) modulated through the degree to which the ionizable residue is "buried" within the protein chain [19, 102].

$$pK_a^{Calc} = pK_a^{Model} + \Delta pK_a; \ \Delta pK_a = pK_a^{Calc} - pK_a^{Model} \qquad (2)$$

To start the procedure, *a*-ARM identifies the experimental pH (i.e., crystallization conditions available in the initial PDB structure file) making the pH selection automatic [A]. Alternatively, the user may type the pH value [M]. After the assignment of the working pH a preliminary preparation of the $PDB_{(i)}^{ARM}$ file is performed that consists in adding the coordinates of the missing heavy atoms of chain residues, as well as the hydrogen atoms. This is carried out by using the PDB2PQR [21, 22] software. Then, both the pK_a^{Calc} and the burying percentage are computed via the PROPKA software [102]. These information is then analyzed for the correct assignment of the protonation states of the ionizable residues.

More specifically, the parameter used to identify the state of the ionizable residues is the side-chain ionization equilibrium. Such equilibrium is estimated by inserting both the pK_a^{Calc} value and the established working pH in the Henderson–Hasselbalch equation [109], which describes the relationship between the pH and the pK_a and the equilibrium concentrations of dissociated [A⁻] and non-

Fig. 6 Step 2. Automatic generation of chromophore cavity. The process initiates with the $PDB_{(i)}^{ARM}$ structure and ends with the `cavity` file that contains the residues forming the chromophore cavity. The parallelograms represent input or output data, the continuous line squares refer to processes or actions, and the dashed lines mean software executions. For the a-ARM [106] protocol, the [A] mark symbolizes fully automation. The code does not require the user's interaction during its execution. Adapted with permission from [106]. Copyright 2019 American Chemical Society

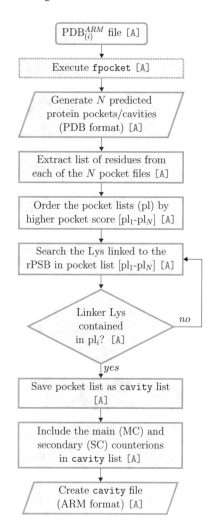

dissociated acid [HA], respectively [93, 109, 116]:

$$pH = pK_a^{Calc} + \log \frac{[A^-]}{[HA]} \tag{3}$$

The charges of the positive and negatively ionizable residues are then deduced from Eq. 3 using the following approximated rules [93]:

$$\ulcorner Q^- \urcorner = \frac{(-1)}{1 + 10^{-(pH - pK_a^{Calc})}}; [93] \qquad \text{for Asp and Glu} \tag{4}$$

and

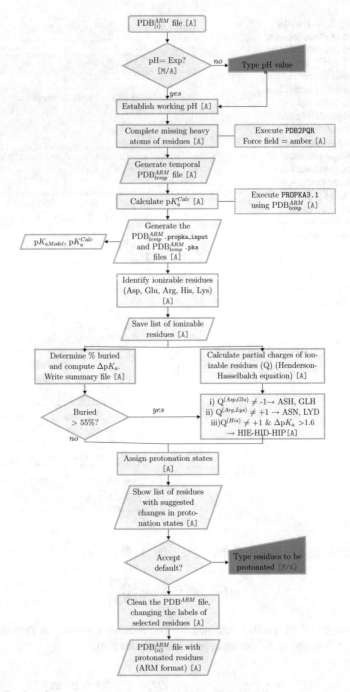

Fig. 7 (Continued)

◀ **Fig. 7 Step 3. Automatic assignment of ionization states.** The process initiates with the PDB$_{(i)}^{ARM}$ structure with standard protonation states and ends with a modified PDB$_{(i)}^{ARM}$ that contains the residues with nonstandard protonation states. The parallelograms represent input or output data, the continuous line squares refer to processes or actions, and the dashed lines mean software executions. For the a-ARM [106] protocol, the [A] mark symbolizes fully automation, the [M] mark represents manual decision with automatic action. The red filled boxes represent tasks in which the user may interact with the program. Adapted with permission from [106]. Copyright 2019 American Chemical Society

$$\ulcorner Q^+ \urcorner = \frac{(+1)}{1 + 10^{+(pH - pK_a^{Calc})}} ; [93] \qquad \text{for Arg, Lys and His.} \qquad (5)$$

where $\ulcorner Q^+ \urcorner$ and $\ulcorner Q^- \urcorner$ are integers obtained by rounding the decimals using the "round half to even" convention.

- *Step 4. Automatic Counterion Placement.*

a-ARM employs a novel procedure for the automatic, systematic, and congruous generation and placement of the external counterions (i.e., Na$^+$/Cl$^-$), totally removing the user manipulation described in Step 4 of Sect. 3.1.1 (see Fig. 8). The main aim is to identify the type (i.e., Cl$^-$ and/or Na$^+$) and number of counterions needed to setup as neutral the protein environment charge, based on the actual charges of the OS and IS surfaces. The OS and IS surfaces are automatically [A] defined by orienting the protein along its z axis, as schematized in Fig. 9 for the case of bovine rhodopsin (Rh) [101]. From now on, this figure will be used as a reference to illustrate the procedure for the counterion placement. Technically, the protein coordinates found in the PDB$_{(ii)}^{ARM}$ file are first centered at the protein center of mass (xyz^{cm}). Subsequently, the obtained set of coordinates are rotated through the main rotational axis that is aligned with the z-axis, via the Orient utility of the VMD [56] software.

Then, the new set of coordinates is re-centered at the center of mass of the rPSB. The above-described coordinates transformation is performed in order to select the residues belonging to either the OS or IS surfaces. To this aim, an imaginary plane orthogonal to the z-axis, that contains the z coordinate of the NZ protonated Schiff base atom (z^{PSB}) of the rPSB moiety, is defined based on the magnitude of the z value to divide the protein environment into two halves (i.e., IS and OS surfaces). Accordingly, the OS surface contains the residues with z value larger than z^{rPSB}, while the IS contains the residues with z lower or equal to z^{rPSB}. Once the IS and OS surfaces are defined, their respective charges (Q$_{IS}$, Q$_{OS}$) are computed as the difference between the number of negatively charged (Asp, Glu and crystallographic Cl$^-$ anions) and positively charged (Arg, Lys, His, and crystallographic Na$^+$ cations) residues. With this information, the type and number of counterions required to neutralize the net charge of each surface independently are automatically identified.

Finally, to obtain the coordinates of the counterions, the PUTION software optimizes their positions on the basis of the Coulomb's law [88], via the generation of an electrostatic potential grid around all charged residues that excludes points

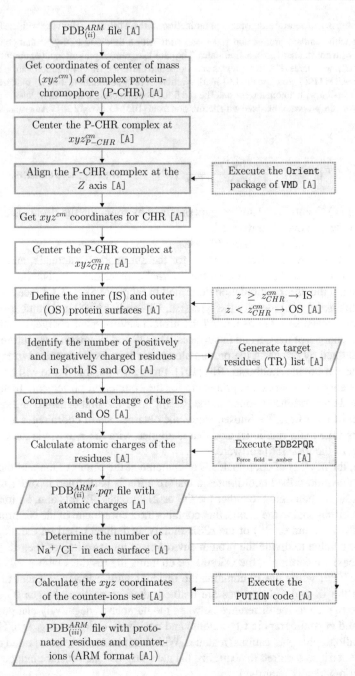

Fig. 8 (Continued)

◄ **Fig. 8   Step 4. Automatic selection of external counterion placement.** The process initiates with the PDB$_{(ii)}^{ARM}$ structure, that contains the residues with nonstandard protonation states, and ends with the PDB$_{(iii)}^{ARM}$ file that contains the external counterions (Na$^+$/Cl$^-$) needed to neutralize both inner (IS) and outer (OS) surfaces. The parallelograms represent input or output data, the continuous line squares refer to processes or actions, and the dashed lines mean software executions. For the a-ARM [106] protocol, the [A] mark symbolizes fully automation. The code does not require the user's interaction during its execution. Adapted with permission from [106]. Copyright 2019 American Chemical Society

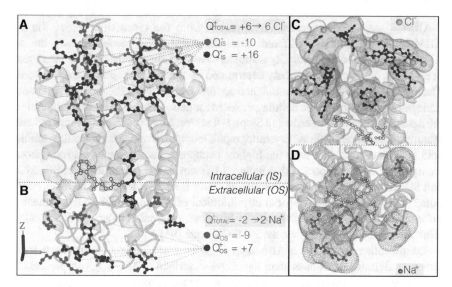

Fig. 9   External counterion placement for bovine rhodopsin. Schematic representation of the procedure for the definition of the number and type of external counterions needed to neutralize the IS (**a**) and OS (**b**) surfaces of bovine rhodopsin. We also illustrate the grid generated by the PUTION code to calculate the coordinates of the Cl$^-$ counterions in the IS (**c**) and the Na$^+$ in the OS (**d**). The negatively and positively charged residues are illustrated as red and blue sticks, respectively, and the Na$^+$ and Cl$^-$ counterions as blue and green spheres, respectively. The net charge of the IS surface is $Q_{IS} = +6$, resulting from 16 positively charged and 10 negatively charged residues, whereas the net charge of the OS is $Q_{OS} = -2$, given by 7 positively charged and 9 negatively charged residues. As a consequence, 6 Cl$^-$ and 2 Na$^+$ must be added to compensate the positive and negative charge of the IS and OS, respectively. "Adapted with permission from [106] Copyright 2019 American Chemical Society."

with distance shorter than 2.0 Å and larger than 8.0 Å from the center of charge of an ionized residue (see Fig. 9c, d).

Therefore, the first counterion is placed on the surface that possesses the highest absolute net charge and the procedure continues alternately by placing the next counterion on the other surface, until both are neutralized. Fig. 9c, d illustrates the external counterions placed for bovine rhodopsin (Rh). The associated coordinates are added to PDB$_{ii}^{ARM}$ to produce the PDB^{ARM} file that is employed as actual input for the second phase of the protocol to build the QM/MM model.

The general driver that links the operations described in Fig. 4, and better explained in Figs. 5, 6, 7, and 8, is a Python3-based package, available as a command-line tool, useful to operate either as a fully automated or as an interactive system for the customized generation of the ARM input. Therefore, the input data required for generating complete PDB^{ARM} and cavity files may be provided via either default (i.e., automatically or [A]) choices or by answering specific questions (i.e., manually [M]) in the command-line of the terminal window.

Based on the selected options, a-ARM is divided in a-ARM$_{default}$ and a-ARM$_{customized}$ approaches. As apparent from the discussion below, the a-ARM$_{customized}$ model is different from the ARM$_{manual}$ model. In fact, the a-ARM$_{customized}$ model variability is subject to well- defined constraints and features, in all cases, a chromophore cavity determined by the Fpocket and not CASTp tool. The a-ARM$_{default}$ is a fully automated approach resulting in the top input preparation speed for the systematic building of ARM models. This is achieved by employing the default parameters described in Steps 1–4 above. More specifically, these models feature: monomeric chain A; crystallographic/comparative water molecules; amino acid side-chain rotamers with the highest occupancy factor; default chromophore cavity obtained with Fpocket that includes Lys covalently linked to the rPSB, MC, and SC; standard/nonstandard protonation states for all the ionizable residues computed at crystallographic pH (or at physiological pH 7.4 in case of no experimental information available), Hid tautomer of histidine; external counterions coordinates (Na^+ and/or Cl^-) predicted via the PUTION code [91].

On the other hand, the a-ARM$_{customized}$ is a semiautomatic approach that, although significantly slower than the above described a-ARM$_{default}$, presents a major degree of accuracy in terms of reproducing the experimental ΔE_{S0-S1} trends, due to customization of certain parameters. In this regard, the user can customize the a-ARM input by selecting in the command-line terminal specific parameters concerning to Steps 1 and 3. For instance, it is possible to select the protein chain, decide the protonation states of ionizable residues by typing their residue number identification and select the tautomer of the histidine. The advantage of these two approaches is that, regardless of the user degree of chemical expertise or computational facilities employed, reproducible inputs, and consequently reproducible QM/MM models are guaranteed when the same initial information (e.g. the PDB file) is selected. This was not the case with the a-ARM$_{manual}$ model.

3.2 QM/MM Model Generator

As illustrated in Fig. 3b, Phase II drives the automatic generation of the QM/MM models. The corresponding procedure is the same for both the ARM and a-ARM protocols. Accordingly, 10 independent QM/MM models are generated, either of the ARM$_{manual}$, a-ARM$_{default}$ or a-ARM$_{customized}$ type according to the employed specific input preparation scheme. These $N = 10$ independent models (replicas) are employed to simulate and explore the possible relative conformational phase space of the cavity residue side-chains and retinal chromophore. Indeed, during the Molecular Dynamics-based (MD) step (see later action 2), 10 parallel MD runs 1 ns long each are performed, each starting with a different, randomly chosen seed to warrant independent initial conditions. In each run, the frame closest to the average of the simulation is then selected as the starting geometry for constructing the corresponding QM/MM model.

Although this is still an approximate solution, previous tests have shown that when considering the limited portion of the flexible a-ARM model (i.e., cavity residues and chromophore), 10 repetitions are an acceptable compromise between computational time and stability of the result (i.e., stability of the computed side-chain conformations and hydrogen bond network). On this subject, [90] have shown that for a set of 3 phylogenetically diverse rhodopsins (Rh, SqRh, and ASR_{13C}), the sensitivity of the computed average vertical excitation energy values as a function of the number of replicas ($N = 1, 5, 10, 15,$ and 20) become substantially stable, with variations lower than $1.0 \, kcal \, mol^{-1}$ beyond 10 replicas. The three steps shown in Fig. 3b correspond to the following actions:

(1) optimization of crystallographic water molecules and addition of hydrogen atoms to polar residues, using DOWSER [151];

(2) molecular mechanics energy minimization and simulated annealing/molecular dynamics relaxation at 298 K on all side-chains of the chromophore cavity; during the MD computation also the retinal chromophore is allowed to move. This MD simulation is repeated $N = 10$ times with randomly chosen seeds, as previously exposed. Notice that, during the molecular dynamics run, the chromophore is represented by using an MM parametrization and partial charges computed as AMBER-like RESP charges which are specific for each employed isomer of the chromophore (e.g., 11-*cis*, all-*trans*, and 13-*cis* rPSB). The default heating, equilibration and production times for the MD are of 50, 150, and 800 fs, respectively, for a total length of 1 ns. The MD is performed employing GROMACS [111].

Finally, each of the previously generated replicas undergoes

(3) QM/MM geometry optimization and single-point energy calculation at the 3-state state-average CASPT2//CASSCF(12,12)/6-31G(d)/AMBER levels, respectively [5]. The QM/MM step makes use of micro-iterations that provide quicker convergence, lower energies, and a more realistic description of chromophore–environment interactions. In addition, suitable level shiftings are employed during the CASSCF and CASPT2 calculations, so as to minimize the possibility of convergence failure via state mixing and intruder state problems. For further details on the QM/MM approach see Sect. 2.1. This part is carried out by employing the interface between [Open]Molcas [6] and TINKER [113]. The CASPT2//CASSCF/6-31G*/MM treatment has been extensively investigated for photobiological studies and its limitations are well understood. As previously documented [90], the rather limited ca. $3–4 \, kcal \, mol^{-1}$ error in excitation energy reported in several studies for this level of theory is somewhat due to error cancellations associated with the limited quality of CASSCF/6-31G* equilibrium geometries. Therefore, the different properties computed by a-ARM are expected to be affected by systematic error cancellations. However, we should stress that, according to the philosophy of the a-

ARM protocol, the main focus of ARM is the ability to reproduce observed trends in vertical excitation energies (i.e., the sign and magnitude of the individual differences with respect to experimental data).

The final output consists of 10 replicas of equilibrated QM/MM models of the type described in Sect. 2.1 and, for each replica, the vertical excitation energy values between S_0 and the first two singlet excited states S_1 and S_2 is provided. A detailed explanation of Phase II of the a-ARM protocol is given in Melaccio et al. [91] and Pedraza-González et al. [106] Finally, it is worth to mention some shortcomings of our simplified model with respect to more sophisticated QM/MM models (see Table 1). For instance, (1) lack of the environment description (membrane + solvent), (2) missing polarizable force field, (3) not flexible backbone and non-cavity side-chains, and (4) approximated ionization states. Our current research is aimed, also, to overcome these points, while maintaining reasonable computational costs, or estimating the errors due to them.

3.3   Benchmark: ARM versus a-ARM

(Most of the content of Sect. 3.3 is adapted with permission from *J. Chem. Theory Comput.* 2019, 15, 3134-3152. Copyright 2019 American Chemical Society). Above we have described the general workflow of the ARM and a-ARM protocols. We have also revised the main features of the algorithms driving the a-ARM two phases, *input file generator* (see Sect. 3.1) and *QM/MM model generator* (see Sect. 3.2). In order to provide information on the accuracy of the ARM models, we now review the results of tests performed to evaluate the predictability of trends of vertical excitation energies (ΔE_{S0-S1}). To this aim, we collated in Fig. 10 computed ARM$_{\text{manual}}$ (yellow circles), a-ARM$_{\text{default}}$ (green triangles) and a-ARM$_{\text{customized}}$ (orange squares) ΔE_{S0-S1} values, forming a set of 25 wild-type and 19 mutant rhodopsins, coming from the animal (A) and microbial (M) families (Type II and Type I rhodopsins, respectively) [28], for which experimental values (blue down triangles) are available.

For each rhodopsin, the ΔE_{S1-S0} values were obtained as the average of the energy difference between the S_0 and S_1 states in the 10 replicas generated for each model (see Sect. 3.2). Consistently with the ARM protocol, the ΔE_{S1-S0} values reported in Fig. 10 are, for all entries, the average of the excitation energies of 10 equilibrium model replicas (see Sect. 3.2). The full benchmark set features values ranging from 458 nm (62.4 kcal mol^{-1}, 2.71 eV) to 575 nm (49.7 kcal mol^{-1}, 2.15 eV). Such a wide range provides information on the method accuracy while the rhodopsin diversity provides information on the transferability and general applicability of the generated models.

In Fig. 10, the computed and observed values are compared after assuming that the observed ΔE_{S0-S1} (down triangles) values can be derived from the experimental λ_{max}^a via the equation $\Delta E_{S0-S1} = hc/\lambda_{\text{max}}^a$. Figure 10 is divided into four different regions: m-set, a-set, Rh−mutants set, and bR−mutants set (see caption of Fig. 10). The m-set and Rh−mutants set are employed to compare the performance of ARM

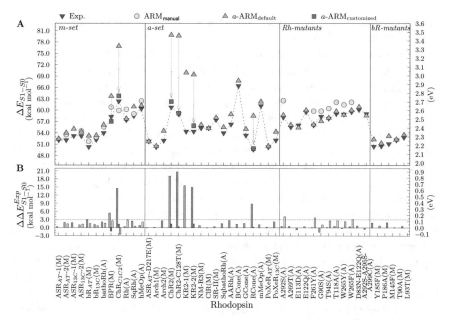

Fig. 10 Benchmarks. a Vertical excitation energies ($\Delta E_{S1\text{-}S0}$) computed with a-ARM$_{\texttt{default}}$ (up triangles) and a-ARM$_{\texttt{customized}}$ (squares), [106] along with reported ARM$_{\texttt{manual}}$ [91] (circles) and experimental data (down triangles). S_0 and S_1 energy calculations were performed at the CASPT2(12,12)//CASSCF(12,12)/AMBER level of theory using the 6-31G(d) basis set. The calculated $\Delta E_{S1\text{-}S0}$ values are the average of 10 replicas. **b** Differences between calculated and experimental $\Delta E_{S1\text{-}S0}$ ($\Delta\Delta E_{S1\text{-}S0}^{\text{Exp}}$). Values presented in kcal mol^{-1} (left vertical axis) and eV (right vertical axis). The m-set corresponds to WT rhodopsins forming the original benchmark set for the ARM protocol; a-set introduces new rhodopsins to the benchmark set of the a-ARM protocol; $Rh-mutants$ set contains mutants of bovine Rhodopsin (Rh) belonging either to the benchmark set of the ARM or a-ARM protocols; and $bR-mutants$ set contains mutants of bR evaluated with the WebARM interface [107] (see Sect. 3.5). ASR: *Anabaena* sensory rhodopsin [1XIO [141]], bR$_A$T: Bacteriorhodopsin all-*trans* [6G7H [98]] and 13-*cis* [1X0S [97]], BPR: Blue proteorhodopsin [4JQ6 [115]], Rh: Bovine rhodopsin [1U19 [101]], ChR$_{C1C2}$: Chimaera channelrhodopsin [3UG9 [64]], SqRh: Squid rhodopsin [2Z73 [94]], hMeOp: Human melanopsin [template 2Z73 [94]], ASR$_{AT}$-D217E: *Anabaena* sensory rhodopsin D217E [4TL3 [23]], Arch1: Archaerhodopsin-1 [1UAZ [26]], Arch2: Archaerhodopsin-2 [3WQJ [71]], ChR2: Channelrhodopsin-2 [6EID [142]], ChR2-C128T: Channelrhodopsin-2 C128T [6EIG [142]], KR2: *Krokinobacter eikastus* rhodopsin 2 [3X3C [65]], NM-R3: Nonlabens marinus rhodopsin-3 [5B2N [52]], ClR: Nonlabens marinus rhodopsin-3 [5G28 [67]], SRII: Sensory rhodopsin II [1JGJ [79]], SqbathoRh: Squid bathorhodopsin [3AYM [95]], AARh: Ancestral archosaur rhodopsin [template 1U19 [101]], BCone: Human blue cone [template 1U19 [101]], GCone: Human green cone [template 1U19 [101]], RCone: Human red cone [template 1U19 [101]], mMeOp: Mouse melanopsin [template 2Z73 [94]], PoXeR: *Parvularcula oceani* Xenorhodopsin [template 4TL3 [23]]. "Adapted with permission from [106]. Copyright 2019 American Chemical Society."

and a-ARM protocols, while the remaining sets focus on the performance of the a-ARM protocol exclusively.

In the following, we compare the ARM and a-ARM protocols, both in terms of trends (Fig. 10a), as well as in terms of individual errors (Fig. 10b) as the $\Delta \Delta E_{S0-S1}^{Exp}$ value. The established error bar for the a-ARM protocol in terms of ΔE_{S1-S0} is reported as +3.0 kcal mol^{-1}, where the calculated value is blue-shifted with respect to the experimental one [91, 106]. We have, however, to warn the reader that the comparison between the ARM$_{manual}$ and a-ARM$_{default}$ models does not appear to be informative since the manual ARM protocol does not have an *input file preparation* phase (see Sect. 3.1.1), thus all the steps needed to generate the ARM input were subject to human decisions and, thus, the reproducibility of ARM$_{manual}$ models is not guaranteed. In practice, the original ARM protocol has a larger level of flexibility and a user may, in principle, take advantage of this by trying different construction parameters to match the experimentally observed value. Instead, a partial comparison between models ARM$_{manual}$ and a-ARM$_{customized}$ is more fair (see m-set), considering that in both cases it is possible to take different decisions to customize the ARM input. Here, as shown in Fig. 10, it is evident that even when the ARM$_{manual}$ models produce values in good agreement with the experimental trend, a slight improvement is obtained when the ARM input is constructed using the *input file generator*. This is mainly attributed to a more proper placement of the external counterions (see Sect. 3.1.2) [106]. Notice, while a-ARM$_{customized}$ models are constructed semiautomatically, they can, in contrast with the ARM$_{manual}$ models, still be reproduced provided the same parameters are selected by different users.

We now focus on analyzing the trends generated using the fully automated a-ARM$_{default}$ approach, starting with the choice of rotamers from a given crystallographic structure. In Fig. 10, two default models are reported for *Anabaena* sensory rhodopsin (ASR), both ASR$_{AT}$ (ASR$_{AT}$-1 and ASR$_{AT}$-2) and ASR$_{13C}$ (ASR$_{13C}$-1 and ASR$_{13C}$-2) conformations, as well as for KR2 (KR2-1 and KR2-2). In the case of

Fig. 11 a-**ARM models for KR2 [PDB ID 3X3C].** Conformational (the occupancy factor of the rotamers Asp-116 and Gln-157 are presented in parentheses) and ionization state variability. "Adapted with permission from [106] Copyright 2019 American Chemical Society."

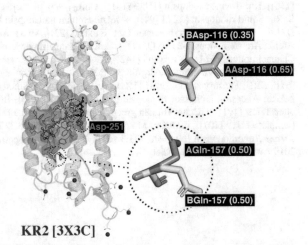

KR2 [3X3C]

ASR, the 1XIO [141] X-ray structure employed as a template to build the ASR_{AT} and ASR_{13C} ARM models contains two possible rotamers for both residues RET-301 (all-*trans* and 13-*cis* rPSB) and Lys-310 (ALys and BLys). These residues correspond to the retinal chromophore and the Lys covalently linked, respectively. In the X-ray structure, each rotamer in each pair (i.e., RET-301 or Lys-310) exhibit an occupancy number of 0.5, that is interpreted as 50% of probability [131, 141]. For this reason, it is not feasible to select the rotamer on the basis of its occupancy number and, as an alternative, a-ARM$_{\text{default}}$ generates four QM/MM models: the all-*trans* configuration using ALys (ASR_{AT}-1) and BLys (ASR_{AT}-2) and the 13-*cis* configuration with ALys (ASR_{13C}-1) and BLys (ASR_{13C}-2). A similar approach was employed for the generation of ARM manual models in previous studies [1, 91, 103, 131, 141]. As observed in Fig. 10, for the same retinal configuration the performance of the models generated with both Lys conformations is similar in terms of ΔE_{S0-S1}. However, for both all-*trans* and 13-*cis* configuration, the model that yields a ΔE_{S0-S1} value closest to the experimental one was selected (i.e., ASR_{AT}-2 and ASR_{13C}-1). Figure 11 illustrates the fact that, in the case of KR2 generated from the template 3X3C [65], two residues exhibit different rotamers. The most relevant case is the MC Asp-116, that presents two conformations, namely, AAsp and BAsp, labeled with occupancy numbers 0.65 and 0.35, respectively. In addition, the residue Gln-157 that is part of the environment subsystem (i.e., fixed during the QM/MM calculations) exhibits two conformations (AGln and BGln) both with occupancy number 0.5. Similar to the case of ASR, according to the occupancy numbers a-ARM$_{\text{default}}$ chooses the rotamer AAsp-116 and generates two models relative to Gln-157: KR2-1, which includes AAsp-116 and AGln-157, and KR2-2, which includes AAsp-116 and BGln-157. As further discussed in Ref. [106], after the analysis of the calculated and observed ΔE_{S0-S1} values KR2-2 is chosen. Considering the method's default error bar, and the choice of models when more than one model is prepared, it is possible to analyze the general trend of a-ARM$_{\text{default}}$. Figure 10A (green triangles) shows that the experimental trend is qualitatively reproduced for most of the models. More specifically, a-ARM$_{\text{default}}$ performs well, showing values within the systematic error (i.e., 3.0 kcal mol^{-1}) in 38 out of 44 cases. Among the identified outliers, BPR and ChR$_{C1C2}$ in the m-set show deviations of 5.4 kcal mol^{-1} and 14.5 kcal mol^{-1}, respectively. In the a-set, ChR2, ChR2-C128T, KR2, and RCone are off the observed value, with deviations between 9 and 21 kcal mol^{-1}. Since the ARM models generated using a-ARM$_{\text{default}}$ for the identified outliers (BPR, ChR$_{C1C2}$, ChR2, ChR2-C128T, and KR2) are unable to provide values inside the experimental trend in ΔE_{S1-S0}, we made use of the a-ARM$_{\text{customized}}$ approach to evaluate a different setup for the protonation states, keeping the other default parameters (e.g., the default chain A, default chromophore cavity, default rotamer choices, etc), as shown in Table 2. Considering that, in certain cases, a-ARM$_{\text{default}}$ does not correctly assign the correct residue charge, to do a more rational assignment of the protonation states with a-ARM$_{\text{customized}}$, we use as an example the four models reported in Table 2. For instance, the default model predicts that, for KR2, the charge of the rPSB is stabilized by a counterion complex comprising two aspartic acid residues, Asp-116 and Asp-251. Based on the protocol outcome and on recent experimental evidence [65, 134], residues Asp-116 and

Asp-251 are identified to be the chromophore MC and SC, respectively. As shown in Table 2, at the crystallographic pH 8.0 the a-ARM$_{\text{default}}$ model suggests that both residues are negatively charged, suggesting the presence of two negative charges around the rPSB chromophore, that would outbalance the single positive charge of the later. In the customized model, Asp-251 is, instead, protonated (i.e., neutral) to counterbalance the charge in the vicinity of the rPSB. These results on KR2 point out a drawback of the a-ARM$_{\text{default}}$ approach of the *input file generator* phase for the assignment of protonation states for residues surrounding the rPSB. However, they also present a case in which the a-ARM$_{\text{customized}}$ approach is useful to explore alternative choices of protonation states based on either chemical reasoning or experimental information. A similar analysis was performed also for the other outliers, and the differences in protonation states are reported in Table 2. For instance, in RCone the protonation states of Glh-83 and Glu-110 were exchanged, leading to a different MC, similarly to bR$_{AT}$ with the protonation states of Asp-85 and Asp-212. In BPR, a different protonation state for Glh-124 lead to a different stabilization of the electronic states. As expected, a certain degree of customization led to a considerable improvement of the computed values. Indeed, the a-ARM$_{\text{customized}}$ models (orange squares in Fig. 10) feature ΔE_{S0-S1} values in agreement with the experimental trend, and $\Delta\Delta E_{S0-S1}$ lower than 3.0 kcal mol^{-1}.

In order to quantify the parallelism between the calculated and observed trends in ΔE_{S1-S0} and, in turn, contrast the performance of the a-ARM$_{\text{default}}$ and a-ARM$_{\text{customized}}$ approaches, the trend deviation factor ($||$Trend Dev.$||$) is introduced. Such factor characterizes the ability of the QM/MM ARM models to reproduce the changes in ΔE_{S1-S0} observed experimentally for a set of rhodopsins, taking as a reference value the $\Delta\Delta E_{S1-S0}$ of a reference rhodopsin (i.e., Rh). To obtain the $||$Trend Dev.$||$ value for both a-ARM$_{\text{default}}$ and a-ARM$_{\text{customized}}$ approaches, the first step is to compute the variation in observed ΔE_{S1-S0} produced for each of the $x = 43$ rhodopsins with respect to the reference Rh, by means of the formula $\delta_{x,\text{Exp}}^{Rh,\text{Exp}} \Delta E_{S1-S0}$. Next, a similar procedure is required but now considering the computed ΔE_{S1-S0} of Rh as the reference to be contrasted with the computed value of the remaining $x = 43$ rhodopsins, with the formula $\delta_{x,\text{Calc}}^{Rh,\text{Calc}} \Delta E_{S1-S0}$. When both $\delta_{x,\text{Exp}}^{Rh,\text{Exp}} \Delta E_{S1-S0}$ and $\delta_{x,\text{Calc}}^{Rh,\text{Calc}} \Delta E_{S1-S0}$ quantities are characterized for each rhodopsin in the set, the difference between them is calculated and the $||$Trend Dev.$||$ value is reported in terms of their mean absolute error (MAE) and mean absolute deviation (MAD). The $||$Trend Dev.$||$, for the benchmark set comprising 43 rhodopsins, expressed as MAE $\pm$ MAD are 2.5 $\pm$ 1.2 kcal mol^{-1} (0.11 $\pm$ 0.05 eV) for a-ARM$_{\text{default}}$, and 0.7 $\pm$ 0.5 kcal mol^{-1} (0.03 $\pm$ 0.02 eV) for a-ARM$_{\text{customized}}$. It is evident that a significant improvement is obtained when considering the a-ARM$_{\text{customized}}$ values for KR2-2, ChR$_{C1C2}$, ChR2, ChR2-C128T, BPR, RCone and bR$_{AT}$ instead of the a-ARM$_{\text{default}}$ values.

We now analyze the effect of the source of the initial (input) rhodopsin structure. Currently, there is only a relatively small number of X-ray crystallographic structure of rhodopsins. When such an experimental structure is not available, QM/MM models must be generated by starting with a structure obtained via comparative modeling. This raises the issue of the actual quality of the comparative modeling and which impact such quality might have on the predicted vertical excitation energies. Figure 12 shows how both a-ARM$_{\texttt{default}}$ and a-ARM$_{\texttt{customized}}$ models show that the $\Delta E_{\text{S1-S0}}$ value has a limited dependence from the origin of the rhodopsin structure. Moreover, out of the 6 outliers found with a-ARM$_{\texttt{default}}$, only one was based on a comparative model structure. These results are encouraging as it seems that X-ray structures-based and comparative model-based a-ARM models exhibit a quality similar to the experimental one. Of course, these conclusions are related to the explored benchmark set of rhodopsins and cannot rigorously be generalized. It is however likely that, if correctly executed, comparative models could provide a sufficiently accurate input structure.

The performance of the a-ARM protocol revised above constitutes the first step toward a fast and cheap computational tool, filling the gap between slow experimental structure determination and fast protein expression and characterization achieved by genetic engineering techniques.

Table 2 Setup of the protonation states for a-ARM$_{\text{default}}$ and a-ARM$_{\text{customized}}$ models. The residues with different protonation states are highlighted. Figures adapted with permission from [106]. Copyright 2019 American Chemical Society

	KR2	RCone	BPR	bR$_{AT}$
	Protonation states:			
a-ARM$_{\text{default}}$	– Asp-251	– Ash-80, – Glh-83, – Glu-110, – Hid-178	– Glh-90, – Glh-124	– Ash-85, – Ash-96, – Ash-115, – Glh-194, – Asp-212
a-ARM$_{\text{customized}}$	– Ash-251	– Ash-80, – Glu-83, – Glh-110, – Hid-178	– Glh-90, – Glu-124	– Asp-85, – Ash-96, – Ash-115, – Glh-194, – Ash-212

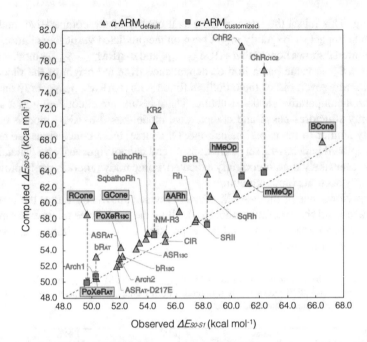

Fig. 12 Performance of a**-ARM in the calculation of vertical excitation energies** (ΔE_{S0-S1}) **for both X-ray crystal structures and comparative models.** Observed *versus* computed values for ΔE_{S0-S1} of wild-type rhodopsins. The symbols refer to: QM/MM models obtained using the a-ARM$_{\text{default}}$ approach (green triangles) and QM/MM models obtained using the a-ARM$_{\text{customized}}$ approach (orange squares). Labels without boxes correspond to X-ray crystal structures, whereas labels in gray boxes correspond to comparative models. S_0 and S_1 energy calculations were performed at the CASPT2//CASSCF(12,12)/Amber level of theory using the 6-31G(d) basis set. The calculated ΔE_{S0-S1} values are the average of 10 replicas

3.4 Advantages and Limitations

(Most of the content of Sect. 3.4 is reproduced/adapted with permission from *J. Chem. Theory Comput.* 2019, 15, 3134-3152. Copyright 2019 American Chemical Society). As previously reported, in spite of the encouraging outcome of the benchmark studies revised above, it is apparent that additional work has to be done for moving to a systematic application of the ARM protocols to larger arrays of rhodopsins:

- *Assignment of the protonation states:* There are two aspects which limit the confidence in the automation of the ionizable state assignment described above. The first is that, due to the fact that the information provided by PROPKA [102] is approximated, the computed pK_a^{Calc} value may, in certain cases, be not sufficiently realistic. The second aspect regards the assignment of the correct tautomer of histidine. This amino acid has charge of +1 when both the δ-nitrogen and ϵ-nitrogen of the imidazole ring are protonated (HIP), while it is neutral when either the δ-nitrogen (HID) or the ϵ-nitrogen (HIE) are deprotonated. a-ARM uses as default the HID tautomer for the automatic assignment [A], or allow the user to choose between the three tautomers for a not automated selection [M]. Therefore, when possible, the user should collect the available experimental data and/or inspect the chemical environment of the ionizable residues including the histidines, and propose the appropriate tautomer [106].

- *Automatic construction of comparative models*: since rhodopsin structural data are rarely available, it would be important to investigate the possibility of building, automatically, the corresponding comparative models. With such an additional tool one could achieve a protocol capable of producing QM/MM models starting directly from the constantly growing repositories of rhodopsin amino acid sequences. This target is currently pursued in our lab.

- *Automatic generation of mutants*: In order to achieve a successful technology for systematically predicting mutant structures, a level of accuracy of the a-ARM models superior to the one currently available is needed. To deal with that, current efforts toward the improvement of the mutations routine are directed to replace SCWRL4 (i.e., a backbone-dependent rotamer library) by a software-based on comparative modeling.

3.5   Web-ARM Interface

The a-ARM protocol described above has been recently implemented at the web level by developing the WebARM interface [107]. Web-ARM is a user-friendly interface written using Python 3 that can be found at the following address: web-arm.org. In this way, the potential user, can access the entire a-ARM protocol features by using an up-to-date browser and, thus, without the need to install software and scripts on his/her computer.

The features of Web-ARM are best described following the four phases into which the interface is divided, as summarized in Fig. 13. During *Phase 1*, the user uploads a rhodopsin structure, in pdb format, and is guided through all the steps necessary to generate input and cavity files. This phase follows the a-ARM protocol as described above (see Sect. 3.1.2) and gives to the user the ability of constructing a-ARM$_{default}$ models, or modify the default choices generating a a-ARM$_{customized}$ model.

The input and cavity files generated during *Phase 1* can be directly transferred to *Phase 3* for submission, or to *Phase 2* to insert point mutations. In the latter phase, the user can provide a list of required mutations, by choosing from drop-down menus which rhodopsin amino acid mutate to what. It is recommended to mutate only cavity amino acids. The mutations are performed as described at the end of Sect. 3.4. For each mutant, the user is subsequently asked to go through *Phase 1*, again, to prepare the corresponding input files.

During *Phase 3*, a *cron* demon takes charge of creating a QM/MM model for each generated input, by running the corresponding 10 repetitions. This follows what is described in Sect. 3.2. The Web-ARM internal driver takes care of performing the

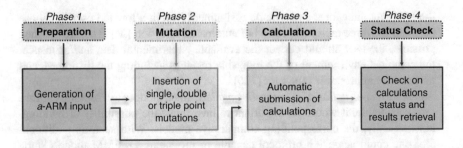

Fig. 13 The four phases of the WebARM interface. The four phases of the Web-ARM protocol

necessary steps, as well as submitting the calculations to the dedicated computational facilities.

Once a QM/MM model has been generated, the user is provided with a summary of all the important data, along with a downloadable file (in compressed format) containing the major output files. Further information and a complete walk-through are provided in a Tutorial that can be accessed/downloaded from the Web-ARM main web page.

Web-ARM is intended as both a research, as well as an teaching tool. In fact, the interface can be employed to systematically screen rhodopsin variants, and thus obtain a qualitative check prior to, e.g., an experimental study. The interface can also be used in teaching and learning activities, e.g., to introduce students to the idea of QM/MM models and corresponding computed data.

4 Applications

In this section, we review few studies based on ARM models of rhodopsins and yielding results which are, in principle, relevant for the development of optogenetics tools [20, 92]. The reader shall be aware of the fact that (unless differently stated) the discussed applications employed the ARM$_{manual}$ models produced using the original ARM protocol. Thus, the preparation of the input to the QM/MM model generator is performed manually according to the criteria already revised in the previous sections. In other words, as explained in Sect. 3, the user must take a number of decisions such as those related to the choice of residue ionization states and external counterion placement. In fact, most of the applications of ARM models presently reported in the literature were performed before the a-ARM protocol was ready. The only exception is the calculation of the TPA spectrum of bovine rhodopsin and IR vision (see Sect. 4.3), in which the a-ARM$_{default}$ model for Rh-WT is employed.

Natural or engineered microbial rhodopsins [28] represent, presently, successful optogenetics tools [87]. For instance, a well-known fluorescent reporter is the rhodopsin *Archaerhodopsin-3* (Arch3) [73] from the archaea *Halorubrum sodomense* found in the Dead Sea. Arch3 is employed to visualize the action potential of neurons with spatial and temporal resolution. Indeed, the fluorescence intensity of Arch3 is sensitive to the cell membrane electrostatics and, more specifically, to the transmembrane electrostatic potential (i.e., voltage) variations associated with the action potential dynamics [36, 43, 50, 53, 82]. However, the Arch3 fluorescence is very dim (*c.a.* 0.001 quantum yield). In fact, three light photons must be absorbed before a detectable photon is emitted. Such photon originates from a photocycle intermediate rather than directly from the DA state (i.e., the thermal equilibrated form of Arch3).

To overcome such limitations, one can search for Arch3 variants [43, 50, 53] or for variants of other microbial rhodopsins and look for [36, 72, 86] enhanced fluorescence intensity resulting from photons emitted after a one-photon absorption process. This requires, as discussed below and schematically illustrated in Fig. 14, to engineer (i.e., through suitably designed mutations) rhodopsin variants featuring: (i) an energy minimum I_f located close to the Franck–Condon (FC) point on the spectroscopic potential energy surface (e.g., usually the S_1 potential energy surface) and (ii) an energy barrier $E_{S1}^{\ddagger}$ preventing the species produced after S_1 relaxation to

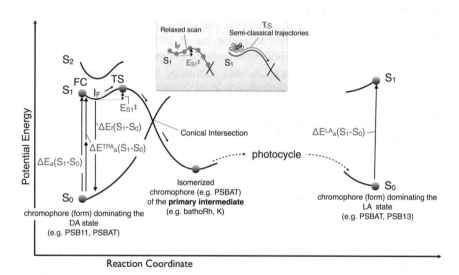

Fig. 14 Properties investigated using ARM models. Schematic diagram displaying the photoiomerization path (including the relevant S_0, S_1, and S_2 energy profiles of a generic rhodopsin. ARM models have been employed to compute the vertical excitation energy for absorption ($\Delta E_a(S_1 - S_0)$), fluorescence ($\Delta E_f(S_1 - S_0)$) and two-photon absorption ($\Delta E_a^{TPA}(S_1 - S_0)$) as well as the excited state reaction barrier ($E_{S1}^{\ddagger}$) associated with emission. In the inset (top center), we schematically illustrate the calculations of an excited state isomerization path providing the $E_{S1}^{\ddagger}$ value via a relaxed scan and of a semiclassical trajectory (this provides, in case of an ultrafast reaction, an estimate of the excited state lifetime τ_S associated to the double-bond isomerization barrier). ARM models have also been employed to investigate the structure and spectroscopy of primary photocycle intermediates (batho and K intermediates) and of photocycle intermediates corresponding to light-adapted states

quickly isomerize and, therefore, decay in the conical intersection region. Features i–ii would yield a fluorescent intermediate/signal, where the $E_{S_1}^{\ddagger}$ value would be one factor controlling the magnitude of the fluorescence quantum yield with λ_{max}^{f} corresponding to the $\Delta E_f(S_1 - S_0)$ energy of the energy minimum.

Another engineering issue is related to the possibility to engineer rhodopsin actuators or reporters that absorb red light. In fact, rhodopsin variants capable to detect red and infrared photons represent important assets for optogenetic applications as red light has higher penetration depth and lower phototoxicity (i.e., is carrying a lower level of energy) [110, 149]. This can be accomplished by engineering rhodopsin variants with lower $\Delta E_a(S_1 - S_0)$ values (see Fig. 14), and therefore corresponding to longer λ_{max}^{a}. The engineering of red-light excitation is also important for the development of optogenetic tools based on bistable (or bichromic) rhodopsins [135]. In fact, bistability is an attractive feature as it provides the basis for engineering photoswitchable fluorescent probes [20].

Bistable rhodopsins are rhodopsins featuring two stable isomeric forms (i.e., characterized by two chromophore isomers such as all-*trans* and 13-*cis*). Certain bistable rhodopsins can be interconverted using light of different wavelengths (photochromism of type II/P) [10]. However, more frequently the light-adapted (LA) state reverts back to the dark-adapted (DA) state thermally (photochromism of type I/T) [10]. The applications of bistable rhodopsins are related to the possibility to use light irradiation to change the rhodopsin isomeric composition passing from a DA state dominated by one form to a LA state dominated by the alternative form, the efficiency of such conversion being proportional to the difference between the two λ_{max}^{a} values. It is thus apparent that achieving bistable rhodopsins featuring one form with a λ_{max}^{a} value significantly shifted to the red may facilitate applications where it is important to switch on-and-off the rhodopsin properties.

Similar to red light absorption, degenerate two-photon absorption (TPA) [140] is important in optogenetics because it allows to provide a way to photoexcite a rhodopsin with infrared (IR) light leading to an improved light penetration with respect to one-photon absorption with visible light and, possibly, maximizing the use of, for instance, bistable rhodopsin tools [135]. In fact, as also illustrated in Fig. 14, photoexcitation to the S_1 state can be also achieved by simultaneous absorption of two photons of the same energy $\Delta E_a^{TPA}(S_1 - S_0)$ corresponding to half the energy necessary for one-photon absorption (OPA). We will see that, a a-ARM$_{default}$ model has been successfully used to compute the TPA cross section (σ_{TPA}) of Rh-WT.

In conclusion, below we revise two combined experimental and theoretical studies [85], in which the engineering of microbial rhodopsins with enhanced fluorescence and λ_{max}^{a} value shifted to the red was the main target [59]. We also revise a study that demonstrates that ARM models can be used to evaluate the TPA spectral band of bovine rhodopsin as a model system for TPA studies [41]. We also show how ARM models can be used to investigate photochemical properties. More specifically, we report on computational studies of the primary photoproduct intermediates in an animal and a microbial rhodopsins including the structure and spectroscopy of both the primary photocycle intermediate and the structure dominating the LA form. Finally, we revise the result of a study where ARM models have been used to investigate the dependence of the absorption spectra of a microbial rhodopsin from the pH [108].

4.1　Search for an Eubacterial Fluorescent Rhodopsin

(Most of the content of Sect. 4.1 is reproduced/adapted with permission from *J. Am. Chem. Soc.* 2019, 141, 1, 262-271. Copyright 2019 American Chemical Society). An attempt to find novel optogenetics fluorescent probes using an eubacterial rhodopsin has been recently reported [85]. Such study focuses on a sensory rhodopsin from the freshwater eubacterium *Anabaena* (ASR) which show bistability and type II photochromism [15, 66, 141] based on the all-*trans* rPSB containing form (ASR$_{AT}$) and the 13-*cis* rPSB containing form (ASR$_{13C}$). ASR$_{AT}$ is the only component of the DA state while ASR$_{13C}$ is the main component of the LA state. It has been demonstrated that ASR$_{AT}$ and ASR$_{13C}$ can be interconverted by irradiation with light of different wavelengths. On the other hand, the unstable LA state converts slowly back to the DA state with half-time above 1 h at 4 °C, presumably due to a relatively large ground state isomerization barrier.

Similar to Arch3, ASR exhibits, in its DA state, a dim fluorescence (i.e., fluorescence quantum yield > 10^{-4}) [80] but the emission appears instantly (within 200 fs) since it is produced directly from the S_1 electronic state of ASR$_{AT}$ (the main form contributing to the DA state). It was observed (see Fig. 15) that certain mutations cause opposite effects on ASR$_{AT}$ fluorescence intensity when compared with the wild-type (WT$_{AT}$). In fact, the double mutation W76S/Y179F displays almost one order of magnitude higher fluorescence, while L83Q displays a decreased, almost negligible, fluorescence [85].

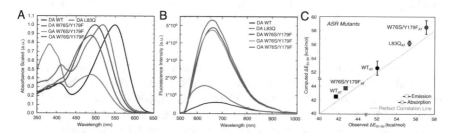

Fig. 15　Steady-state spectra and excited states dynamics of ASR$_{AT}$. a Scaled absorption spectra of the light-adapted W76S/Y179F mutant and dark-adapted (DA) WT, W76S/Y179F, and L83Q mutants. Light adaptation was carried out with either an orange (OA) or green (GA)LED. **b** Steady-state emission of DA WT ASR, DA L83Q ASR, DA,OA, and GA W76S/Y179F mutant of ASR. **c** Comparison between simulated and observed λ^a_{max} (circles) and λ^f_{max} (squares) values for the DA state. Deviation bars for the computed excitation energy values are shown as black segments. Adapted with permission from [85]. Copyright 2019 American Chemical Society

The molecular mechanisms explaining these observations were investigated by assuming that the fluorescence quantum yield Φ_F is proportional to the excited state isomerization barrier $E_{S1}^{\ddagger}$. More rigorously, Φ_F is given by the following equation:

$$\Phi_F = k_r^S \tau_S \tag{6}$$

where k_r^S is the rate constant for radiative deactivation ($1/k_r^S$ is called radiative lifetime and becomes the fluorescence lifetime in case the only radiative decay channel is fluorescence) of S_1 to S_0 with emission of fluorescence and τ_S is the first singlet S_1 excited state lifetime whose expression is

$$\tau_S = \frac{1}{k_r^S + k_{nr}^S} \tag{7}$$

where k_{nr}^S is the rate constant for non-radiative deactivation of S_1 to S_0. When assuming that k_r^S is similar in different but homologue rhodopsins (i.e., in different rhodopsin variants), Φ_F becomes inversely proportional to k_{nr}^S. Finally, by assuming that the k_{nr}^S magnitude is a negative exponential function of the S_1 energy barrier $E_{S1}^{\ddagger}$ controlling access to the S_1/S_0 conical intersection associated with the chromophore isomerization (see Fig. 14), one creates a direct link between such barrier and the critical quantity Φ_F.

The X-ray crystallographic structures of both ASR$_{AT}$ and ASR$_{13C}$ are available [141], making possible the construction of ARM$_{manual}$ models (see Fig. 16) without the need to build a comparative model first. By computing, using such a model, the S_1 relaxed scan along the C13=C14 twisting of the chromophore, it was found [85] that the observed fluorescence intensities can be related to the change in electronic character along the isomerization.

More specifically, the ARM$_{manual}$ models of the WT$_{AT}$, L83Q, and W76S/Y179F of ASR$_{AT}$ were used to produce the S_1 energy profiles along approximated C13=C14 isomerization coordinates. As shown in Figure 15C, the models were able to reproduce the observed trend in absorption and emission λ_{max} (λ_{max}^a and λ_{max}^f, respectively) computed in terms of $\Delta E_a(S_1 - S_0)$.

It is apparent from inspection of Fig. 17a–c, that the steep S_1 potential energy profile of L83Q must lead to the conical intersection more effectively than the flatter WT$_{AT}$ and W76S/Y179F surfaces. More specifically, WT$_{AT}$ and W76S/Y179F display profiles (see Fig. 17b, c) featuring a ca. 3 and 6 kcal mol^{-1} $E_{S1}^{\ddagger}$ values. The experimental τ_S values were 0.48 ps, 0.86 ps, and 5.7 ps for L83Q, WT$_{AT}$ and W76S/Y179F, respectively, and qualitatively in line with the computed barrier heights. Thus, an approximate mechanism could be proposed based on the idea that Φ_F is a function of the S_1 lifetime and, in turn, of the $E_{S1}^{\ddagger}$ value.

Since the paths of Fig. 17a–c do not consider kinetic energy effects, the above conclusion was supported by using the ARM models to probe the reaction dynamics with a 200 fs single semiclassical trajectory released from the FC region without initial velocities. Such trajectories would approximate the motion of the center of the

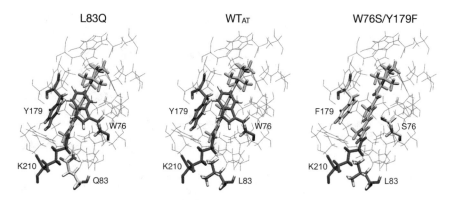

Fig. 16 Cavity residues of the ARM models of ASR$_{AT}$ and two of its mutants. The amino acids of the chromophore cavity are represented by a thin framework representation. The variable cavity residues 76, 83, and 179 are shown in tube representation. The rPSB$_{AT}$ chromophore and lysine side-chain (K210) covalently linked to the chromophore are also shown in tube representation. Reprinted with permission from [85]. Copyright 2019 American Chemical Society

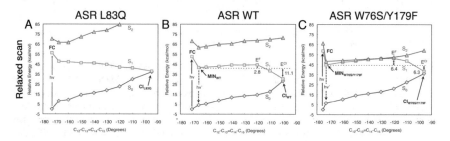

Fig. 17 C_{13}=C_{14} **photoisomerization paths along** S_1. (a)–(c). CASPT2//CASSCF/AMBER energy profiles along S_1 (squares) isomerization paths. S_0 (diamonds) and S_2 (triangles) profiles along the S_1 path are also given. The S_1 is computed in terms of a relaxed scan along the C_{12}-C_{13}=C_{14}-C_{15} dihedral angle. Adapted with permission from [85]. Copyright 2019 American Chemical Society

vibrational wave-packet [37]. The authors assumed that during such a short time, the trajectories describe the average evolution of population on the lowest excited states (S_1 and S_2) [83]. A barrier would slow down the isomerization on the excited state causing an increase in τ_S and, in turn, an emission enhancement. The results given in Fig. 18a–c indicates a behavior consistent with the observed fluorescence lifetime and with the isomerization paths. For instance, L83Q reaches the relevant S_1/S_0 conical intersection decay channel in ca. 100 fs (see Fig. 18a). This is consistent with the lack of barrier along the S_1 potential energy profile of Fig. 17a.

The above results were discussed on the basis of a theoretical framework (see more details in recent publications [46, 84]) based on the analysis of the electronic structure of the first three electronic states of a gas-phase model of the rhodopsin chromophore with five conjugating double bonds (PSB5) [80, 84]. The S_0, S_1, and

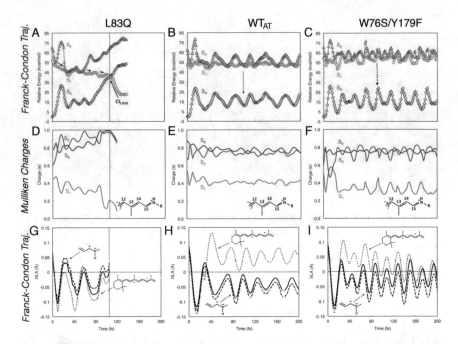

Fig. 18 Trajectory computation on S_1 for L83Q, WT and W76S/Y179F, respectively. a–c QM/MM FC trajectories computed at two-root state-averaged-CASSCF/AMBER level of theory and corrected at the CASPT2 level. S_0 (diamonds), S_1 (squares) and S_2 (triangles) CASPT2//CASSCF/AMBER energy profiles along the FC trajectories. **d–f** Mulliken charge variations of the =CH-CH=CH-CH=NH_2 moiety of the chromophore. **g–i** Evolution of the total BLA (full line) and of the BLA of two specific moieties (dotted and dashed lines) for L83Q, WT, and W76S/Y179F, respectively. Adapted with permission from [85]. Copyright 2019 American Chemical Society

S_2 electronic characters dominating a planar constrained PSB5 molecule are labeled as 1Ag, 1Bu, and 2Ag (see the resonance formulas in the all-*trans* rPSB chromophore representation in Fig. 19). These labels indicate the electronic terms of a homologous all-*trans* polyene with C_{2h} symmetry. S_1 has a 1Bu character characterized by a positive charge spread toward the H_2C=CH—end of the PSB5 framework (i.e., charge-transfer (CT) character). This is qualitatively different from the 1Ag closed-shell character of S_0 as 1Ag displays the positive charge localized on –C=NH_2. In contrast, the second singlet excited state (S_2) has 2Ag character which is associated to a diradical (DIR), rather than charge-transfer, structure. Thus, S_2 features, similar to S_0, a positive charge mostly located on –C=NH_2.

As illustrated by the resonance formulas in Fig. 19a, the mentioned 1Ag, 1Bu, and 2Ag charge distributions of PSB5 are assumed to reflect the electronic characters of the less symmetric rPSB chromophore to follow how the electronic character changes along the S_1 PESs (e.g., along a reaction path or trajectory).

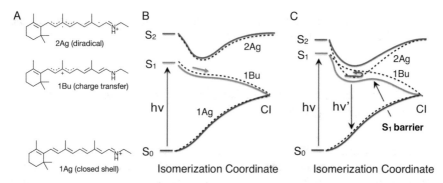

Fig. 19 Schematic S_0, S_1, and S_2 energy profiles along the S_1 PES path driving the chromophore excited state isomerization. **a** Resonance (Lewis) formulas corresponding to the electronic structure of the three relevant electronic states of the retinal chromophore. **b** An S_1 PES dominated by a 1Bu character is associated with a barrier-less path. **c** A mixed 1Bu/2Ag character is associated with the presence of a barrier along the path. The dashed energy profiles represent the energy of diabatic states corresponding to "pure" 1Bu and 2Ag electronic characters. Notice that the transient excited state has been also been identified as a locally excited state in previous literature. [57, 75] Adapted with permission from [85]. Copyright 2019 American Chemical Society

Figure 17a–c shows that the average S_2-S_1 energy gap along the S_1 path decreases in the order L83Q>WT>W76S/Y179F. In particular, in W76S/Y179F the molecular population exiting from the FC point and carrying a 1Bu character (i.e., the S_1 charge of $=CH–CH=CH–CH=NH_2$ is less than in S_0 and S_2), relaxes through S_1 PES regions with increased DIR character ($MIN_{W76S/Y179F}$ in Figs. 17c and 19f) indicating character mixing. The same population then reaches regions (ca. $-120°$) where the CT and DIR characters have close weights (i.e., similar charge distribution in S_1 and S_2). This happens at a lesser extent in WT_{AT} and L83Q. In fact the first features a larger CT weight in S_1 and in the second the charge becomes less than $+0.2$.

The results above suggest that in W76S/Y179F the S_1 barriers are generated via avoided crossings between states with pure CT and DIR characters as illustrated in Fig. 19 (right). The dynamic description of the same mixing process is provided by the calculation of FC trajectories. Thus, in the L83Q case (Fig. 18a, e) the S_2 energy profile only crosses the S_1 energy profile in the 20–50 fs time segment and then becomes destabilized leading to a fully reactive event. In fact, the molecule remains dominated by a CT reactive character without being trapped in S_1. On the other hand, a trapping event is detected in WT_{AT} (see Fig. 18b, e) and W76S/Y179F (see Fig. 18c, f) as a consequence of the existence of a barrier, and therefore of a fluorescent S_1 intermediate (I_F in Fig. 14). In Fig. 18g–i, it is demonstrated that the CT or DIR character is also dynamically reflected in the bond-length alternation (BLA) geometry. It is evident, and consistent with the resonance formulas of Fig. 19, that the β-ionone containing fragment has a larger BLA value consistently with a similar length for the initial formal single and double bonds. Such BLA value is instead lower (actually negative) when the CT character dominates and therefore the lengths of the double and single bond invert.

4.2 Search for Red-Shifting Mutations in a Sodium Pumping Eubacterial Rhodopsin

(Most of the content of Sect. 4.2 is adapted with permission from *Nat Commun.* 2019, 10, 1993. Open access under a CC BY license (Creative Commons Attribution 4.0 International License)). As schematically displayed in Fig. 14, the λ_{max}^a value of a rhodopsin in its DA state is determined by the gap between S_0 and S_1. Such ΔE_{S1-S0} value must depend on the steric and electrostatic interaction between the chromophore and protein environment. The former modulates the geometry (e.g., planarity) of the chromophore so that, for instance, a mutation twisting the chromophore backbone about a single bond effectively shortens the π-conjugation leading to a blue-shifting effect. In contrast, the protein electrostatics modulate the ΔE_{S1-S0} value by interacting with the distinct electronic structures of the S_0 and S_1 states which, in turn, depend on their electronic characters. In fact, as shown in Fig. 19, the S_0 state the rPSB chromophore has a positive charge substantially localized on the C15=N moiety while in a S_1 state with CT character the charge is delocalized along the chromophore chain toward the β-ionone ring. Thus, a negatively charged residue placed near the positively charged C15=N moiety selectively stabilizes the S_0 state, whereas the energy of the S_1 state is stabilized by a negatively charged residue placed in the vicinity of the β-ionone ring.

The color-tuning effect caused by the polarity of amino acid residues located in the vicinity of the retinal chromophore has been reported for channelrhodopsin (ChR2) from Chlamydomonas reinhardtii which is the most popular optogenetic tool [143]. A highly red-shifted variant of ChR2 has been found in Volvox carteri (VChR1) and in Chlamydomonas noctigama (Chrimson). Recently, the x-ray crystallographic structure of Chrimson was solved, providing molecular insights into the mechanism causing its red-shifted absorption [99]. It was concluded that the residues surrounding the chromophore impact the λ_{max}^a of Chrimson by modulating (i) the protonation state of the counterion, (ii) the distribution of polar residues around the β-ionone ring and (iii) the steric interaction with the highly rigid chromophore cavity. However, the above electronic character discussion indicates that also the distribution of the polar residues near C15=N should affect the λ_{max}^a value.

Below we review a combined computational and experimental study carried out with ARM_{manual} models and focusing on polar residues located both near the β-ionone ring and the C15=N moiety of the chromophore of Na^+ pumping rhodopsin. *Krokinobacter rhodopsin 2* (KR2) is the first identified outward (from the cytoplasm to extracellular milieu) Na^+ pumping rhodopsin [58]. In fact, KR2 was reported to be able to inhibit the neuronal action potential without the unnecessary intracellular Cl^- accumulation and pH change, sometimes demonstrated to cause unexpected cellular activity. However, since it has an absorption in the relatively short wavelength region ($\lambda^a_{\max} = 525$ nm), color tuning to a more red-shifted region is required for *in vivo* applications.

By changing the polarity of the residues near the β-ionone ring and C15=N moiety it was possible to construct a KR2 variant displaying a $\lambda^a_{\max}$ value 40 nm red-shifted with respect to the wild-type form (WT$_{KR2}$). In order to investigate the mechanism allowing such a result, ARM$_{\texttt{manual}}$ models were built for WT$_{KR2}$ and its P219T, P219G, S254A, and P219T/S254A mutants constructed by taking the WT$_{KR2}$ X-ray crystallographic structure (PDB code: 3X3C) as the template. The computed ΔE_{S1-S0} values were compared with the corresponding observed $\lambda^a_{\max}$ (see Fig. 20a). As reported for ASR$_{AT}$ in Fig. 15, the trend of energy shifts of the mutants relative to WT$_{KR2}$ were reproduced supporting the validity of the models.

The models allowed to disentangle the electrostatic and steric effects responsible for the observed decreasing ΔE_{S1-S0} values (i.e., red-shifting) along the mutant series. To do so, the ΔE_{S1-S0} values for the corresponding chromophore in isolated condition (i.e., removing the protein part from the model while keeping the chromophore geometry fixed at its equilibrium geometry in the protein environment) were computed. As shown in Fig. 20a, it was found that the gas-phase rPSB and full rhodopsin trends are opposite. Since the gas-phase rPSB trend must be the result of geometrical changes sterically induced by the rhodopsin cavity, it was concluded that the electrostatic interaction is responsible for the observed trend. More specifically (see Fig. 20b, c), while the blue-shifting changes in the chromophore geometry were mainly located in the C9=C10–C11=C12–C13=C14–C15=N moiety for stretching deformations and on the C13=C14–C15=N moiety for torsional deformations, the red-shifting electrostatic contribution imposed by the environment and cavity sub-

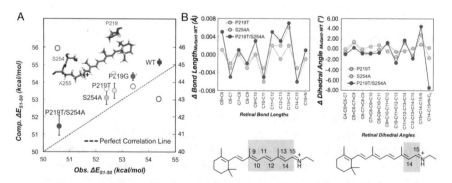

Fig. 20 $\lambda^f_{\max}$ **of KR2 and its mutants. a** Comparison between simulated and observed $\lambda^a_{\max}$ values for protein (full circles) and gas-phase (open circles) models for the DA state of WT$_{KR2}$ and its selected mutants dominated by the all-*trans* rPSB chromophore. The chromophore structure and the WT$_{KR2}$ mutation sites are displayed in tube representation. **b** Difference in bond lengths of the mutants relative to WT$_{KR2}$. The moiety undergoing the largest change is highlighted at the bottom. **c** Difference in torsional angle of the mutants relative to WT$_{KR2}$. The moiety undergoing the largest change is highlighted at the bottom. Adapted with permission from [59] open access under a CC BY license (Creative Commons Attribution 4.0 International License)

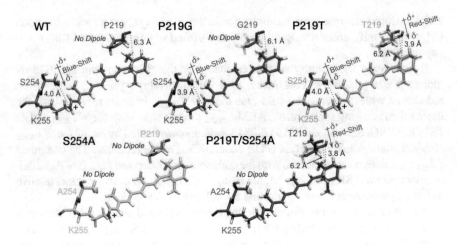

Fig. 21 Comparison between the ARM structures of KR2 mutants. Retinal chromophore and mutated side-chains S254 and P219 for WT_{KR2} and four selected mutants. The mutant structures also show, in transparent representation, the WT_{KR2} side-chains. Adapted with permission from [59] open access under a CC BY license (Creative Commons Attribution 4.0 International License)

systems dominated and quenched the geometrical effects resulting in a net red-shifted change.

In Fig. 21, the above electrostatic effects are interpreted in terms of changes in dipole moment at the 219 and 254 residues located close to the β-ionone ring and –C=N-moieties, respectively. For instance, the P219T mutation creates a new dipole moment having the partially negatively charged oxygen atom of the side-chain facing the β-ionone ring, thereby selectively stabilizing the S_1 state relative to S_0 state. In contrast, S254A removes the dipole moment stabilizing S_0 relative to S_1. Using the same ARM models, it was also possible to investigate the role played by each protein amino acid residues in decreasing the ΔE_{S1-S0} value relative to WT_{KR2}. In fact, one can computationally set the MM point charges of each residue to zero and then recompute the vertical excitation energy now called ΔE_{off}. The differences between the ΔE_{S1-S0} and ΔE_{off} values revealed an effect based on two components: (i) a direct component due to a change in number, magnitude, and position of the point charges in the mutated site and (ii) an indirect component which originates from the reorganization of the local environment and hydrogen bond network induced by the same mutation. The second effect is due to the fact that, in the mutants, conserved residues and water molecules change in position or conformation thus displaying different ΔE_{S1-S0} and ΔE_{off} values with respect to the WT_{KR2}. Such indirect effects contribute to the total excitation energy changes significantly. For instance, when comparing the P219T, P219G, and S254A mutants, double mutant (P219T/S254A), and WT_{KR2}, the data show that the amino acid substitutions at residues 219 and 254 result in a red-shift in absorption relative to WT_{KR2} and contribute to the strong red-shifting observed in the double mutant. However, such changes are accompanied by variations in the effects of the conserved residues. In fact, the effect of the conserved S254 residue in P219T and the conserved P219 residue in S254A are not close to the one found in WT_{KR2}.

4.3 Calculation of the TPA Spectrum of Bovine Rhodopsin and IR Vision

(Most of the content of Sect. 4.3 is reproduced/adapted with permission from *J. Phys. Chem. Lett.* 2019, 10, 20, 6293-6300. Copyright 2019 American Chemical Society). ARM models can be employed to calculate the TPA spectra of rhodopsins. This has been shown in the literature, where a calculation of the TPA electronic spectrum of Rh (bovine rhodopsin) has been reported. To achieve the quality necessary for such a computation, the Rh model (in this case produced with the automated *a*-ARM protocol discussed in Sect. 3.1.2), was improved by using a multistate multi-configurational second-order perturbation level of QM theory (XMCQDPT2) and a correlation-consistent atomic basis set. This level is superior to the single-state CASPT2 method used for ARM models. Also, notice that, to be able to carry out a consistent comparison with the TPA spectrum, the standard OPA electronic spectrum (e.g., the one providing the λ_{max}^a values discussed in the previous subsections) was recomputed using an 8-root state-average (with equal weights as it is always the case for all state-average calculations mentioned in this chapter) wave function while, usually, the simulation of the OPA properties of ARM models are evaluated at the 3-root SA wave function level and 6-31G(d) level for the atomic basis.

After constructing the a-$ARM_{default}$ model starting from the Rh X-ray crystallographic structure, the resulting OPA and TPA spectra (λ_{max}^a and transition intensities) were computed and compared with the available experimental data. The reported results are displayed in Fig. 22. The computed OPA $\lambda_{max}^a = 475$ nm for the first transition, which differs from the 498 nm experimental value by less than 3.0 kcal mol^{-1}, is displayed as a stick spectrum and, to allow comparison with the TPA spectrum, placed at double the wavelength (950 nm) required for OPA excitation, namely, the calculated wavelength of photons that would be absorbed in a degenerate TPA transition. Similarly, the λ_{max}^a predicted for the weaker OPA transition of $\lambda_{max}^a = 339$ nm and placed at 678 nm in the figure diagram, differs from the experimental data by only 0.2 kcal mol^{-1}.

The TPA cross section (σ_{TPA}) is related to the imaginary part of the second hyperpolarizability. The authors used a sum-over-state (SOS) approach based on the mentioned approximation employing an 8-root state-average wave function, and therefore 8 electronic states. As anticipated above, they assumed that the two simultaneously absorbed photons are degenerate, namely, of the same wavelength. The SOS calculation predicts an average TPA σ_{TPA} of 472 GM at λ_{max}^a of 950 nm, for the first band corresponding to the $S_0 \rightarrow S_1$ transition. The calculated σ_{TPA} value is comparable with the value reported for the standard ChR2 optogenetics tool, which

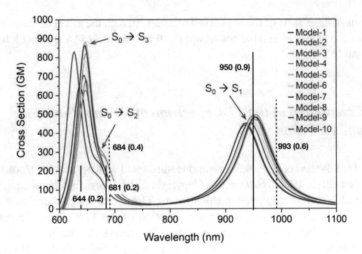

Fig. 22 One- and two-photon absorption spectra of bovine rhodopsin. Computed λ^{a}_{max} corresponding to photons with average half-energy value of the OPA λ^{a}_{max} (full vertical lines and bold numbers with oscillator strength in parenthesis) and TPA spectral bands for the first three transition of Rh obtained from the ten QM/MM models generated by the ARM protocol at the MS-MC-PT2 level of theory (XMCQDPT2/cc-pVTZ//CASSCF/6-31G(d)/AMBER models). The dashed vertical lines refer to standard 3-root state-average CASPT2/6-31G(d) level of theory used with ARM models to compute their $S_0 \rightarrow S_1$ and $S_0 \rightarrow S_2$ λ^{a}_{max} values. Adapted with permission from [41]. Copyright 2019 American Chemical Society.

is ~260 GM at 920 nm and, therefore, a vertical excitation energy of 30 kcal mol^{-1}. This suggests that ARM models may be useful to assist the design of novel optogenetics tool with large σ_{TPA} values.

The documented first TPA transition corresponds to the OPA transition responsible for the ultrafast 11-*cis* to all-*trans* photoisomerization of the 11-*cis* rPSB chromophore of Rh. Therefore, in agreement with the results by Palczewska et al., the results support the perception of the IR light by Rh and the hypothesis that OPA and TPA result in the same photoisomerization process, producing all-*trans* rPSB which, in turn, activates the Rh photocycle and leads to visual perception. This appears to be a straightforward conclusion when considering the computational and experimental evidence in favor of a barrier-less nature of the S_1 double-bond isomerization path [37, 121].

OPA and TPA calculations of the a-ARM$_{default}$ model indicate that while the second excitation ($S_0 \rightarrow S_2$) is weak in the OPA spectrum (see the oscillator strengths in Figure 22), it is, as expected, more intense in TPA (22). In fact, the calculated average λ^{a}_{max} related to the $S_0 \rightarrow S_2$ transition, is located at 678 (2 × λ^{a}_{max}, OPA = 339) nm with the average cross section of 231 GM. The calculated average λ^{a}_{max} TPA related to the $S_0 \rightarrow S_3$ transition is located at 644 (2 × λ^{a}_{max}, OPA = 322) nm and shows an average σ_{TPA} of 771 GM. Similar to the $S_0 \rightarrow S_2$ transition, the $S_0 \rightarrow S_3$ OPA transition is weak (i.e., it has a small oscillator strength) but it is predicted to be strongly allowed in TPA. Furthermore, while both $S_0 \rightarrow S_2$ and $S_0 \rightarrow S_3$ transitions

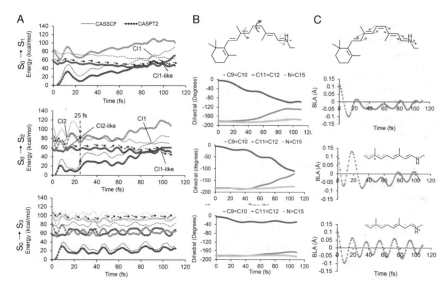

Fig. 23 Geometrical progression along the trajectories triggered via TPA absorption. a Energy profiles describing the excited state relaxation occurring after the population of the S_1, S_2, and S_3 state. **b** Corresponding progression along three relevant torsional deformations. (C) Progression along the corresponding collective stretching deformation called bond-length alternation (BLA) and defined as the difference between the average single bond length and average double-bond length). Adapted with permission from [41]. Copyright 2019 American Chemical Society

are predicted to have the same intensity in OPA (i.e., same oscillator strength), the latter has a higher σ_{TPA} (771 GM vs. 231 GM).

The dynamics triggered by the TPA population of the S_2, and S_3 potential energy surfaces were investigated by computing the corresponding FC trajectories and comparing them with that reported for the reactive S_1 state (see Fig. 23). As discussed above, FC trajectories provide an approximate description of the motion of the center of the excited state population and, in the present context, are used to provide qualitative information on the excited state reactivity. The top panel of Fig. 23a displays the S_1 FC trajectory energy profiles of Rh computed at the 2-root state-average CASSCF/6-31G(d)/AMBER level and corrected by 3-root state-average CASPT2 level via single-point calculations, starting at the FC point on S_1 and reaching the S_1/S_0 conical intersection on a ca. 100 fs time scale (CI1 and CI1-like points in CASSCF and CASPT2 profiles, respectively) [80, 84]. In order to demonstrate that the reactivity is maintained also when populating the S_2 state, we look at the trajectory obtained at the 3-root state-average level starting at the FC point on S_2 as displayed in the middle panel of Fig. 23a. The results indicate that the reactive state S_2 is populated after ~15 *fs* (CI2 and CI2-like in CASSCF and CASPT2 profiles, respectively) upon decay from S_2 to S_1. The now S_1 population reaches the reactive S_1/S_0 conical intersection on a ca. 100 *fs* time scale (CI1 and CI1-like points in CASSCF and CASPT2 profiles, respectively). A similar analysis has been done for S_3 which displays a large σ_{TPA} value. As shown in the bottom panel of Fig. 23a, this state appears less reactive as no decay of the S_3 state (computed using a 4-root state-average CASSCF wave function) is detected within 200 fs.

The same trajectories provide information on the reaction coordinate spanned during the excited state relaxation. Such progression is reported in Fig. 23b, c for the relevant torsional and stretching (BLA) deformations respectively. It can be clearly seen that the reaction coordinate is dominated by a bicycle-pedal motion [146] centered on the C11=C12 and C9=C10 torsions with no contribution along C15=N. Such a reaction coordinate is clearly conserved for excitations to the S_2 state via TPA (see the central panel in Fig. 23b). Such motion does not seem to be activated after the population of S_3 (see bottom panel of Fig. 23b). It might be worth noting here that internal conversion from S3 to S2 might be efficient simply through efficient vibronic coupling. The initially unreactive nature of S_3 can also be seen in the BLA coordinate of Fig. 23c. In fact, the comparison of the central and bottom panels of such figure clearly show how the BLA is not completely inverted in S_3 which shows oscillation around a zero BLA value (i.e., around similar single and double-bond lengths) reflecting a limited reactive character.

4.4 Modeling Primary Photoproducts, Reaction Paths, and Bistable States in an Animal and a Microbial Rhodopsin

In Fig. 23a, we have shown that a-ARM$_{default}$ models can be employed to investigate the excited state reactivity of an animal rhodopsin via FC trajectory computations. Such computations, when continued for a time long enough to reach the S_0 valley, generates a model of the structure of the primary stable photocycle intermediate (see also Fig. 14). For Rh such intermediate is called bathorhodopsin (bathoRh) and its model is displayed in Fig. 24. It can be seen that the FC trajectory provides information useful for computing the structure of a S_1/S_0 (minimum energy conical intersection) MECI which is the critical structure driving the decay to S_0 (the photochemical funnel) and therefore the isomerization.

The comparative analysis of the Rh, MECI, and bathoRh structures provide information on the reaction coordinate in terms of the main variations in the chromophore and protein geometry. Such variations are, in part, displayed in Fig. 24. The same structures also provide information on the electronic character changes occurring along the same reaction coordinate and showing that the initial charge transfer character is lost immediately after decay to S_0 (see the bottom of the figure where the S_1 and S_0 charges of a specific chromophore fragment are given and reflect the change in electron distribution along the reaction path). Of course, the quality of the bathoRh model can be evaluated by comparison of the observed and predicted λ_{max}^a values and, if possible, by comparing the X-ray crystallographic structure with predicted structure.

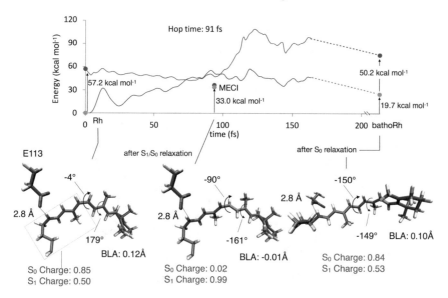

Fig. 24 CASPT2//CASSCF/AMBER energy profiles along the S_1 photoisomerization path of Rh. Energy profiles along the S_0 and S_1 states along the FC trajectory describing the OPA triggered progression along the S_1 state (see top panel in Fig. 23a). The a-ARM$_{default}$ used for this calculation is the same employed above for the discussed TPA computations. The geometrical progression is illustrated in terms of the structure representative of the DA state featuring and 11-*cis* rPSB chromophore, the conical intersection (MECI) controlling the decay from S_1 to S_0 and the geometry of the primary photoproduct bathoRh. Both the MECI and the bathoRh are geometrically optimized, and therefore do not correspond to trajectory snapshots. The energies of MECI and bathoRh are relative to the ground state Rh. These are unpublished data from the authors but an earlier version of this study was reported in Ref. [121]. The charges in S_0 and S_1 are given as positive electron charge

The Rh calculation shows that ARM models can be used to track the S_1 reaction paths, predict the structure of the MECI and that of the first primary photoproduct (photocycle intermediate). We now discuss and additional study in which not only the primary photoproduct K is modeled but, if suitable structural information is given, the structure of the meta-stable LA form of a bistable microbial rhodopsin. Indeed, we discuss the S_1 reaction path computation of a light-driven ion-pumping rhodopsin [58] from a deep-ocean marine bacterium, *Parvularcula oceani* [78] (PoXeR) as an example of the application of an ARM models to the study of a rhodopsin with a LA state with a lifetime of tens of seconds. The study has the target of inferring the character of the K intermediate featuring a 13-*cis* rPSB chromophore. Such intermediate eventually gets back to the DA state thermally (see Fig. 14) and shows a $\tau = 91$ s.

P. oceani is a α-proteobacterium found at a depth of 800 m in the southeastern Pacific ocean. Given the small amount of light available, PoXeR performs its function as inward light-driven H$^+$ pump efficiently and using only one photon to complete

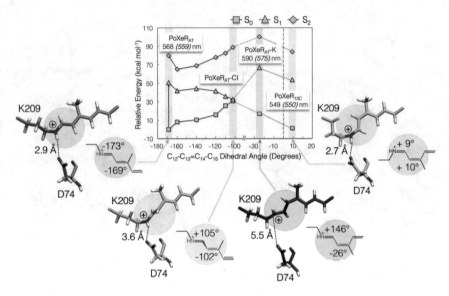

Fig. 25 S_0, S_1 **and** S_2 **energy profiles along the** S_1 **isomerization path of PoXeR**$_{AT}$. The geometries corresponding to PoXeR$_A T$, PoXeR$_A T$-CI, PoXeR$_A T$-K and PoXeR$_{13C}$ were obtained through a ground state relaxed scan along the reactive C_{12}-C_{13}=C_{14}-C_{15} dihedral angle (i.e., taking the torsion about C13=C14 as the driving coordinate). All energies are reported relative to that of the ground state of PoXeR$_{AT}$. The computed and observed (italic in parenthesis) wavelength values are shown for the full protein-retinal complex. In each state, the isomerizing bond is highlighted

one entire DA → LA → DA photocycle. As opposed to the ASR photocycle, which features the long lived (hours) meta-stable form ASR$_{13C}$, the 13-*cis* to all-*trans* thermal back-conversion (PoXeR$_{13C}$ → PoXeR$_{AT}$) takes place in minutes rather than hours. This is why a single photon is sufficient to achieve the full rotation in the PoXeR photocycle which, in total, lasts minutes. In other words PoXeR exibits, effectively, Type I photochromism. This is different from the Type II photochromism of ASR which requires a second photon to quickly reconvert ASR$_{13C}$ to the DA dominating ASR$_{AT}$ form.

PoXeR shares a 51% of its amino acid sequence with ASR. This allows to employ the ASR x-ray crystallographic structure as a template for performing comparative modeling and use the resulting structure to build an ARM$_{manual}$ model of PoXeR$_{AT}$ and of the meta-stable form PoXeR$_{13C}$. The PoXeR$_{AT}$ model can be employed to compute the S_1 isomerization path leading to the production of K (in microbial rhodopsins K has the same role of bathoRh in animal rhodopsins). PoXeR$_{AT}$, PoXeR$_{13C}$ and the the primary photocycle intermediate PoXeR$_{AT}$-K, feature different λ^a_{max} values. These values, provide a trend in experimental quantities that acceptable ARM$_{manual}$ models must reproduce.

The results of the above-mentioned study are reported in Fig. 25. From inspection of the figure, it is evident that the trend in observed λ^a_{max} values is reproduced by the three models which display a blue-shifted error in the case of PoXeR$_{AT}$ and

PoXeR$_{AT}$-K. The reaction path displays a limited barrier at ca. $-140°$ which is reminiscent of the barrier documented for ASR (see Fig. 17b above) and found at ca. $-120°$ twisting about C13=C14.

The results also provide information on the detailed structural changes occurring during the isomerization reaction as well as on the distorted (i.e., nonplanar) conformation of PoXeR$_{AT}$-K when compared to the PoXeR$_{AT}$ reactant. It must also be noticed that the planarity is roughly recovered at PoXeR$_{13C}$ which features $+10°$ and $+9°$ twisted C13=C14 and C15=N double bonds. Notice that both PoXeR$_{AT}$ and PoXeR$_{13C}$ are pre-twisted of ca. $10°$ in the same direction indicating unidirectional isomerization motion.

4.5 Modeling the pH Sensitivity of the λ_{max}^a in ASR

(Most of the content of Sect. 4.5 is reproduced/adapted with permission from *J. Chem. Theory Comput.* 2019, 15, 8, 4535-4546. Copyright 2019 American Chemical Society).

We end the application section by revising a study [108] of the sensitivity of the vertical excitation energy ΔE_{S1-S0} value to pH in ASR$_{AT}$ carried out with ARM$_{manual}$ models. These models have been used to obtain the excitation energies on a selected subset of "protonation microstates" of this rhodopsin DA state. As explained in Sect. 3.1.1, the ionization state of each ionizable protein residue has to be provided during the building of the input for construction of an ARM model. One usually employs approximate methodologies like the one provided by the PROPKA software [102]. However, the software usually provides only one of the possible protonation microstates which are then assumed to be a largely dominating microstate. However, it is then possible that the dominating microstate changes when changing the pH or an even more complex situation may arise where more microstates in equilibrium among them may contribute to the description of the protein ionization.

The authors of the mentioned study have tested a certain number of dominating microstates with respect to a chosen reference microstate (REF) supposed to dominate at pH 7. It is then possible to roughly mimic the effect of a change of pH by modifying the dominating protonation microstate which will feature a different protonation in one or more of several titratable residues with respect to the reference. The result of the comparison of the investigated microstates, each one represented by a distinct ARM$_{manual}$ model, is shown in Fig. 26. As expected, the largest shift (>120 nm) is due to the protonation of the retinal counterion, D75, keeping the retinal as a protonated Schiff base. This shift is out of the range reported experimentally (the maximal observed variation is 10 nm), thus D75 is always kept deprotonated. The deprotonation of D217 (15.4 Å from the retinal) induces a blue-shift of 13 nm and it is in qualitative agreement with specific experimental observations [119]. Closer to the retinal (14.2 Å), the deprotonation of D198 causes a much larger 41 nm λ_{max}^a blue-shift. Simultaneous deprotonation of both D198 and D217 results in a blue-shift which is 32 nm with respect to the reference value, hence less than the one obtained

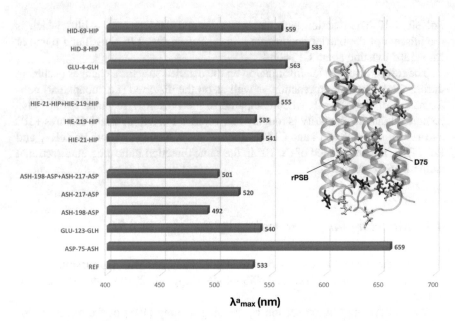

Fig. 26 λ_{max}^{a} **values of several ASR$_{AT}$ protonation microstates**. The REF model corresponds to the most likely protonation microstate at pH = 7. The 3-letter code for each amino acid has been chosen to unambiguously denote each change of the protonation state on the y-axis, as follows: initial protonation state–amino acid number–final protonation state. The structure on the right provides a schematic representation of the ASR model, featuring 18 titratable residues between pH = 3 and pH = 8: aspartic acids in red; glutamic acids in yellow, histidines in blue. Adapted with permission from [108]. Copyright 2019 American Chemical Society

by the single deprotonation of D198. The smallest red-shift (2 nm) is due to protonation of H219. Accordingly, H219 could be a good candidate for explaining the ASR red-shift at acidic pH. This tiny effect is probably related to the large distance between H219 and the retinal, more than 19 Å. The smaller distance between H21 and the retinal (18 Å results in a slightly larger red-shift (8 nm) upon H21 protonation. However, protonation of both H21 and H219 results in a 22 nm red-shift, larger than the ones resulting from each protonation taken separately. In this configuration, H21 and H219 form an interacting pair in the ASR structure. Because of the large electrostatic repulsion between two, close, positively charged moieties, this doubly protonated situation has very little chance to occur in the considered pH range. In other words, the corresponding protonation microstate is unlikely to be populated. These ARM$_{manual}$ models result in observable λ_{max}^{a} changes and can thus be used to predict the changes themselves. However, the author stresses that such information needs to be complemented with the probability of such a process to take place at a given pH. For instance, it may occur that the large blue-shift caused by D198 deprotonation actually does not matter if this reaction is not likely to happen, because the D198 pKa value could be significantly far from the pH range under study. Rephrased in a probabilistic language, the analysis above do not take into account the relative populations of protonated and deprotonated residues at a given pH. This must be evaluated with different technologies exploiting full MM parametrized models such as CpHMD (Constant pH Molecular Dynamics) simulations [54].

5 Future Applications: Use a-ARM rhodopsin model building and Forget original ARM protocol

Most of the applications presented in Sect. 4 were performed on the basis of QM/MM models generated with the first version of the ARM protocol [90]. As further described in Sect. 3.1.1, the main pitfalls of this version are related with the input file preparation phase, that is manually achieved by following a procedure composed of four different steps and yielding models whose construction was only partially automated. In fact, such a procedure requires researchers to make a series of decisions, based on their chemical/physical knowledge and intuition. Doubtless, the great flexibility to operate in each of the four steps may impair the reproducibility of the final QM/MM models, making difficult, if not impossible, to get the same model from two different researchers operating independently and without exchanging information. In this regard, although the pool of results reported in Sects. 4.1, 4.2, 4.4, and 4.5 are reliable, in the sense that they were thoughtfully produced, their reproducibility in other labs and without step-by-step instructions, cannot not be guaranteed.

As an alternative, it is expected that the use of QM/MM models constructed with the updated version of the protocol (a-ARM [106, 107]) that implements a Python-based module for the automatic input preparation (see Sect. 3.1.2) will give rise to reproducible results. Indeed, a first step toward demonstrating that the performance of automatic a-ARM models is consistent with that of manual ARM is to compare their ability to reproduce trends in maximum absorption wavelength. As observed in Fig. 10, this is achieved via benchmark calculations for a set of phylogenetically diverse rhodopsins variants (i.e., wild-type and mutants). In this figure, it is evident that, similar to what observed for manual ARM, the automatic a-ARM version results in excellent agreement with experimental data. Subsequently, we compared the $\lambda^a_{\max}$ computed with both versions of the protocol for each of the studied rhodopsins (e.g., ASR_{AT} and ASR_{13C} in Sect. 4.1; $KR2_{AT}$ in Sect. 4.2; $PoXeR_{AT}$ and $PoXeR_{13C}$ in Sect. 4.4) as an attempt to evaluate whether the automatic a-ARM QM/MM models would be suitable for the applications previously summarized. In doing so, it is noticeable that both versions produce consistent values, as shown in Table 3.

This is a positive indication that conclusions derived from each of the applications reported in Sect. 4 can be transferred when using models produced with a-ARM. Nevertheless, such a hypothesis must be verified by reproducing a complete photochemical study beyond the simple calculation of $\lambda^a_{\max}$. To this aim, we attempted to recreate one of the studies of Sect. 4 by using the corresponding a-ARM models. In this regard, we selected the "search for an eubacterial fluorescent rhodopsin" as a case study (see Sect. 4.1) due to the relevant findings/conclusions derived from both

Table 3 Comparison between spectroscopic data calculated by using ARM_{manual} and $a\text{-}ARM_{automatic}$ QM/MM models. Experimental maximum absorption wavelength, λ_{max}^{a}, in nm, and first vertical excitation energy, ΔE_{S0-S1}^{a}, in kcal mol^{-1}

Rhodopsin variant	PDB ID	RET-C	Experimental		ARM_{manual} [a]		$a\text{-}ARM_{automatic}$ [a]	
			λ_{max}^{a}	ΔE_{S0-S1}^{a}	λ_{max}^{a}	ΔE_{S0-S1}^{a}	λmax^{a}	ΔE_{S0-S1}^{a}
ASR-WT$_{AT}$	1XIO	all-$trans$	550[85]	52.0	543[85]	52.6$_{1.0}$	547[d]	52.3$_{0.2}$
ASR-L83Q$_{AT}$	1XIO[b]	all-trans	517[85]	55.3	508[85]	56.2$_{0.4}$	513[d]	55.7$_{0.4}$
ASR-W76S/Y179F$_{AT}$	1XIO[b]	all-trans	488[85]	58.6	488[85]	58.5$_{1.0}$	504[d]	56.8$_{2.0}$
KR2-WT$_{AT}$	3X3C	all-trans	525[65]	54.5	517[59]	55.2$_{0.2}$	511[106]	55.9$_{0.4}$
PoXeR-WT$_{AT}$	4TL3[c]	all-trans	568[58]	50.3	559[e]	51.1$_{0.9}$	566[106]	50.5$_{0.5}$
PoXeR$_{13C}$	4TL3[c]	13-cis	549[58]	52.1	550[e]	51.9$_{1.0}$	525[106]	54.4$_{0.2}$

[a] Average value of 10 replicas, along with the corresponding standard deviation given as subindex
[b] X-ray structure template model for mutations using $a\text{-}ARM$
[c] X-ray structure template model for comparative model
[d] Original data produced for this work
[e] Original data produced for this work. See Sect. 4.4

Table 4 Overview of structural features and both experimental and computational data for a-ARM QM/MM models of *Anabaena* sensory rhodopsin (ASR)

Anabaena sensory rhodopsin (ASR)			
General Information			
PDB ID:	1XIO[141]	Chromophore:	Retinal (RET)
RET Configuration:	*all-trans* (AT)	Lysine Linker:	A-Lysine 210 (K210)
Proton acceptor:	Aspartic 75 (D75)	Proton Donor:	—
Experimental Spectroscopic Information			
Absorption:			
WT:	ΔE_{S1-S0}: 52.0 kcal mol^{-1} (2.25 eV)	λ_{max}^{a}: 550 nm[1]	
L83Q:	ΔE_{S1-S0}: 55.3 kcal mol^{-1} (2.40 eV)	λ_{max}^{a}: 517 nm[1]	
W76S/Y179F:	ΔE_{S1-S0}: 58.6 kcal mol^{-1} (2.54 eV)	λ_{max}^{a}: 488 nm[1]	
Emission:			
WT:	ΔE_{S1-S0}: 42.4 kcal mol^{-1} (1.83 eV)	λ_{max}^{f}: 674 nm[85]	
W76S/Y179F:	ΔE_{S1-S0}: 43.5 kcal mol^{-1} (1.88 eV)	λ_{max}^{f}: 658 nm[85]	
pKa Analysis at crystallographic pH 5.6			
Protonated residues:	ASH 198 ASH 217 GLH 36 HID 8		
Counterion Distribution:	Inner surface: 7Cl^{-}	Outer surface: 1Na^{+}	
Cavity Residues:	11^{b}, 43^{b},47^{b},73,75,76,79,80,83,86^{b},109,112,113,116,119,131,132,134^{b}, 135,136,137^{b},139,176,179,180,183,198,202,203^{b},206^{b},209,210,211^{b},214^{b}		
Computational Results			
Absorption^{a}:			
WT:	ΔE_{S1-S0}: 52.3 kcal mol^{-1} (2.27 eV)	λ_{max}^{a}: 547 nm	f_{Osc}: 1.29
Error:	+0.3 kcal mol^{-1}		
Standard Deviation (DEV.ST):	0.16 kcal mol^{-1}		
L83Q:	ΔE_{S1-S0}: 55.7 kcal mol^{-1} (2.41 eV)	λ_{max}^{a}: 513 nm	f_{Osc}: 1.02
Error:	+0.4 kcal mol^{-1}		
Standard Deviation (DEV.ST):	0.42 kcal mol^{-1}		
W76S/Y179F:	ΔE_{S1-S0}: 56.8 kcal mol^{-1} (2.46 eV)	λ_{max}^{a}: 504 nm	f_{Osc}: 1.06
Error:	−1.8 kcal mol^{-1}		
Standard Deviation (DEV.ST):	1.98 kcal mol^{-1}		

(continued)

Table 4 (continued)

Anabaena sensory rhodopsin (ASR)			
Emission calculated from $MINS_1^c$:			
WT:	ΔE_{S1-S0}^f: 35.4 kcal mol^{-1} (1.53 eV)	λ_{max}^f: 808 nm	Error: -7.0 kcal mol^{-1}
W76S/Y179F:	ΔE_{S1-S0}^f: 34.9 kcal mol^{-1} (1.51 eV)	λ_{max}^f: 820 nm	Error: -8.6 kcal mol^{-1}
Emission calculated from FC Trajectory^d:			
WT:	ΔE_{S1-S0}^f: 35.1 kcal mol^{-1} (1.52 eV)	λ_{max}^f: 815 nm	Error: -7.3 kcal mol^{-1}
W76S/Y179F:	ΔE_{S1-S0}^f: 35.4 kcal mol^{-1} (1.53 eV)	λ_{max}^f: 809 nm	Error: -8.1 kcal mol^{-1}

^{a} Average maximum absorption wavelength of the N-independent replicas ($N = 10$), calculated at the single-state CASSCF/AMBER//CASPT2/6-31G(d) level
^{b} Residues added to the customized cavity
^{c} Maximum emission wavelength computed from the $MINS_1$ structure, of the replica with λ_{max}^a closest to the average value, obtained at the state-average 3-roots SA3-CASSCF/AMBER//CASPT2/6-31G(d) level
^{d} Maximum emission wavelength calculated as the average ΔE_{S1-S0}^f values along the FC trajectories of Fig. 29b, c, starting 15 fs after the initial relaxation (state-average 2-roots SA2-CASSCF/AMBER//CASPT2/6-31G(d) level)

experimental and computational photochemical studies. In the following, we report the computational results obtained by using, as a reference test case, the a-ARM models for ASR_{WT} and its variants L83Q and W76S/Y179F.

We first assess the performance of automatic a-ARM QM/MM models for the reproduction of experimental trends in spectroscopic data (e.g., absorption and emission wavelength). Furthermore, in order to fully discuss the reproducibility of the set of reference results, we provide complete information on the features of these models (i.e., retinal configuration; rotamer for linker lysine; residues forming the chromophore cavity; protonation states of ionizable residues; and number/type of external counterions in each protein surface) obtained with the *input file generator* (see Sect. 3.1.2). Table 4 summarizes information on both model features and spectroscopic data for the a-ARM models. The main discrepancies between the automatic a-ARM models described in this table and the manual ARM models reported in Sect. 4.1, are related to a different choice of the protonation states that, in the latter case, were set up by visual inspection (i.e., ASH 217, HID 219, HIE 69, and HID 8; see Ref. [85]). As expected, this choice has an impact also on the number of external counter-ions (i.e., Inner surface: 5 Cl$^-$, Outer surface: 3 Na$^+$) that were manually placed.

As observed in Table 4, for the a-ARM models the difference between calculated and experimental λ_{max}^a for the three ASR_{AT} variants is lower than 3.0 kcal mol^{-1}, that is considered the expected error bar when using the a-ARM protocol [106]. The model with the largest deviation with respect to experimental λ_{max}^a is the double mutant W76S/Y179F, however, it is noticeable that the standard deviation is higher

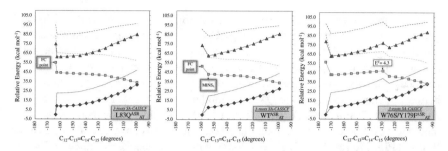

Fig. 27 $C_{13}=C_{14}$ **photoisomerization paths along** S_1, **using the automatic** a**-ARM models described in** Table 4. Energy profiles along S_1 (squares) isomerization paths. S_0 (diamonds) and S_2 (triangles) profiles along the S_1 path are also given. The S_1 is computed in terms of a relaxed scan along the C_{12}-$C_{13}=C_{14}$-C_{15} dihedral angle, using the 3-roots State-Averaged SA3-CASSCF/AMBER/6-31G(d) corrected at 3-roots CASPT2/6-31G(d) level of theory. Original data produced for this work

than the other cases and the experimental value falls within this error bar. On the other hand, the experimental order for the emission wavelength, λ_{max}^{f}, for the WT and W76S/Y179F is not respected when these values are calculated from the MINS$_1$ structure. However, when they are extracted from the FC trajectories (see Sect. 4.1) the order is reproduced. These results are consistent with Ref. [85], where it is explained that the values calculated from the MINS$_1$ lack the kinetic energy of the molecule, which is, instead, considered during the FC trajectories calculations.

We now turn our attention to the excited state reaction paths, from now on called Relaxed Scan (RS). Ideally, the RS should be computed from the MINS$_1$ geometry. However, for systems in which a MINS$_1$ is not located along the S_1 PES (e.g., ASR single mutant L83Q), an extrapolated path can be computed by using as input geometry the FC point. Therefore, RS plotted in Fig. 27 were computed from the MINS$_1$ geometry for both WT (Fig. 27, middle) and double mutant W76S/Y179F (Fig. 27, right), whereas for the case of the single mutant L83Q (Fig. 27, left) the FC point was used. These plots should be, in principle, compared with those reported in Fig. 17. The discontinuity observed around $-125°$ for the double mutant W76S/Y179F (Fig. 27c) is attributed to a change of conformation on the C15=N coordinate, which is not constrained during the geometry optimization. As observed in Fig. 27, the WT profile seems to present a small energy barrier and the double mutant W76S/Y179F presents a significant barrier, whereas the energy profile for the single mutant L83Q is flat up to the CI. These results are consistent with experimental observations reported in Ref. [85], where L83Q is not fluorescent, WT has a dim fluorescence and W76S/Y179F presents an enhanced fluorescence. The discrepancies observed in the behavior of S_1, S_2, and S_3 energy profiles when comparing both manual (Fig. 17) and automatic a-ARM models (Fig. 27) could be attributed to the use of a different methodology, which implies the setup of many variables such as the number of state-averaged roots to compute the constrained optimizations and the type and number of constraints imposed over the rPSB. Whereas the RS produced for the a-ARM models of Fig. 27

were generated via constrained 3-roots SA-CASSCF/AMBER//CASPT2/6-31G(d) geometry optimizations along the C_{12}-C_{13}=C_{14}-C_{15} dihedral angle (constrained three dihedral angles), in Ref. [85] the authors have, regrettably, specified neither the number of roots nor the type and number of constraints imposed over the rPSB for the geometry optimizations. Therefore, with the aim of performing a fair comparison between manual and automatic QM/MM models, we computed the RS for the WT and W76S/Y179F manual ARM QM/MM model this time using exactly the same methodology employed to compute the corresponding RS presented in Fig. 27b for WT and Fig. 27c for W76S/Y179F automatic a-ARM models, as shown in Fig. 28. It is evident that when using the same methodology, both manual and automatic QM/MM models present a similar behavior in terms of energy profiles. Remarkably, the discontinuity observed for the double mutant when using a-ARM models is also present when using manual ARM models.

Finally, we analyze the FC trajectories computed by using as input the ground state a-ARM QM/MM models described above for the three ASR$_{AT}$ variants. The energy profiles produced for a simulation time of 500 fs, plotted in Fig. 29 and to be compared with those shown in Fig. 18, are in good agreement with experimental data on decay lifetime. As observed, for a-ARM-based FC trajectories the single mutant L83Q exhibits a decay time of c.a. 120 fs, whereas the WT and the double mutant W76S/Y179F are not reactive during the simulation time. These results are in agreement with experimental time-resolved data, since the decay time is above around 270 fs for the former, while it is around 500 fs for the latter two [85]. However, it is evident a discrepancy between the S_1 and S_2 profiles when comparing FC trajectories for manual and automatic models. In this regard, we should mention that FC trajectories reported in Fig. 18 of Sect. 4.1 were produced by using a slightly different methodology with respect to that currently employed as a standard in a-ARM for producing Fig. 29. Specifically, and again due to an "operator" choice, whereas in the latter only the side-chain atoms of the residues forming the chromophore cavity were allowed to relax during the calculations, to be consistent with the philosophy of the ARM protocol [91, 106], in the former further degrees of freedom were considered to account for the delicate nature of the system by allowing to relax both the backbone and side-chain atoms of the residues forming the chromophore cavity during the calculations. Such a difference in methodology could explain that in the case of a-ARM based FC trajectories (Fig. 29), there is not an evident coupling between states S_1 and S_2. In order to test this hypothesis, and make a fair comparison between manual and automatic QM/MM models, we computed the FC trajectory for the WT manual ARM QM/MM model this time using exactly the same methodology employed to compute the trajectory presented in Fig. 29b for WT a-ARM model, as shown in Fig. 30. It is evident that when using the same methodology, both manual and automatic QM/MM models present a similar behavior in terms of energy paths (Fig. 30 top), Mulliken charges (Fig. 30 middle) and BLA (Fig. 30 bottom). Nevertheless, a further investigation (e.g., via benchmark calculations) is needed to justify whether or not the backbone atoms should be relaxed during the FC trajectory computations. Indeed, one of the current priorities on the development of a-ARM is the automation of the whole pool of the analysis reported in this section (i.e., location of MINS$_1$,

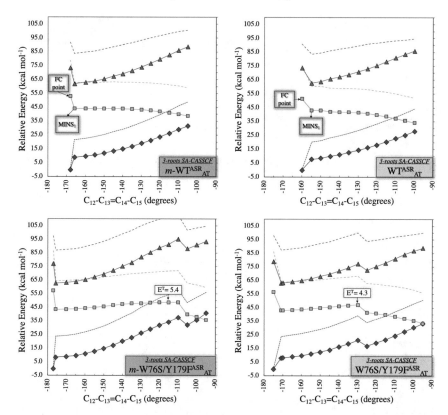

Fig. 28 $C_{13}=C_{14}$ **photoisomerization paths along** S_1**, using the manual ARM models reported in Ref.** [85] **(see** Sect. 4.1) **(left) and the automatic** a-ARM **models described in** Table 4 **(right).** Energy profiles along S_1 (squares) isomerization paths. S_0 (diamonds) and S_2 (triangles) profiles along the S_1 path are also given. The S_1 is computed in terms of a relaxed scan along the C_{12}-$C_{13}=C_{14}$-C_{15} dihedral angle, using the 3-roots State-Averaged SA3-CASSCF/AMBER/6-31G(d) corrected at 3-roots CASPT2/6-31G(d) level of theory. Original data produced for this work

computation of FC trajectories and calculation of photoisomerization paths along S_1), as a unique pipeline for the High-Throughput Screening of Rhodopsin Variants with Enhanced Fluorescence. In that work, current efforts are being performed to establish a congruent and standard methodology to guarantee full reproducibility of the obtained results.

From the computational analysis presented above, it is possible to confirm that the data produced with the new version of the ARM protocol is consistent with that reported when using the original version of the protocol. However, the discrepancies found for the different analysis performed when using ARM or a-ARM methodologies, exhibit the importance of the development of an automatic and standardized tool. Such a task is work in progress.

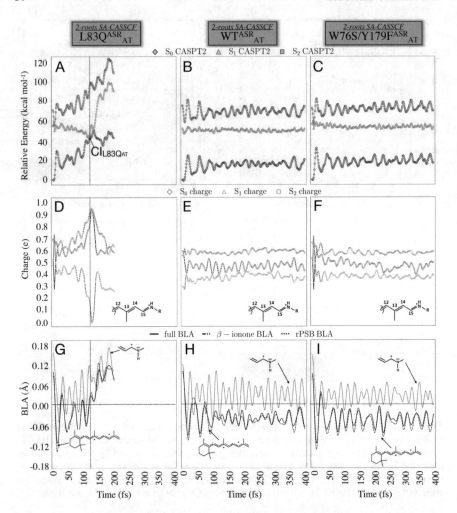

Fig. 29 Trajectory computation on S_1 for $L83Q_{AT}$, WT_{AT} and $W76S/Y179F_{AT}$, respectively, using as initial structure the automatic a-ARM models described in Table 4. **a–c** a-ARM QM/MM FC trajectories computed at two-root state-averaged-CASSCF/AMBER level of theory and corrected at the 3-root CASPT2 level. The backbone of the protein is fixed whereas the sidechain of the residues forming the chromophore cavity are allowed to relax during the computation. S_0 (diamonds), S_1 (squares), and S_2 (triangles) energy profiles along the FC trajectories. **d–f** Mulliken charge variations of the =CH-CH=CH-CH=NH$_2$ moiety of the chromophore. **g–i** Evolution of the total BLA (full line) and of the BLA of two specific moieties (dotted and dashed lines) for L83Q, WT, and W76S/Y179F, respectively. Original data produced for this work

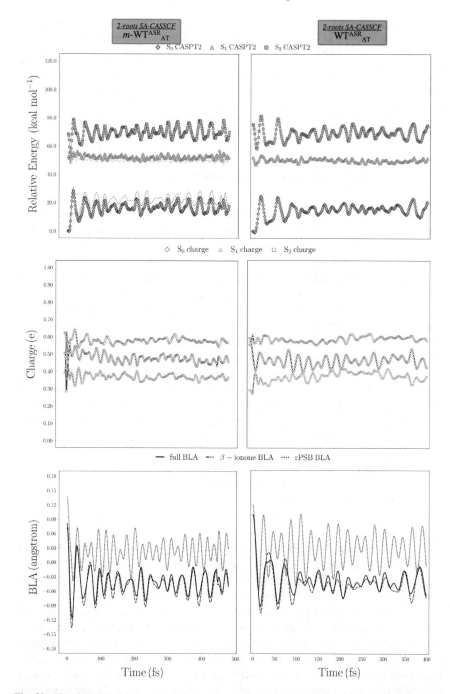

Fig. 30 (Continued)

◄ **Fig. 30 Trajectory computation on S_1 for WT$_{AT}$, using as initial structure the manual ARM models reported in Ref. [85] (see Sect. 4.1) (left) and the automatic a-ARM models described in** Table 4 **(right)**. (top) QM/MM FC trajectories computed at two-root state-averaged-CASSCF/AMBER level of theory and corrected at the 2-roots CASPT2 level; the backbone of the protein is fixed whereas the side-chain of the residues forming the chromophore cavity are allowed to relax during the computation. S_0 (diamonds), S_1 (squares), and S_2 (triangles) energy profiles along the FC trajectories. (middle) Mulliken charge variations of the =CH–CH=CH–CH=NH$_2$ moiety of the chromophore. (bottom) Evolution of the total BLA (full line) and of the BLA of two specific moieties (dotted and dashed lines). Original data produced for this work

6 Concluding Remarks and Perspectives

QM/MM models are by now established as powerful computational tools for treating chemical reactivity and other electronic processes in large molecules [11, 120, 126]. In their simplest form that corresponds, substantially, to the two-layer structure of the ARM models, they can be applied whenever one needs to simulate an electronic event (for instance, photoexcitation and bond breaking) occurring in a localized active-site under the influence of the interaction with a surrounding larger molecular environment. This has lead to tools for helping to build QM/MM models with different characteristics and according to the specific instructions provided by the user [127, 148]. Accordingly, an increasing number of applications in several branches of chemistry are being reported.

In spite of their obvious success, QM/MM models are not straightforward to build consistently and are thus often not easy to be reproduced in different laboratories due to different user choices during the building. In fact, as also discussed above, the building is not usually driven by "black-box" protocols and a user must, not only exercise great care during the selection of the appropriate QM and MM methods, frontier definition, and type of interaction between the corresponding QM and MM layers, but also understand the details of the structure preparation. For proteins, these details may include type, number, and location of counterions and protonation states of ionizable residues. Such complexity, which also expands the time necessary for model building, impairs important applications in protein design and in comparative protein studies.

With this chapter, we have tried to revise some recent work on the automation of the building protocol of QM/MM models focusing on a single class of light-responsive proteins. We argue that automation, in contrast to user manipulation, opens up new perspectives for comparative protein function studies even in a field as complex as photobiology where one has to describe the consequences of light absorption at both the electronic and geometrical levels. In fact, automation appears to be an unavoidable prerequisite for the production of sizable arrays (from hundreds to thousands) of congruous rhodopsin models: namely, models that can be used for investigating property trends, for designing novel optogenetic tools via *in silico* screening of mutant rhodopsins [89, 106] or for following evolutionary steps along the branches of a phylogenetic tree [81].

The ARM approach reveals that the price that one has to pay for automation is that of a lesser flexibility in the QM/MM model construction which, in principle, shall be the result of a completely defined building protocols involving minimal user choices. This translates in the production of specialized models valid within a single protein family and providing only specific information and above we have shown that a number of steps in such direction have been completed for rhodopsins. However, we have also shown that most of the applications reported until now are based on models whose input has still to be manually prepared. It is therefore clear that the next step along this research line would be the evaluation and application of rhodopsin QM/MM models prepared with a fully automated protocol such as a-ARM. Notice that the comparison between the experimentally observed and simulated quantities of the QM/MM models generated with the ARM and a-ARM protocol discussed above, confers to the corresponding models a significance which goes beyond that of having simply reproduced experimental trends or predicting new trends in, for instance, λ_{max}^a. More specifically, after having made sure that the QM/MM models are able to reproduce the trend in λ_{max}^a and λ_{max}^f values, the authors have used such models for mechanistic studies and, ultimately, for extracting rules useful for engineering new rhodopsins displaying novel properties.

Finally, we stress that the structure of the ARM protocols discussed in this chapter could, in principle, be replicated for other biologically or technologically important photoresponsive proteins (e.g., the natural photoactive yellow proteins [117] or the synthetic rhodopsin mimics [18] or the receptors that incorporate Light–Oxygen–Voltage (LOV) motifs and tetrapyrrole-binding phytochromes) [122]. Therefore, our research effort can also be considered the first step toward a more general photobiology tool applicable outside the rhodopsin area.

Acknowledgements Thanks to all current and past group members that have contributed to the ARM protocol: Federico Melaccio, Alessio Valentini, Leonardo Barneschi, Yoelvis Orozco-Gonzalez, Nicolas Ferré, Samira Gholami, Fabio Montisci, Silvia Rinaldi, Marco Cherubini, Xuchun Yang, Michael Stenrup, and Hoi-Ling Luk. A special thank goes to Dr. Samer Gozem of Georgia State University for a critical reading of the manuscript. The authors acknowledge a MIUR (Ministero dell'Istruzione, dell'Università e della Ricerca) Grant "Dipartimento di Eccellenza 2018-2022". The research has been partially supported by the following grants MIUR (PRIN 2015) and NSF CHE-CLP-1710191. The authors are also grateful for a USIAS 2015 fellowship from the University of Strasbourg (France) which has allowed further development of the method.

References

1. Agathangelou D, Orozco-Gonzalez Y, del Carmen Marín M, Roy P, Brazard J, Kandori H, Jung KH, Léonard J, Buckup T, Ferré N, Olivucci M, Haacke S (2018) Effect of point mutations on the ultrafast photo-isomerization of anabaena sensory rhodopsin. Faraday Discuss 207:55–75
2. Altun A, Yokoyama S, Morokuma K (2008) Mechanism of spectral tuning going from retinal in vacuo to bovine rhodopsin and its mutants: multireference ab initio quantum mechanics/molecular mechanics studies. J Phys Chem B 112(51):16883–16890

3. Altun A, Yokoyama S, Morokuma K (2008) Spectral tuning in visual pigments: an oniom (qm: Mm) study on bovine rhodopsin and its mutants. J Phys Chem B 112(22):6814–6827

4. Andersson K, Malmqvist PÅ, Roos BO (1992) Second-order perturbation theory with a complete active space self-consistent field function. J Chem Phys 96(2):1218–1226. https://doi.org/10.1063/1.462209, http://scitation.aip.org/content/aip/journal/jcp/96/2/10.1063/1.462209

5. Andruniów T, Ferré N, Olivucci M (2004) Structure, initial excited-state relaxation, and energy storage of rhodopsin resolved at the multiconfigurational perturbation theory level. Proc Natl Acad Sci USA 101(52):17908–17913

6. Aquilante F, Autschbach J, Carlson RK, Chibotaru LF, Delcey MG, De Vico L, Fdez Galván I, Ferré N, Frutos LM, Gagliardi L, Garavelli M, Giussani A, Hoyer CE, Li Manni G, Lischka H, Ma D, Malmqvist PÅ, Müller T, Nenov A, Olivucci M, Bondo Pedersen T, Peng D, Plasser F, Pritchard B, Reiher M, Rivalta I, Schapiro I, Segarra-Martí J, Stenrup M, Truhlar DG, Ungur L, Valentini A, Vancoillie S, Veryazov V, Vysotskiy VP, Weingart O, Zapata F, Lindh R (2016) Molcas8: new capabilities for multiconfigurational quantum chemical calculations across the periodic table. J Comput Chem 37(5):506–541. https://doi.org/10.1002/jcc.24221, https://doi.org/10.1002/jcc.24221

7. Battaglia S, Lindh R (2019) Extended dynamically weighted caspt2: the best of two worlds. arXiv preprint arXiv:191104996

8. Bja O, Spudich EN, Spudich JL, Leclerc M, DeLong EF (2001) Proteorhodopsin phototrophy in the ocean. Nature 411(6839):786–789

9. Bogomolni R, Spudich J (1987) The photochemical reactions of bacterial sensory rhodopsin-i. flash photolysis study in the one microsecond to eight second time window. Biophys J 52(6):1071–1075

10. Bouas-Laurent H, Dürr H (2001) Organic photochromism. Pure Appl Chem 73:639–665. https://doi.org/10.1351/pac200173040639

11. Boulanger E, Harvey J (2018) QM/MM methods for free energies and photochemistry. Curr Opin Struct Biol 49:72–76

12. Bravaya K, Bochenkova A, Granovsky A, Nemukhin A (2007) An opsin shift in rhodopsin: Retinal s0–s1 excitation in protein, in solution, and in the gas phase. J Am Chem Soc 129(43):13035–13042

13. Breed MF, Harrison PA, Blyth C, Byrne M, Gaget V, Gellie NJC, Groom SVC, Hodgson R, Mills JG, Prowse TAA, Steane DA, Mohr JJ (2019) The potential of genomics for restoring ecosystems and biodiversity. Nat Rev Genet 20:615–628. https://doi.org/10.1038/s41576-019-0152-0

14. Campomanes P, Neri M, Horta BAC, Röhrig UF, Vanni S, Tavernelli I, Rothlisberger U (2014) Origin of the spectral shifts among the early intermediates of the rhodopsin photocycle. J Am Chem Soc 136(10):3842–3851

15. Cheminal A, Leonard J, Kim S, Jung KH, Kandori H, Haacke S (2013) Steady state emission of the fluorescent intermediate of anabaena sensory rhodopsin as a function of light adaptation conditions. Chem Phys Lett 587:75–80

16. Chen F, Song Y, Li X, Chen J, Mo L, Zhang X, Lin Z, Zhang L (2019) Genome sequences of horticultural plants: past, present, and future. Hortic Res 6:112. https://doi.org/10.1038/s41438-019-0195-6

17. Cornell WD, Cieplak P, Bayly CI, Gould IR, Merz KM, Ferguson DM, Spellmeyer DC, Fox T, Caldwell JW, Kollman PA (1995) A second generation force field for the simulation of proteins, nucleic acids, and organic molecules. J Am Chem Soc 117(19):5179–5197. https://doi.org/10.1021/ja00124a002

18. Crist RM, Vasileiou C, Rabago-Smith M, Geiger JH, Borhan B (2006) Engineering a rhodopsin protein mimic. J Am Chem Soc 128(14):4522–4523

19. Davies MN, Toseland CP, Moss DS, Flower DR (2006) Benchmarking pk_a prediction. BMC Biochem 7(1):18–30

20. Deisseroth K (2011) Optogenetics. Nat Methods 8(1):26–29

21. Dolinsky TJ, Nielsen JE, McCammon JA, Baker NA (2004) Pdb2pqr: An automated pipeline for the setup of poisson-boltzmann electrostatics calculations. Nucleic Acids Res 32:W66–W667

22. Dolinsky TJ, Czodrowski P, Li H, Nielsen JE, Jensen JH, Klebe G, Baker NA (2007) Pdb2pqr: Expanding and upgrading automated preparation of biomolecular structures for molecular simulations. Nucleic Acids Res 35:W522–W525

23. Dong B, Sánchez-Magraner L, Luecke H (2016) Structure of an inward proton-transporting anabaena sensory rhodopsin mutant: mechanistic insights. Biophys J 111:963–972. https://doi.org/10.1016/j.bpj.2016.04.055

24. Doyle SA (2009) High throughput protein expression and purification: methods and protocols. Springer

25. Dundas J, Ouyang Z, Tseng J, Binkowski A, Turpaz Y, Liang J (2006) CASTp: computed atlas of surface topography of proteins with structural and topographical mapping of functionally annotated residues. Nucleic Acids Res 34(suppl_2):W116–W118

26. Enami N, Yoshimura K, Murakami M, Okumura H, Ihara K, Kouyama T (2006) Crystal structures of archaerhodopsin-1 and -2: common structural motif in archaeal light-driven proton pumps. J Mol Biol 358(3):675–685. https://doi.org/10.1016/j.jmb.2006.02.032

27. Engqvist MKM, McIsaac RS, Dollinger P, Flytzanis NC, Abrams M, Schor S, Arnold FH (2014) Directed evolution of gloeobacter violaceus rhodopsin spectral properties. J Mol Biol

28. Ernst OP, Lodowski DT, Elstner M, Hegemann P, Brown LS, Kandori H (2013) Microbial and animal rhodopsins: Structures, functions, and molecular mechanisms. Chem Rev 114(1):126–163

29. Ferré N, Ángyán JG (2002) Approximate electrostatic interaction operator for qm/mm calculations. Chem Phys Lett 356(3–4):331–339

30. Ferré N, Olivucci M (2003) The amide bond: pitfalls and drawbacks of the link atom scheme. THEOCHEM 632:71–82

31. Ferré N, Olivucci M (2003) Probing the rhodopsin cavity with reduced retinal models at the caspt2//casscf/amber level of theory. J Am Chem Soc 125(23):6868–6869

32. Ferré N, Cembran A, Garavelli M, Olivucci M (2004) Complete-active-space self-consistent-field/amber parameterization of the lys296-retinal-glu113 rhodopsin chromophore-counterion system. Theor Chem Acc 112(4):335–341

33. Filatov M, Liu F, Kim KS, Martínez TJ (2016) Self-consistent implementation of ensemble density functional theory method for multiple strongly correlated electron pairs. J Chem Phys 145(24):244104

34. Filatov M, Martínez TJ, Kim KS (2017) Description of ground and excited electronic states by ensemble density functional method with extended active space. J Chem Phys 147(6):064104

35. Finley J, Malmqvist P, Roos BO, Serrano-Andrés L (1998) The multi-state caspt2 method. Chem Phys Lett 288(2–4):299–306

36. Flytzanis NC, Bedbrook CN, Chiu H, Engqvist MK, Xiao C, Chan KY, Sternberg PW, Arnold FH, Gradinaru V (2014) Archaerhodopsin variants with enhanced voltage-sensitive fluorescence in mammalian and caenorhabditis elegans neurons. Nat Commun 5:4894–4902

37. Frutos LM, Andruniów T, Santoro F, Ferré N, Olivucci M (2007) Tracking the excited-state time evolution of the visual pigment with multiconfigurational quantum chemistry. Proc Natl Acad Sci 104(19):7764–7769

38. Fujimoto K, Jun-ya H, Hayashi S, Shigeki K, Nakatsuji H (2005) Mechanism of color tuning in retinal protein: Sac-ci and qm/mm study. Chem Phis Lett 414:239–242

39. Fujimoto K, Hayashi S, Hasegawa JY, Nakatsuji H (2007) Theoretical studies on the color-tuning mechanism in retinal proteins. J Chem Theory Comput 3(2):605–618

40. Geerke DP, Thiel S, Thiel W, van Gunsteren WF (2007) Combined qm/mm molecular dynamics study on a condensed-phase sn2 reaction at nitrogen: the effect of explicitly including solvent polarization. J Chem Theor Comput 3(4):1499–1509

41. Gholami S, Pedraza-González L, Yang X, Granovsky AA, Ioffe IN, Olivucci M (2019) Multi-state multi-configuration quantum chemical computation of the two-photon absorption spectra of bovine rhodopsin. J Phys Chem Lett 10(20):6293–6300

42. Gómez-Consarnau L, Raven JA, Levine NM, Cutter LS, Wang D, Seegers B, Arístegui J, Fuhrman JA, Gasol JM, Sañudo-Wilhelmy SA (2019) Microbial rhodopsins are major contributors to the solar energy captured in the sea. Sci Adv 5(8):eaaw8855

43. Gong Y, Li JZ, Schnitzer MJ (2013) Enhanced archaerhodopsin fluorescent protein voltage indicators. PLoS One 8(6):e66959–e66969

44. Govorunova EG, Sineshchekov OA, Li H, Spudich JL (2017) Microbial rhodopsins: diversity, mechanisms, and optogenetic applications. Annu Rev Biochem 86:845–872

45. Gozem S, Huntress M, Schapiro I, Lindh R, Granovsky A, Angeli C, Olivucci M (2012) Dynamic electron correlation effects on the ground state potential energy surface of a retinal chromophore model. J Chem Theory Comput 8:4069–4080

46. Gozem S, Luk HL, Schapiro I, Olivucci M (2017) Theory and simulation of the ultrafast double-bond isomerization of biological chromophores. Chem Rev 117(22):13502–13565

47. Granovsky AA (2011) Extended multi-configuration quasi-degenerate perturbation theory: the new approach to multi-state multi-reference perturbation theory. J Chem Phys 134:214113–214127

48. Guareschi R, Valsson O, Curutchet C, Mennucci B, Filippi C (2016) Electrostatic versus resonance interactions in photoreceptor proteins: the case of rhodopsin. J Phys Chem Lett 7(22):4547–4553

49. Hickey LT, Hafeez N, A, Robinson H, Jackson SA, Leal-Bertioli SCM, Tester M, Gao C, Godwin ID, Hayes BJ, Wulff BBH (2019) Breeding crops to feed 10 billion. Nat Biotechnol 37:744–754. https://doi.org/10.1038/s41587-019-0152-9

50. Hochbaum DR, Zhao Y, Farhi SL, Klapoetke N, Werley CA, Kapoor V, Zou P, Kralj JM, Maclaurin D, Smedemark-Margulies N, L SJ, Boulting GL, Straub C, Cho YK, Melkonian M, Wong GKS, Harrison DJ (2014) All-optical electrophysiology in mammalian neurons using engineered microbial rhodopsins. Nat Methods 11(8):825–833

51. Horbatenko Y, Lee S, Filatov M, Choi CH (2019) Performance analysis and optimization of mixed-reference spin-flip time-dependent density functional theory (mrsf-tddft) for vertical excitation energies and singlet-triplet energy gaps. J Chem A J Phys Chem A 123(37):7991–8000

52. Hosaka T, Yoshizawa S, Nakajima Y, Ohsawa N, Hato M, DeLong EF, Kogure K, Yokoyama S, Kimura-Someya T, Iwasaki W, Shirouzu M (2016) Structural mechanism for light-driven transport by a new type of chloride ion pump, Nonlabens marinus rhodopsin-3. J Biol Chem 291:17488–17495

53. Hou JH, Venkatachalam V, Cohen AE (2014) Temporal dynamics of microbial rhodopsin fluorescence reports absolute membrane voltage. Biophys J 106(3):639–648

54. Huang Y, Chen W, Wallace JA, Shen J (2016) All-atom continuous constant ph molecular dynamics with particle mesh ewald and titratable water. J Chem Theory Comput 12(11):5411–5421

55. Huix-Rotllant M, Filatov M, Gozem S, Schapiro I, Olivucci M, Ferré N (2013) Assessment of density functional theory for describing the correlation effects on the ground and excited state potential energy surfaces of a retinal chromophore model. J Chem Theory Comput 9:3917–3932

56. Humphrey W, Dalke A, Schulten K (1996) Vmd: visual molecular dynamics. J Mol Graph 14(1):33–38

57. Huntress M, Gozem S, Malley KR, Jailaubekov AE, Vasileiou C, Vengris M, Geiger JH, Borhan B, Schapiro I, Larsen D, Olivucci M (2013) Towards an understanding of the retinal chromophore in rhodopsin mimics. J Phys Chem B 117:10053–10070

58. Inoue K, Ito S, Kato Y, Nomura Y, Shibata M, Uchihashi T, Tsunoda SP, Kandori H (2016) A natural light-driven inward proton pump. Nat Commun 7:13415

59. Inoue K, Marín MdC, Tomida S, Nakamura R, Nakajima Y, Olivucci M, Kandori H (2019) Red-shifting mutation of light-driven sodium-pump rhodopsin. Nat Commun 10(1):1993

60. Jing Z, Liu C, Cheng SY, Qi R, Walker BD, Piquemal JP, Ren P (2019) Polarizable force fields for biomolecular simulations: recent advances and applications. Annu Rev Biophys 48:371–394

61. Kandori H (2019) Retinal proteins: photochemistry and optogenetics. Bull Chem Soc Jpn 93(1):76–85
62. Kandori H, Shichida Y, Yoshizawa T (2001) Photoisomerization in rhodopsin. Biochem (Mosc) 66(11):1197–1209
63. Karasuyama M, Inoue K, Nakamura R, Kandori H, Takeuchi I (2018) Understanding colour tuning rules and predicting absorption wavelengths of microbial rhodopsins by data-driven machine-learning approach. Sci Rep 8(1):15580
64. Kato HE, Zhang F, Yizhar O, Ramakrishnan C, Nishizawa T, Hirata K, Ito J, Aita Y, Tsukazaki T, Hayashi S, Hegemann P, Maturana AD, Ishitani R, Deisseroth K, Nureki O (2012) Crystal structure of the channelrhodopsin light-gated cation channel. Nature 482:369
65. Kato HE, Inoue K, Abe-Yoshizumi R, Kato Y, Ono H, Konno M, Hososhima S, Ishizuka T, Hoque MR, Kunitomo H, Ito J, Yoshizawa S, Yamashita K, Takemoto M, Nishizawa T, Taniguchi R, Kogure K, Maturana AD, Iino Y, Yawo H, Ishitani R, Hideki K, Nureki O (2015) Structural basis for na+ transport mechanism by a light-driven na+ pump. Nature 521(7550):48–53
66. Kawanabe A, Furutani Y, Jung KH, Kandori H (2007) Photochromism of anabaena sensory rhodopsin. J Am Chem Soc 129(27):8644–8649
67. Kim K, Kwon SK, Jun SH, Cha JS, Kim H, Lee W, Kim JF, Cho HS (2016) Crystal structure and functional characterization of a light-driven chloride pump having an ntq motif. Nat Commun 7:12677–12687
68. Kim SY, Waschuk SA, Brown LS, Jung KH (2008) Screening and characterization of proteorhodopsin color-tuning mutations in escherichia coli with endogenous retinal synthesis. Biochim Biophys Acta 1777:504–513
69. Kircher M, Kelso J (2010) Highthroughput dna sequencing-concepts and limitations. Bioessays 32(6):524–536
70. Klapoetke NC, Murata Y, Kim SS, Pulver SR, Birdsey-Benson A, Cho YK, Morimoto TK, Chuong AS, Carpenter EJ, Tian Z (2014) Independent optical excitation of distinct neural populations. Nat Methods 11(3):338
71. Kouyama T, Fujii R, Kanada S, Nakanishi T, Chan SK, Murakami M (2014) Structure of archaerhodopsin-2 at 1.8 å resolution. Acta Crystallogr D 70(10):2692–2701
72. Kralj JM, Hochbaum DR, Douglass AD, Cohen AE (2011) Electrical spiking in escherichia coli probed with a fluorescent voltage-indicating protein. Science 333(6040):345–348
73. Kralj JM, Douglass AD, Hochbaum DR, Maclaurin D, Cohen AE (2012) Optical recording of action potentials in mammalian neurons using a microbial rhodopsin. Nat Methods 9(1):90–95
74. Lam SD, Das S, Sillitoe I, Orengo C (2017) An overview of comparative modelling and resources dedicated to large-scale modelling of genome sequences. Acta Crystallographica Section D 73(8):628–640. https://doi.org/10.1107/S2059798317008920, https://doi.org/10.1107/S2059798317008920
75. Laricheva E, Gozem S, Rinaldi S, Melaccio F, Valentini A, Olivucci M (2012) Origin of fluorescence in 11-cis locked bovine rhodopsin. J Chem Theory Comput 8:2559–2563
76. Le Guilloux V, Schmidtke P, Tuffery P (2009) Fpocket: an open source platform for ligand pocket detection. BMC Bioinform 10(1):168
77. Le Guilloux V, Schmidtke P, Tuffery P (2009) Fpocket: an open source platform for ligand pocket detection. Bioinformatics 10(1):168–179
78. Li S, Tang K, Liu K, Yu CP, Jiao N (2014) Parvularcula oceanus sp. nov., isolated from deep-sea water of the southeastern pacific ocean. Antonie van Leeuwenhoek 105(1):245–251
79. Luecke H, Schobert B, Lanyi JK, Spudich EN, Spudich JL (2001) Crystal structure of sensory rhodopsin ii at 2.4 angstroms: insights into color tuning and transducer interaction. Science 293(5534):1499–1503
80. Luk HL, Melaccio F, Rinaldi S, Gozem S, Olivucci M (2015) Molecular bases for the selection of the chromophore of animal rhodopsins. Proc Natl Acad Sci USA 112:15297–15302
81. Luk HL, Bhattacharyya N, Montisci F, Morrow JM, Melaccio F, Wada A, Sheves M, Fanelli F, Chang BSW, Olivucci M (2016) Modulation of thermal noise and spectral sensitivity in lake baikal cottoid fish rhodopsins. Sci Rep 6:38425

82. Maclaurin D, Venkatachalam V, Lee H, Cohen AE (2013) Mechanism of voltage-sensitive fluorescence in a microbial rhodopsin. Proc Natl Acad Sci USA 110(15):5939–5944

83. Manathunga M, Yang X, Luk HL, Gozem S, Frutos LM, Valentini A, Ferré N, Olivucci M (2016) Probing the photodynamics of rhodopsins with reduced retinal chromophores. J Chem Theory Comput 12(2):839–850

84. Manathunga M, Yang X, Orozco-Gonzalez Y, Olivucci M (2017) Impact of electronic state mixing on the photoisomerization time scale of the retinal chromophore. J Phys Chem letters 8(20):5222–5227

85. Marín MdC, Agathangelou D, Orozco-Gonzalez Y, Valentini A, Kato Y, Abe-Yoshizumi R, Kandori H, Choi A, Jung KH, Haacke S, Olivucci M (2019) Fluorescence enhancement of a microbial rhodopsin via electronic reprogramming. J Am Chem Soc 141(1):262–271

86. McIsaac RS, Engqvist MK, Wannier T, Rosenthal AZ, Herwig L, Flytzanis NC, Imasheva ES, Lanyi JK, Balashov SP, Gradinaru V, Arnold FH (2014) Directed evolution of a far-red fluorescent rhodopsin. Proc Natl Acad Sci USA 111(36):13034–13039

87. McIsaac RS, Bedbrook CN, Arnold FH (2015) Recent advances in engineering microbial rhodopsins for optogenetics. Curr Opin Struct Biol 33:8–15

88. McLaughlin S (1989) The electrostatic properties of membranes. Annu Rev Biophys Biomol Struct 18(1):113–136

89. Melaccio F, Ferré N, Olivucci M (2012) Quantum chemical modeling of rhodopsin mutants displaying switchable colors. Phys Chem Chem Phys 14:12485–12495

90. Melaccio F, del Carmen Marín M, Valentini A, Montisci F, Rinaldi S, Cherubini M, Yang X, Kato Y, Stenrup M, Orozco-Gonzalez Y, Ferré N, Luk HL, Kandori H, Olivucci M (2016) Towards automatic rhodopsin modeling as a tool for high-throughput computational photobiology. J Chem Theor Comput 12:6020–6034

91. Melaccio F, Marín MdC, Valentini A, Montisci F, Rinaldi S, Cherubini M, Yang X, Kato Y, Stenrup M, Orozco-González Y, Ferré N, Luk HL, Kandori H, Olivucci M (2016) Toward automatic rhodopsin modeling as a tool for high-throughput computational photobiology. J Chem Theor Comput 12(12):6020–6034

92. Miesenböck G (2011) Optogenetic control of cells and circuits. Annu Rev Cell Dev Biol 27:731–758

93. Moore DS (1985) Amino acid and peptide net charges: a simple calculational procedure. Biochem Educ 13(1):10–11

94. Murakami M, Kouyama T (2008) Crystal structure of squid rhodopsin. Nature 453(7193):363–367

95. Murakami M, Kouyama T (2011) Crystallographic analysis of the primary photochemical reaction of squid rhodopsin. J Mol Biol 413(3):615–627

96. Nass Kovacs G, Colletier JP, Grünbein ML, Yang Y, Stensitzki T, Batyuk A, Carbajo S, Doak RB, Ehrenberg D, Foucar L, Gasper R, Gorel A, Hilpert M, Kloos M, Koglin JE, Reinstein J, Roome CM, Schlesinger R, Seaberg M, Shoeman RL, Stricker M, Boutet S, Haacke S, Heberle J, Heyne K, Domratcheva T, Barends TRM, Schlichting I (2019) Three-dimensional view of ultrafast dynamics in photoexcited bacteriorhodopsin. Nat Commun 10(1):3177

97. Nishikawa T, Murakami M, Kouyama T (2005) Crystal structure of the 13-*cis* isomer of bacteriorhodopsin in the dark-adapted state. J Mol Biol 352(2):319–328

98. Nogly P, Weinert T, James D, Carbajo S, Ozerov D, Furrer A, Gashi D, Borin V, Skopintsev P, Jaeger K, Nass K, Bath P, Bosman R, Koglin J, Seaberg M, Lane T, Kekilli D, Brünle S, Tanaka T, Wu W, Milne C, White T, Barty A, Weierstall U, Panneels V, Nango E, Iwata S, Hunter M, Schapiro I, Schertler G, Neutze R, Standfuss J (2018) Retinal isomerization in bacteriorhodopsin captured by a femtosecond x-ray laser. Science eaat0094

99. Oda K, Vierock J, Oishi S, Rodriguez-Rozada S, Taniguchi R, Yamashita K, Wiegert JS, Nishizawa T, Hegemann P, Nureki O (2018) Crystal structure of the red light-activated channelrhodopsin chrimson. Nat Commun 9(1):3949

100. Okada T, Fujiyoshi Y, Silow M, Navarro J, Landau EM, Shichida Y (2002) Functional role of internal water molecules in rhodopsin revealed by x-ray crystallography. Proc Natl Acad Sci 99(9):5982–5987

101. Okada T, Sugihara M, Bondar AN, Elstner M, Entel P, Buss V (2004) The Retinal Conformation and its Environment in Rhodopsin in Light of a New 2.2 Å crystal structure. J Mol Biol 342(2):571–583
102. Olsson MH, Søndergaard CR, Rostkowski M, Jensen JH (2011) Propka3: Consistent treatment of internal and surface residues in empirical pk_a predictions. J Chem Theor Comput 7(2):525–537
103. Orozco-Gonzalez Y, Manathunga M, del Carmen Marín M, Agathangelou D, Jung KH, Melaccio F, Ferré N, Haacke S, Coutinho K, Canuto S, Olivucci M (2017) An average solvent electrostatic configuration protocol for qm/mm free energy optimization: implementation and application to rhodopsin systems. J Chem Theor Comput 13(12):6391–6404
104. Pagano K, Paolino M, Fusi S, Zanirato V, Trapella C, Giuliani G, Cappelli A, Zanzoni S, Molinari H, Ragona L, Olivucci M (2019) Bile acid binding protein functionalization leads to a fully synthetic rhodopsin mimic. J Phys Chem Lett 10:2235–2243
105. Park JW, Shiozaki T (2017) Analytical derivative coupling for multistate caspt2 theory. J Chem Theor Comput 13(6):2561–2570
106. Pedraza-González L, De Vico L, Marín MdC, Fanelli F, Olivucci M (2019) a-arm: automatic rhodopsin modeling with chromophore cavity generation, ionization state selection and external counter-ion placement. J Chem Theor Comput 15(5):3134–3152. https://doi.org/10.1021/acs.jctc.9b00061
107. Pedraza-González L, Marín MdC, Jorge AN, Ruck TD, Yang X, Valentini A, Olivucci M, De Vico L (2020) Web-arm: a web-based interface for the automatic construction of qm/mm models of rhodopsins. J Chem Inf Model 60(3):1481–1493. https://doi.org/10.1021/acs.jcim.9b00615, https://doi.org/10.1021/acs.jcim.9b00615
108. Pieri E, Ledentu V, Sahlin M, Dehez F, Olivucci M, Ferré N (2019) Cphmd-then-qm/mm identification of the amino acids responsible for the anabaena sensory rhodopsin ph-dependent electronic absorption spectrum. J Chem Theor Comput 15:4535–4546
109. Po HN, Senozan N (2001) The henderson-hasselbalch equation: its history and limitations. J Chem Edu 78(11):1499–1503
110. Prigge M, Schneider F, Tsunoda SP, Shilyansky C, Wietek J, Deisseroth K, Hegemann P (2012) Color-tuned channelrhodopsins for multiwavelength optogenetics. J Biol Chem 287(38):31804–31812. https://doi.org/10.1074/jbc.M112.391185, http://www.jbc.org/content/287/38/31804.abstract
111. Pronk S, Páll S, Schulz R, Larsson P, Bjelkmar P, Apostolov R, Shirts MR, Smith JC, Kasson PM, Van Der Spoel D, Hess B, Lindahl E (2013) Gromacs 4.5: a high-throughput and highly parallel open source molecular simulation toolkit. Bioinformatics 29(7):845–854
112. Pushkarev A, Béjà O (2016) Functional metagenomic screen reveals new and diverse microbial rhodopsins. ISME J 10(9):2331
113. Rackers JA, Wang Z, Lu C, Laury ML, Lagardére L, Schnieders MJ, Piquemal JP, Ren P, Ponder JW (2018) Tinker 8: software tools for molecular design. J Chem Theor Comput 14(10):5273–5289
114. Raja R, Lee YS, Streicher K, Conway J, Wu S, Sridhar S, Kuziora M, Liu H, Higgs BW, Brohawn PZ, Bais C, Jallal B, Ranade K (2017) Integrating genomics into drug discovery and development: challenges and aspirations. Pharm Med (Cham, Switz) 31:217–233. https://doi.org/10.1007/s40290-017-0192-8
115. Ran T, Ozorowski G, Gao Y, Sineshchekov OA, Wang W, Spudich JL, Luecke H (2013) Cross-protomer interaction with the photoactive site in oligomeric proteorhodopsin complexes. Acta Crystallogr D 69(10):1965–1980
116. Reijenga J, Van Hoof A, Van Loon A, Teunissen B (2013) Development of methods for the determination of pk_a values. Anal Chem Insights 8:53–71
117. Reis J, Woolley G (2016) Photo control of protein function using photoactive yellow protein. Methods Mol Biol 1408:79–92
118. Roos BO, Taylor PR, Siegbahn PEM (1980) A complete active space scf method (casscf) using a density matrix formulated super-ci approach. Chem Phys 48(2):157–173

119. Rozin R, Wand A, Jung KH, Ruhman S, Sheves M (2014) ph dependence of anabaena sensory rhodopsin: retinal isomer composition, rate of dark adaptation, and photochemistry. J Phys Chem B 118(30):8995–9006
120. Ryde U (2016) Qm/mm calculations on proteins. Methods Enzymol 577:119–158
121. Schapiro I, Ryazantsev MN, Frutos LM, Ferré N, Lindh R, Olivucci M (2011) The ultrafast photoisomerizations of rhodopsin and bathorhodopsin are modulated by bond alternation and hoop driven electronic effects. J Am Chem Soc 133:3354–3364
122. Schmidt D, Cho YK (2015) Natural photoreceptors and their application to synthetic biology. Trends Biotechnol 33:80–91
123. Seib KL, Dougan G, Rappuoli R (2009) The key role of genomics in modern vaccine and drug design for emerging infectious diseases. PLoS Genet 5(10):1–8. https://doi.org/10.1371/journal.pgen.1000612, https://doi.org/10.1371/journal.pgen.1000612
124. Sekharan S, Sugihara M, Buss V (2007) Origin of spectral tuning in rhodopsin. it is not the binding pocket. Angewandte Chemie Int Ed 46(1-2):269–271
125. Senn HM, Thiel W (2009) QM/MM methods for biomolecular systems. Angew Chem 48(7):1198–1229
126. Senn HM, Thiel W (2009) QM/MM methods for biomolecular systems. Angew Chem Int Ed 48(7):1198–1229
127. Sherwood P, de Vries AH, Guest MF, Schreckenbach G, Catlow CRA, French SA, Sokol AA, Bromley ST, Thiel W, Turner AJ (2003) Quasi: a general purpose implementation of the qm/mm approach and its application to problems in catalysis. J Mol Struct: THEOCHEM 632(1–3):1–28
128. Sinicropi A, Andruniow T, De Vico L, Ferré N, Olivucci M (2005) Toward a computational photobiology. Pure Appl Chem 77(6):977–993
129. Snyder JW Jr, Curchod BF, Martínez TJ (2016) Gpu-accelerated state-averaged complete active space self-consistent field interfaced with ab initio multiple spawning unravels the photodynamics of provitamin d3. J Phys Chem Lett 7(13):2444–2449
130. Stein CJ, Reiher M (2019) Autocas: a program for fully automated multiconfigurational calculations. J Comput Chem
131. Strambi A, Durbeej B, Ferré N, Olivucci M (2010) Anabaena sensory rhodopsin is a light-driven unidirectional rotor. Proc Natl Acad Sci USA 107(50):21322–21326
132. Teller DC, Okada T, Behnke CA, Palczewski K, Stenkamp RE (2001) Advances in determination of a high-resolution three-dimensional structure of rhodopsin, a model of g-protein-coupled receptors (gpcrs). Biochemistry 40(26):7761–7772
133. Tomasello G, Olaso-González G, Altoè P, Stenta M, Serrano-Andrés L, Merchán M, Orlandi G, Bottoni A, Garavelli M (2009) Electrostatic control of the photoisomerization efficiency and optical properties in visual pigments: on the role of counterion quenching. J Am Chem Soc 131(14):5172–5186
134. Tomida S, Ito S, Inoue K, Kandori H (2018) Hydrogen-bonding network at the cytoplasmic region of a light-driven sodium pump rhodopsin KR2. Biochim Biophys Acta—Bioenerg. https://doi.org/10.1016/j.bbabio.2018.05.017
135. Tsukamoto H, Terakita A (2010) Diversity and functional properties of bistable pigments. Photochem Photobiol Sci 9:1435–1443. https://doi.org/10.1039/C0PP00168F, http://dx.doi.org/10.1039/C0PP00168F
136. Udvarhelyi A, Olivucci M, Domratcheva T (2015) Role of the molecular environment in flavoprotein color and redox tuning: QM cluster versus qm/mm modeling. J Chem Theor Comput 11:3878–3894
137. Valsson O, Campomanes P, Tavernelli I, Rothlisberger U, Filippi C (2013) Rhodopsin absorption from first principles: bypassing common pitfalls. J Chem Theor Comput 9(j):2441–2454
138. Van Keulen SC, Solano A, Rothlisberger U (2017) How rhodopsin tunes the equilibrium between protonated and deprotonated forms of the retinal chromophore. J Chem Theor Comput 13(9):4524–4534
139. Virshup AM, Punwong C, Pogorelov TV, Lindquist BA, Ko C, Martínez TJ (2009) Photodynamics in complex environments: ab initio multiple spawning quantum mechanical/molecular mechanical dynamics. J Phys Chem B 113(11):3280–3291

140. Vivas MG, De Boni L, Mendonça CR (2018) Chapter 8 - Two-Photon Spectroscopy of Organic Materials. In: Gupta V (ed) Molecular and laser spectroscopy, Elsevier, pp. 165–191. https://doi.org/10.1016/B978-0-12-849883-5.00008-5, http://www.sciencedirect.com/science/article/pii/B9780128498835000085

141. Vogeley L, Sineshchekov OA, Trivedi VD, Sasaki J, Spudich JL, Luecke H (2004) Anabaena sensory rhodopsin: a photochromic color sensor at 2.0 å. Science 306(5700):1390–1393

142. Volkov O, Kovalev K, Polovinkin V, Borshchevskiy V, Bamann C, Astashkin R, Marin E, Popov A, Balandin T, Willbold D, Büldt G, Bamberg E, Gordeliy V (2017) Structural insights into ion conduction by channelrhodopsin 2. Science 358:eaan8862

143. Volkov O, Kovalev K, Polovinkin V, Borshchevskiy V, Bamann C, Astashkin R, Marin E, Popov A, Balandin T, Willbold D, Büldt G, Bamberg E, Gordeliy V (2017) Structural insights into ion conduction by channelrhodopsin 2. Science 358:1018

144. Walczak E, Andruniów T (2015) Impacts of retinal polyene (de) methylation on the photoisomerization mechanism and photon energy storage of rhodopsin. Phys Chem Chem Phys 17(26):17169–17181

145. Wanko M, Hoffmann M, Frauenheim T, Elstner M (2006) Computational photochemistry of retinal proteins. J Comput-Aided Mol Des 20(7–8):511–518

146. Warshel A (1976) Bicycle-pedal model for the first step in the vision process. Nature 260:679–683. https://doi.org/10.1038/260679a0

147. Warshel A (1981) Calculations of enzymatic reactions: calculations of pk_a, proton transfer reactions, and general acid catalysis reactions in enzymes. Biochemistry 20:3167–3177

148. Weingart O, Nenov A, Altoè P, Rivalta I, Segarra-Martí, Dokukina I, Garavelli M (2018) Cobramm 2.0–a software interface for tailoring molecular electronic structure calculations and running nanoscale (qm/mm) simulations. J Mol Model 24(9):271

149. Wen L, Wang H, Tanimoto S, Egawa R, Matsuzaka Y, Mushiake H, Ishizuka T, Yawo H (2010) Opto-current-clamp actuation of cortical neurons using a strategically designed channelrhodopsin. PLoS One 5(9):1–13. https://doi.org/10.1371/journal.pone.0012893, https://doi.org/10.1371/journal.pone.0012893

150. Yu JK, Liang R, Liu F, Martínez TJ (2019) First principles characterization of the elusive i fluorescent state and the structural evolution of retinal protonated schiff base in bacteriorhodopsin. J Am Chem Soc 141(45):18193–18203

151. Zhang L, Hermans J (1996) Hydrophilicity of cavities in proteins. Proteins: Struct, Funct, Bioinform 24(4):433–438

Photo-Active Biological Molecular Materials: From Photoinduced Dynamics to Transient Electronic Spectroscopies

Irene Conti, Matteo Bonfanti, Artur Nenov, Ivan Rivalta, and Marco Garavelli

Abstract We present an overview of a methodology for the simulation of the photo-response of biological (macro)molecules, designed around a Quantum Mechanics / Molecular Mechanics (QM/MM) subtractive scheme. The resulting simulation work-flow, that goes from the characterization of the photo-active system to the modeling of (transient) electronic spectroscopies is implemented in the software COBRAMM, but is completely general and can be used in the framework of any specific QM/MM implementation. COBRAMM is a smart interface to existing state-of-the-art theoretical chemistry codes, combining different levels of description and different algorithms to realize tailored problem-driven computations. The power of this approach is illustrated by reviewing the studies of two fundamental problems involving biological light-sensitive molecules. First, we will consider the photodynamics of the retinal molecule, the pigment of rhodopsin, a visual receptor protein contained in the rod cells of the retina. Retinal, with its light-induced isomerization, triggers a cascade of events leading to the production of the nerve impulse. Then, we will review some studies focusing on the interaction of DNA systems with ultraviolet (UV) light, a

---

I.C., M.B., A.N. and I.R. equally contributed to this work.

I. Conti · M. Bonfanti · A. Nenov · I. Rivalta · M. Garavelli (✉)
Dipartimento di Chimica Industriale Toso Montanari, Universita di Bologna, Viale del Risorgimento 4, 40136 Bologna, Italy
e-mail: marco.garavelli@unibo.it

I. Conti
e-mail: irene.conti@unibo.it

M. Bonfanti
e-mail: matteo.bonfanti@unibo.it

A. Nenov
e-mail: artur.nenov@unibo.it

I. Rivalta
e-mail: i.rivalta@unibo.it

I. Rivalta
Univ Lyon, Ens de Lyon, CNRS, Université Lyon 1, Laboratoire de Chimie, UMR 5182, 69342 Lyon, France

© Springer Nature Switzerland AG 2021
T. Andruniów and M. Olivucci (eds.), *QM/MM Studies of Light-responsive Biological Systems*, Challenges and Advances in Computational Chemistry and Physics 31,
https://doi.org/10.1007/978-3-030-57721-6_2

problem that has become one of the benchmark for the development of nonlinear spectroscopy, because of the ultrashort excited state lifetimes that arise from very efficient radiationless excited state decay and consent self-protection of DNA against UV damage.

Keywords Transient electronic spectroscopy · Excited state dynamics · Retinal photo-isomerization · DNA excited state decay · COBRAMM · Computational spectroscopy

1 Introduction

The interaction with light plays an essential role in biology. In many biological (macro)molecules, the absorption of light triggers extremely complex phenomena, that have been optimized by a long natural evolution and show unique characteristics and performances. In this chapter, we will consider two representative and very diverse examples. In the rhodopsin protein, a visual receptor contained in the rod cells of the retina, the photoexcitation is used to initiate an isomerization reaction that is the first step in a sequence of events, leading to the production of the nerve impulse. In DNA, on the other hand, the absorption of light is undesired since it can lead to the formation of free radicals and subsequently to harmful reactions, thus very efficient photoprotection mechanisms have been developed to quench the excitations by a fast internal conversion.

A detailed molecular understanding of these phenomena is incredibly challenging, as they involve several electronic states of molecular systems embedded in a complex environment and in wide ranges of energy and time. The relevant processes to consider are very different and include absorption, emission, internal conversion, intersystem crossing, chemical reactions. To disentangle the photophysical and photochemical complexity is then required to accurately describe the evolution in time, with a controlled selection of the initial conditions and a high resolution in energy and time.

In this sense, an invaluable contribution has come from time-resolved spectroscopy. Since experiments with microsecond long pulses were first realized in the 50s [152, 170], a great effort has been put in achieving ever shorter pulses. Nowadays, the advanced techniques that are available for generating laser pulses in the nanosecond-femtosecond scale, and even down to the attoseconds, have given the opportunity to study a wide range of physical phenomena, ranging from processes as slow as protein conformational changes to others as fast as electron dynamics.

In the broad variety of spectroscopic experiments, two-dimensional electronic spectroscopy (2DES) plays a prominent role [31, 188]. In this technique, a sequence of ultrashort light pulses with femtosecond duration is employed to obtain a wealth of very detailed information. This is often crucial to disentangle contributions that overlap in less advanced laser spectroscopies [39, 89, 94, 100, 145].

2D optical spectroscopy was initially used to investigate vibrational dynamics using infrared (IR) light. In this context, the technique was useful to determine the coupling between vibrational modes [84] and to study the dynamics on the picosecond timescale [74, 110]. More recently, the 2D technique has been extended to the range of visible light, with the signal that is resonant with electronic transitions [33, 49]. 2DES has proved particularly useful to investigate the structure and the dynamics of multi-chromophoric systems [34, 140], since it is best suited to the task of mapping how an excitation is transferred from one chromophore to another. In more detail, 2DES has many advantages over conventional 1D femtosecond transient absorption (TA) spectroscopy, for example, it allows to directly observe and quantify the couplings between different chromophores from the cross peaks appearing in the 2DES maps, it can monitor in real time energy relaxation and energy transfer between different electronic levels, it allows to single out the various contributions to the linewidth broadening of each optical transition [31].

The methodology is well established in the near-IR and in the visible range, where it has been used with great success to study a broad variety of problems, ranging from photosynthetic complexes [34, 205] to semiconductors [197, 213]. The extension to ultraviolet (UV), although is made more challenging by a number of technical difficulties, has also been realized and opens up the possibility of studying a multitude of molecular targets [31] and benchmarking theoretical models using chromophores of small molecular size [30, 150].

Understanding time-resolved experimental spectra in general, and 2D spectra, in particular, is not always immediate nor straightforward. In this context, the comparison with theory can be crucial to unambiguously identify spectral signatures and to determine the origin of their time evolution, e.g., by disentangling the various deactivation pathways that contribute to the lifetime of a given signal. However, simulations are not trivial, as they require high-level methods and the combination of different approaches. In more detail, a very accurate modeling of the electronic excitations of the chromophore(s) is required, described within a faithful representation of the environment, that in the context of the biological system can be the solvent alone or also some protein embedding. Furthermore, to account for the evolution of the system between the pulses, an extensive characterization of the excited state potential energy surfaces (PES) or an explicit dynamical simulation may also be needed.

In this chapter, we will describe a complete simulation protocol, that goes from the characterization of the photo-active biological system to the modeling of (transient) electronic spectroscopies. This protocol has been built in the last few years around a Quantum Mechanics / Molecular Mechanics (QM/MM) subtractive scheme, initially developed in our software COBRAMM [5, 6], a smart interface to existing state-of-the-art theoretical chemistry codes. COBRAMM has grown over the years and now includes the tools and capabilities to carry out the complete simulation protocol in an effortless and unique workflow [222].

Other combined classical and quantum mechanical approaches are commonly used to describe ultrafast photoprocesses in biological and artificial photosystems, where the total QM/MM energy can either be described within a subtractive or an additive scheme [27, 35, 68, 102, 117, 193], with advantages and disadvantages in either approach. In both schemes, the whole system is divided into a QM inner part, a MM outer part and a link region that requires separate treatment. In the subtractive scheme, after computing the energy of the whole system, the high layer energy is substituted by its QM energy. Extended subtractive schemes are applied in the IMOMM [70] and ONIOM [54] methods implemented in the Gaussian set of routines, the COBRAMM QM/MM code [5, 222] and as an option in GROMACS. Instead, in the additive scheme the total energy is a sum of the outer system MM energy, the QM energy for the inner system plus link and terms coupling the inner and outer layers. The latter scheme is, for example, available in the ChemShell suite of programs [138], in GROMACS [1], MOLCAS [1], and in NWChem [215].

In Sect. 2, we will give a schematic account of the general methodology. In the following two sections, we will describe two illustrative cases: the photoisomerization dynamics of the retinal molecule (Sect. 3) and the simulation of spectroscopic experiments for some DNA systems (Sect. 4). In Sect. 5, we will conclude and present some perspectives regarding the future development of the QM/MM simulation scheme.

2 Methodology

The strategy that we have chosen to study photo-active biological molecules are based on a QM/MM scheme for representing the chromophore and its environment. Usually, a high level ab initio method is necessary for the quantum mechanics (QM) part, since a reliable description of the electronic transitions are required to reach spectroscopic accuracy. Furthermore, biological systems need to be studied in their native condensed-phase environment and this requires an explicit consideration of the electrostatic embedding of the chromophore. Starting from these QM/MM calculations, we can then model spectroscopic experiments using different frameworks that depend on the type of experiment: linear (absorption, emission) and transient (pump-probe, 2D).

A "static" picture, obtained by characterizing some reference points of the PES, suffices for the simplest applications, e.g., for the simulation of linear absorption spectra, or when strong approximations are applicable in the spectroscopy modeling. However, in the most complex cases, nuclear dynamics need to be explicitly considered. In the cases that will be addressed here, this has been done with non-adiabatic trajectories computed with Tully's Fewest Switches Surface Hopping (FSSH) [209, 211]. Trajectories have fundamental limitations deriving from the underlying classical approximation on the nuclear motion, but have considerable practical advantages: they are computationally cheap, can be computed on-the-fly by using only first derivatives of the potential and appeal to chemical intuition by describing the dynamics in terms of hops between adiabatic electronic surfaces.

We schematically describe the general framework that we used to describe the systems with four steps (see Fig. 1):

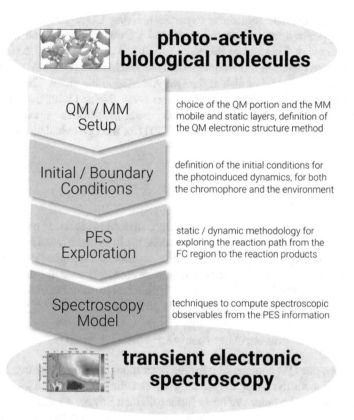

Fig. 1 Schematic representation of the modeling strategy

1. **QM/MM setup**: including the definition of the QM portion, the molecular mechanics (MM) mobile and static layers, and the choice of the QM electronic structure method;
2. **initial/boundary conditions**: the definition of appropriate boundary conditions (for static applications) or initial conditions (for dynamics application), for both the chromophore and the environment;
3. **PES exploration**: choice of the static/dynamics methodology adopted to explore the reaction path, from the Franck–Condon (FC) region to the photoproducts;
4. **spectroscopic modeling**: definition of the techniques to simulate the spectroscopic observables from the PES information obtained along the reaction path.

In the applications that we will present, this complete workflow has been realized with the help of COBRAMM [222]. COBRAMM is a software package that interfaces widely known software for molecular modeling, combining different levels of description (molecular mechanics / electronic structure) and different algorithms (optimizations, reaction path computations, vibrational analysis, adiabatic/non-adiabatic molecular dynamics). Using COBRAMM, one can then select the preferred features of the interfaced software packages and use them in a synergistic way. This allows the setup of tailored problem-driven computations, not only at the quantum mechanical/molecular mechanical (QM/MM) level which will be described here, but also at the QM and MM levels alone. The calculations that will be described in the following have been obtained with the interfaces to Amber [37] (for the MM calculations), Gaussian [69] (for the nuclear geometry optimizations), Molpro [10], and Molcas [223] (for the QM calculations).

2.1 QM/MM Setup

The QM/MM methodology implemented in COBRAMM is based on a subtractive scheme [193], in which we identify two models: the full system S and a smaller, inner portion I—the *high layer*—that we want to describe at the higher level. When the inner part is connected to the rest by one or more covalent bonds, the dangling bonds left in I are saturated by hydrogen atoms. Within this partitioning of the system, the global QM/MM energy $E_{QM/MM}(S)$ is then given by

$$E_{QM/MM}(S) = E_{MM}(S) + E_{QM}(I) - E_{MM}(I) \tag{1}$$

and requires one QM computation for the inner model, $E_{QM}(I)$, and two MM computations of both the inner model $E_{MM}(I)$ and the full system $E_{MM}(S)$. In our implementation, the outer region is further divided in two, a *medium layer* and a *low layer*. The former contains MM atoms that are allowed to move in geometry optimizations or molecular dynamics simulations. The latter consists of atoms that are kept frozen, and thus, define the boundary conditions of the calculation. Further details on the QM/MM implementation are available in Ref. [222].

In modeling a biological photo-active system, the choice of the QM/MM layers is intuitive and straightforward. The chromophore—understood as the part of the molecule in which the relevant electronic excitations are localized—is described at the full QM level. The remainder—e.g., the solvent, the protein environment of biological pigments, the sugar, and phosphate skeleton of nucleic acids—can be treated at the lower MM level. Within the MM part, the medium layer is usually composed by a variable portion of the covalently bound structure and the close solvation shell (possibly including the counterions). The size of the mobile portion varies according to the problem considered. As a rule of thumb, when doing a structure relaxation or a geometry optimization the movable part can be kept smaller than in dynamical simulations, where stronger deformations are expected. However, the setup needs to be defined in accordance with the specific characteristic of the problem under study. As an example, when considering the ultrafast dynamics of a chromophore, we can freeze the protein environment, excluded the closest residues, since the typical motions of a protein take place on a much longer timescale, and thus, should have a negligible effect [166].

The assignment of the high/medium/low layers a priori is not so straightforward for complex biological systems if the photo-active moiety is, for example, inside a protein environment [200]. It is essential to verify that some important quantum effects of the medium or low layers are not missed: a benchmark procedure to distinguish the "photoreaction center", including atoms that are directly involved in the photoreaction, and the "spectator" region, in which the atoms do not directly participate in the reaction, is necessary.

For example, for some processes in solution, few solvent molecules directly interacting with the chromophore (e.g., via strong H-bonds) should be comprised in the high layer. As a general rule of thumb, the first few solvation shells should be included in the medium layer, in order to leave appropriate flexibility to the high layer portion. The bulk solvent, instead, influences the photoreaction just via long-range interactions and is normally included in the fixed low layer.

Similarly, inside proteins, the photo-process in the active site could involve protein residues (or water molecules) directly interacting with the chromophore, and should be included (when possible) in the high layer. The other residues within photoreaction active site should, in principle, be part of the medium layer, unless (like in opsins) the photoreaction is known to involve space-saving mechanisms. The rest of the system could be considered as an electrostatic background that may or may not facilitate the photoinduced processes.

Moreover, regarding the high-medium interface, in case of involvement of covalent bonds between QM and MM portions, the placement of the bond(s) cut(s) must be considered carefully, avoiding artifacts due to erroneous charge distributions. In general, the cut(s) should be sufficiently distant from the chromophoric part, and cuts through double or higher order bonds, as well as cuts through bonds of different atom types (e.g., a C–N bond) must be avoided. Senn and Thiel suggest cutting an 'innocent' C–C bond ca. 3–4 bonds away from the active component [193].

Further evolution of these hybrid methods is the adaptive QM/MM which reclassifies atoms/groups as QM or MM on-the-fly continuously and smoothly during dynamics simulations, which is not allowed in conventional QM/MM [62, 88, 163]. The present method can be used to combine multilevel methods with sampling schemes for systems with atoms or groups of atoms entering or leaving the active zone (the QM region) during the simulation.

Particular importance should be given to the choice of the QM electronic structure technique, as it greatly affects the accuracy of the electronic transition. An inadequate QM description can often lead not only to a significant quantitative mismatch, but even to a completely wrong qualitative picture of the spectroscopic features. Single-reference methods are usually good enough for linear absorption, since doubly excited states and charge transfer states do not contribute significantly to the overall signal. In this context, Time-Dependent Density Functional Theory (TDDFT) is particularly convenient, since one can adjust the QM level by choosing the functional that shows the best agreement with the experimental results.

If one wants to consider excited state dynamics, computationally expensive methods are usually needed, as Multireference Configuration Interaction (MRCI) and Multireference Perturbation Theory (MRPT). In recent years, it has now become possible to do optimizations, or even run MD simulations, at the level of MRPT, which is the gold standard for many problems involving electronic transitions in correlated organic systems. This has been made possible by the combination of advancements in the electronic structure, the growing available computational power, and one of the COBRAMM main development, i.e., the trivially parallel implementation of numerical first derivatives for computing energy gradients and non-adiabatic couplings (NAC).

TDDFT and Algebraic-Diagrammatic Construction (ADC) perform relatively well in simulating non-adiabatic dynamics involving excited states but they have some problems with a certain family of molecules: the order of the states is sometimes incorrect (e.g., adenine [183]), some states are missing (e.g., doubly excited states), charge transfer transitions can be overestimated (e.g., in retinal [57]). Furthermore, these methods do not describe correctly conical intersection topology (a problem common to SS-CASPT2 as well), and this prevents the prediction of quantum yields. Nevertheless, some functionals give better results than others on these particular systems, confirming that the development of new functionals is still a useful work.

In general, multi-configurational methods with an additional treatment based on perturbation theory, e.g., CASPT2 or RASPT2, are the recommended QM approaches, when applicable, for the chromophoric units in photoreactive centers. These approaches should guarantee accuracy below 0.2 eV for computations of electronic transition energies with respect to experiments and a qualitative agreement when comparing transition dipole moments with experimental absorption spectra. The biggest bottleneck of explicitly correlated multi-configurational methods is the lack of analytical gradients. Only recently analytical gradients and non-adiabatic couplings have been introduced for the extended multistate variant of the CASPT2 method, however, their use is still limited by a large amount of memory required.

Relevantly, comparison against available experimental data have to consider the details of the experimental conditions (e.g., chemical environments, aggregation phenomena, measured observable) and the choice of computation parameters (e.g., active space, basis-set, state average level, etc.) should be carefully benchmarked, trying to push the computational costs to the limit to set the best available theoretical reference. For the excited state MD simulations, the multi-configurational approaches usually access sampling within 1 ps for a small chromophore (e.g., a pyrimidine nucleobase), reducing to few tens of femtoseconds for larger molecules. Active space-based approaches, in fact, suffer from a factorial scaling with the active space size, which until recently restrained their application to small and medium-size molecules. Recent implementations of density matrix renormalization group (DMRG) [59] and variational 2-electron reduced-density-matrix-driven (v2RDM) [66], that allow to handle active spaces with many tens of orbitals are becoming more and more popular.

Single-reference wavefunction-based methods—like algebraic diagrammatic construction (ADC) [60] and coupled-cluster (CC)—and TDDFT represent an excellent and cheaper alternative to multi-configurational approaches. However, these methods (a) exhibit weakness in the description of charge transfer states; (b) are blind to doubly excited states; (c) fail at describing the topology of the S1/S0 conical intersection seam. For these reasons, one should carefully benchmark single-reference methods against more accurate multi-configurational results. When the comparison is good, single-reference methods then allow treating much larger system sizes, i.e., multi-chromophoric systems, and longer sampling time, up to tens of ps.

Still, the availability of high-level benchmark calculations and the possibility of one-to-one comparisons with spectroscopic experiments are now encouraging the adoption of computationally cheaper QM methods, even semiempirical methods as OMx-MRCI [212, 214], using careful "calibration" strategies [222]. This is an interesting and a very promising route to address problems that extend to larger dimensions and longer timescales, provided that the cheap method gives at least qualitatively good results.

2.2   *Boundary and Initial Conditions*

When biological systems are considered, the definition of the starting geometry (or geometries) for the simulation already presents a considerable challenge. Organic molecules are, in fact, intrinsically "floppy", and the condensed-phase environment is characterized by an extremely large and chaotic conformational space. The situation is complicated by the fact that spectroscopic signals may be extremely sensitive to some of the geometrical parameters, as the stacking of aromatic rings, the torsion angles of conjugated double bonds, or the distance of the chromophore from charged residues and counterions.

In MD simulations, the starting geometry is obviously important as it defines the initial conditions of the trajectory, and thus, needs to be sampled according to the physical properties—e.g., temperature and pressure—of the system. However, it should be noted that within our QM/MM scheme, the choice of the starting geometry is determinant also in static calculations, since the setup of the frozen low layer defines the boundary conditions of the simulation.

In general, two types of problems need to be considered regarding the sampling of initial/boundary conditions. On the one hand, the biological system may be characterized by the presence of more than one stable conformation, and each conformer may present different types of spectral signatures. For this reason, an initial conformational analysis may be necessary and samples should be selected among the most relevant stable structures. On the other hand, the nuclear motions of both the chromophore and the environment is associated with a certain level of disorder, due to quantum effects and thermal agitation, and this is determinant for the lineshape of the spectroscopic signals. Although lineshapes can in many cases be modeled with a phenomenological broadening, they carry interesting information, e.g., they can show a vibronic structure, or can give an insight into the environmental disorder or the lifetime of the electronic excitation. It would be then desirable to describe lineshapes within the theoretical model, by using an appropriate sampling of the nuclear configuration space.

To describe the lineshape within the QM/MM framework, the phase-space sampling has to take into account two different sources of broadening of the electronic transitions:

- the thermal and quantum fluctuations of the chromophore vibrations,
- the conformational disorder of the environment, at the given temperature and pressure of the global system.

The conformational disorder of the environment can be simply described by running an equilibration dynamics for the full system, and then sampling the resulting canonical ensemble. However, given the length of the equilibration dynamics, usually the necessary simulation is out of reach with the QM/MM setup, and hence, it is done at a lower MM level. The quantum fluctuations of the chromophore, instead, need to be described with an inherently quantum sampling scheme. One of the most common choice is to adopt a Wigner sampling scheme, based on the assumption of a harmonic approximation of the potential around the chromophore ground-state minimum.

The algorithm that we have refined over the years to take into account the signal broadening is then constructed as follows [151]. First, the whole system consisting of the chromophore and the environment is equilibrated by performing a NPT molecular dynamics simulation. Depending on the flexibility of the chromophore, one or more snapshots are then selected to represent the relevant conformations in the equilibrated environment. For a small and rigid system, the lowest energy snapshot is usually enough. In complex cases that are characterized of well separated stable conformers, a more rigorous cluster analysis may be necessary. The selected chromophore + environment conformation(s) are then optimized at the QM/MM level, to let the closest atoms adjust to the higher level of description of the QM part. At this point, a vibrational analysis is needed to compute the harmonic expansion of the potential and to obtain displaced geometries of the chromophore according to the Wigner distribution [16]. During this step, we have usually adopted the practice of freezing soft vibrational modes, for which the harmonic approximation breaks down and the Wigner sampling gives unreasonable results. Other authors have developed alternative strategies, e.g., by adjusting the chromophore to a higher temperature corresponding to the zero-point energy of its vibrational modes [126]. In practice, the Wigner sampling in COBRAMM is realized through an interface with a stand-alone script, part of the JADE program [61], which considers temperature effects. This is usually enough to describe the most important source of signal broadening, while the conformational disorder of the environment can be considered phenomenologically by an arbitrary broadening of the transition lines. Optionally, one can remove the bias that comes from the use of a single or a few samples of the equilibration dynamics, by additionally propagating with MD in the microcanonical ensemble the MM part, at QM fixed. This last step reintroduces some broadening coming from the disorder of the environment. When computing linear absorption spectra, we work in a "snapshot approximation", in which we do not perform any dynamics of the state, thus all the static disorder that we introduce in the sampling contributes to the inhomogeneous broadening of the lineshapes. Signal broadening due to spectral diffusion (i.e., homogeneous broadening) and finite excited state lifetime must be added phenomenologically. However, when simulating pump-probe spectra with trajectories, we are considering a "dynamically evolving" static disorder that describes, at least in an approximate way, homogeneous broadening effects.

This procedure described above is feasible only when the environment is relatively simple, e.g., is made by a solvent droplet. In more complex cases, e.g., when the chromophore is embedded in a protein, it becomes unavoidable to resort to an experimentally refined molecular structure [166]. In these cases, the availability of a high resolution structural study then becomes an essential condition.

2.3 PES Exploration

When one wants to simulate a transient spectroscopy experiment, it is required to explore the excited state PES beyond the FC region. For this purpose, a hierarchy of methods is available in COBRAMM, ranging from a simple static characterization of the PES to the simulation of the dynamics after photoexcitation with mixed quantum-classical MD.

The most basic option is to map the significant points of the PES: excited state minima, transition states (ST), and conical intersections (CI). For this purpose, it is crucial to rely on a robust and efficient technique for determining stationary points of the PES. The "hybrid" strategy that is implemented in COBRAMM is based on the use of the optimization algorithm of Gaussian [69], taking advantage of the "external" option that allows the use of energy and gradient from an independent source.

For minima and TS, COBRAMM uses the standard Gaussian algorithms, that require only the gradient of the energy. When QM analytical first derivatives are available, the gradient is constructed according to the QM/MM subtractive scheme, Eq. 1, using contributions that come from the external QM and MM codes and are then combined internally by COBRAMM. Otherwise, COBRAMM has an internal implementation for computing numerical first derivatives by finite difference, allowing to determine stationary points of the PES at any level, including MRPT.

In the case of CI, the optimizations are performed employing Bearpark's gradient projection method [22]:

$$\mathbf{g} = 2(E_1 - E_2)\frac{\mathbf{X}_1}{||\mathbf{X}_1||} + \mathbf{P}_{\perp X_1, X_2}(\nabla_R E_2) \tag{2}$$

E_1 and E_2 are the adiabatic energies, $\mathbf{X}_1$ and $\mathbf{X}_2$ are the gradient difference and the derivative coupling vectors, $\nabla_R E_2$ is the gradient on the higher state, while $\mathbf{P}_{\perp X_1, X_2}(\nabla_R E_2)$ is its projection on the $3N - 8$ complement orthogonal to the intersection space spanned by $\mathbf{X}_1$ and $\mathbf{X}_2$. The effective gradient $\mathbf{g}$ is computed inside COBRAMM and forwarded to the Gaussian optimizer With our algorithm, we can take into consideration two embedding effects: the electrostatic interaction of the QM part with the field of the MM charges appearing in the QM calculation, and the full system gradient that enters in the $\mathbf{X}_1$ vector. The only missing term is the part of the non-adiabatic coupling vector that refers to the MM atoms, but this can be assumed to be negligible if the QM layer is large enough.

This static approach has the advantage of giving a complete characterization of the PES and a comprehensive picture of the possible dynamical paths. Static information can even be utilized in modeling the transient spectroscopy (see next section) assuming some mechanistic model for the population decay. However, as the size of the system grows, the configuration space becomes larger and more structured, and it may become difficult or even impossible to identify and characterize all the relevant features of the PES. Furthermore, the static treatment cannot answer the question whether an energetically accessible channel is really populated or not during the photodynamics of the system.

In these conditions, it is convenient to turn to molecular dynamics. The first, simplest level of description is intermediate between a static and a dynamical picture. It consists of finding the photochemical minimum energy path (MEP) that starts from the FC point. In practice, one can initialize the system in the minimum geometry of the ground state, and then follow the excited state gradient until the CI seam is met and a hop to a lower adiabatic PES is to be expected. This idealized path, that represents the 0 K overdamped classical MD, neglects temperature and dynamical effects but can already give a picture of the process by indicating the portion of the configuration space that is most likely visited during the dynamics.

However, the most complete and thorough picture of the transient evolution comes from a realistic dynamical description of the process. In this context, the typical framework that we adopt is a mixed quantum-classical approximation [210], that neglects the quantum character of the nuclear motion but allows us to work in a trajectory-based picture, in which the system evolves on the adiabatic PES and is subjected to instantaneous hops from one surface to another [211]. This is much more costly than the static or pseudo-static approaches outlined above, since it necessitates a proper sampling of the initial conditions and then requires to run at least few tens of trajectories, with the computational cost coming from the large number of single-point QM calculations that are necessary to compute the forces necessary to the trajectory integrator.

There are different ways of handling the non-adiabatic hops. In a very simplified picture, one can rely on the fact that non-adiabatic couplings are large only in the vicinity of CI, and thus, can assume that hops occur from the higher to the lower surface every time the system gets close to the intersection seam. This procedure is, for example, very useful to investigate the photoreaction mechanism and to follow the relaxation path, although it cannot accurately determine the quantum yield of the processes involved. To obtain more accurate results, it is necessary to use an advanced stochastic description of the non-adiabatic hops, as in the case of Tully's Fewest Switches Surface Hopping (FSSH) [209, 211].

COBRAMM implements all the necessary tools for the aforementioned strategies. The interface with Gaussian intrinsic reaction coordinate (IRC) algorithm is exploited to construct photochemical MEPS. Adiabatic MD trajectories can be computed with an internal implementation of the velocity Verlet integration of Hamilton's equations of motion. Non-adiabatic jumps can be enforced with a number of algorithms. In particular, Tully's FSSH is implemented in COBRAMM with the popular decoherence correction of Granucci and Persico [81]. Further details on the FSSH implementation are available in Ref. [222].

2.4 Spectroscopic Modeling

The methodologies that we adopt for the simulation of experimental results cover a broad range of techniques, ranging from steady-state linear spectroscopies, both in absorption and emission, to transient spectroscopies with a pump-probe or a 2DES setup.

The simplest case is the simulation of absorption spectra that requires a very limited characterization of the ground state PES. Assuming Franck–Condon approximation and a classical description of the nuclei, the absorption intensity is computed by summing over all the vertical transitions at a fixed nuclear configuration $\mathbf{R}$ starting from electronic state g, with a weighting given by the corresponding oscillator strength $f_{e \leftarrow g}(\mathbf{R})$:

$$I(\omega, \mathbf{R}) \propto \sum_e f_{e \leftarrow g}(\mathbf{R}) \, \delta(E^{(e)}(\mathbf{R}) - E^{(g)}(\mathbf{R}) - \hbar\omega) \tag{3}$$

where $E^{(e)}(\mathbf{R})$ and $E^{(g)}(\mathbf{R})$ are the excited and ground state adiabatic energies, computed at the nuclear configuration $\mathbf{R}$.

To take into account the different contributions to the broadening of the spectral lines, as described in Sect. 2.2, we need to sample the nuclear configuration $\mathbf{R}$ from a classical or *quasi*-classical distribution, and repeat the computation of the adiabatic energies $E^{(e)}$, $E^{(g)}$ and of the oscillator strengths $f_{e \leftarrow g}$ for each of the chosen configuration. Consequently, Eq. 3 becomes our working equation

$$\bar{I}(\omega) \propto \sum_i \sum_e f_{e \leftarrow g}(\mathbf{R}_i) \, \delta(E^{(e)}(\mathbf{R}_i) - E^{(g)}(\mathbf{R}_i) - \hbar\omega) \tag{4}$$

where the sum over i extends on the set of sampled nuclear geometries R_i. In practice, the Dirac delta of the vertical transition in Eqs. 3–4, are substituted with Gaussian functions, that take into account with a phenomenological broadening a part of the lineshape width.

Emission is modeled in a very similar fashion, only reversing the role of the ground and excited states and substituting the oscillator strength with the appropriate rate coefficients

$$W_{se} \propto \sum_i A_{g \leftarrow e}(\mathbf{R}_i) \, \delta(E^{(e)}(\mathbf{R}_i) - E^{(g)}(\mathbf{R}_i) - \hbar\omega) \tag{5}$$

where W_{se} is the rate of spontaneous emission and $A_{g \leftarrow e}(\mathbf{R}_i)$ is Einstein's A coefficient computed from the transition dipole moment at the nuclear configuration $\mathbf{R}_i$) [14]. This is made possible by the assumption that the emission process takes place from an excited state minima after the system has fully equilibrated. Under these conditions, we can apply the same framework of the absorption modeling, by choosing a configuration distribution that describes the nuclear geometries at equilibrium in the excited states, and summing up the vertical transitions that start from the excited state and end in the ground state.

Pump-probe spectroscopy can still be considered as an extension of linear spectroscopy techniques, in which the pump excites the system in a given electronic state and then the probe measures the absorption and emission from that target state after a variable delay t_d from the pump. In practice, we can start from configurations sampling the ground state distribution, as in the absorption simulation. For each of these samples i, we excite the electronic state to the target state determined by the energy of the pump. We then let the system evolve with NVE molecular dynamics for a time t_d, to simulate the evolution of the system during the delay. At each instant of time, the transient spectrum is then calculated as a superposition of stimulated emission to lower electronic states and photoinduced absorption to higher electronic states, again assuming simple vertical transitions

$$
\bar{I}(\omega, t_d) \propto \sum_i \left[\sum_f f_{f \leftarrow e}(\mathbf{R}_i(t_d)) \, \delta(E^{(f)}(\mathbf{R}_i(t_d)) - E^{(e)}(\mathbf{R}_i(t_d)) - \hbar\omega) \right.
$$
$$
\left. - \sum_i f_{g \leftarrow e}(\mathbf{R}_i(t_d)) \, \delta(E^{(e)}(\mathbf{R}_i(t_d)) - E^{(g)}(\mathbf{R}_i(t_d)) - \hbar\omega) \right]
\tag{6}
$$

where e denotes a state from the manifold of excited states visited by the system during the relaxation dynamics at the instance t_d.

When excited state dynamics simulations are too costly or the dynamics long (ps timescales and beyond), we resort to a drastic approximation, that we refer to as the "static picture", in which we treat the chromophores as uncoupled from the bath of vibrations, and thus, as a closed quantum system having only electronic degrees of freedom, and we adopt an entirely phenomenological description of the lineshape. In this context we perform electronic structure calculations at equilibrium structures encountered during the PES exploration to simulate the spectral signatures of long-lived intermediates. The assumption behind this practice is that the population that does exhibit an ultrafast (i.e., on sub-ps time scale) decay, must get trapped in an excited state local minimum, possibly also dissipating partially the initially acquired vibrational energy. Therefore, one can assume that the electronic structure of the local minimum would dominate the time-resolved spectrum for pump-probe delay times in the order of the excited state lifetime. Within this framework and by neglecting the nuclear configuration sampling, Eq. 6 simplifies to

$$
\bar{I}(\omega, \mathbf{R}) \propto \sum_f f_{f \leftarrow e}(\mathbf{R}) \, \delta(E^{(f)}(\mathbf{R}) - E^{(e)}(\mathbf{R}) - \hbar\omega)
$$
$$
- f_{g \leftarrow e}(\mathbf{R}) \, \delta(E^{(e)}(\mathbf{R}) - E^{(g)}(\mathbf{R}) - \hbar\omega)
\tag{7}
$$

where $\mathbf{R}$ denotes the coordinates of the local minima, intermediately populated during the excited state decay. While being rather harsh, this approximation allows to push the limits of the electronic structure methods used. To account for the finite pulsewidths used in the experiments, the calculated transient signals in Eqs. 6 and 7, can be broadened in time by a convolution with a Gaussian function.

The simulation of 2DES is much more challenging, as the 2D spectra refer to the third-order nonlinear polarization of the sample, generated by a sequence of three ultrashort laser pulses interacting with the sample in precisely controlled time intervals t_1 and t_2, and emitted in specific phase-matched direction. Thereby, a 2D spectrum (or map) is obtained for a fixed "waiting time" t_2 (analogous to the time delay between the pump and the probe pulse in 1D pump-probe (PP) spectroscopy) by a 2D Fourier transformation of the emitted signal $S(t_1,t_2,t_3)$ as the function of the time interval t_3 after the interaction with the third pulse in the sequence and the delay time t_1, i.e., $S(\omega_1,t_2,\omega_3)$. In this way, a correlation is established between the wavelength of excitation and probing, thus increasing the spectral resolution without sacrificing temporal resolution. The conventional 1D PP signal can be formally seen (and mathematically obtained) as the marginal of the 2D signal along ω_1. Within several controlled approximations (rotating wave approximation (RWA), impulsive limit) the nonlinear polarization emitted in a given phase-matched direction becomes proportional to the third-order nonlinear response $R^{(}3)(t_1, t_2, t_3)$ of the sample [3, 144].

Similarly, as for the linear spectroscopy, in our strategy we focus on the accuracy of the electronic description, rather than on the nuclear motion [149, 188]. Again, we adopt the "static picture" and rely on electronic structure calculations at equilibrium structures along the PES exploration. The third-order nonlinear response $R^{(}3)(t_1, t_2, t_3)$ along the phase-matching direction consists of three contributions, ground state bleach (GSB), stimulated emission (SE), and excited state absorption (ESA), each one associated with a different combination of electronic transitions excited by the incident pulse sequence (also known as Feynman diagrams). Within the "static picture" approximation and by assuming a "long" (ps) dynamics to neglect coherences during t2, these three contributions read

$$R_{GSB}^{(3)}(t_1, t_3) = +\left(\frac{\imath}{\hbar}\right)^3 \sum_{f,e} \mu_{eg}^4 \exp\left(-\imath\xi_{eg}t_3 - \imath\xi_{ge}t_1\right)$$

$$R_{SE}^{(3)}(t_1, t_2, t_3) = +\left(\frac{\imath}{\hbar}\right)^3 \sum_{f,e,e'} \mu_{ge'}^2\mu_{eg}^2 \exp\left(-\imath\xi_{e'g}t_3 - \imath\xi_{ge}t_1\right) \mathbb{G}_{e\to e'}(t_2) \qquad (8)$$

$$R_{ESA}^{(3)}(t_1, t_2, t_3) = -\left(\frac{\imath}{\hbar}\right)^3 \sum_{f,e,e'} \mu_{fe'}^2\mu_{eg}^2 \exp\left(-\imath\xi_{fe'}t_3 - \imath\xi_{ge}t_1\right) \mathbb{G}_{e\to e'}(t_2)$$

where $\xi_{ab} = \omega_{ab} - \imath\gamma_{ab}$ contains the adiabatic transition energies $\omega_{ab} = \varepsilon_a - \varepsilon_b$ and the phenomenological dephasing constants γ_{ab}. μ_{ij} denote transition dipole moments between the manifolds of ground (g) and excited (e, e', f) states. $\mathbb{G}_{e\to e'}(t_2)$ is a function describing the electronic dynamics of the system during the time interval t_2.

This function is usually computed by solving some rate equations. The transition $e \rightarrow e'$ could describe both evolution in the same electronic state and population transfer to another electronic state. The constants γ_{ab} sum up all the different contributions that determine the lineshape of the 2D signals, including the broadening due to the finite lifetime of the excited state and all the sources already described in Sect. 2.2.

This above "static" approach can be thought as the spectroscopic counterpart of the MEP protocol as it provides information about the change of the *electronic* structure along a 0 K overdamped path but has no information about the timescale on which these changes occur. This information is reconstructed based on the energy barriers connecting the minima to the conical intersection seam and is implicitly incorporated in the spectra simulations by summing only over surviving states (e') at a given "waiting time" t_2.

To overcome the strong limitations of the "static" picture, we need to go beyond the use of a phenomenological dephasing function, using a function that includes effects coming from the intramolecular vibrational dynamics, the interaction with the environment and from the excited state lifetimes. We have attempted to re-introduce the vibrational dynamics with a methodology based on a simple model, in which the electronic degrees of freedom are linearly coupled to a Gaussian bath, known as cumulant expansion of Gaussian fluctuations (CGF)[3]. In practice, a composite line-shape function (also called phase function) is introduced in the expression of the 2D signal, Eq. 8. This lineshape function includes two types of terms: (1) a contribution coming from the vibrational progressions in the spectra due to undamped molecular vibrations that are represented with multidimensional uncoupled displaced harmonic oscillator (DHO) model [144] and (2) a contribution that accounts for the homogeneous broadening due to the coupling to a low frequency modes continuum, which is described by the semi-classical Brownian oscillator (OBO) model [115]. Thereby vibrational coherences are described through memory-conserving expressions for the phase function during all three periods t_1 to t_3 at various levels of sophistication, whereas population dynamics is treated incoherently (i.e., in the Markovian limit) on the basis of rate models. The time-dependent population in each state obtained as a solution of the rate equations, acts as time-dependent weighting factors of the state specific signatures [3, 150, 164]. The CGF formulation is exact for adiabatic dynamics, while non-adiabatic effects can be included at various levels of sophistication [2]. Generalizations beyond the harmonic approximation [7], and weak system-bath coupling have also been reported recently [162].

3 Retinal Systems

Manipulation of the photophysical and photochemical properties of retinal chromophores by modification of their chemical formula (protonation states, functionalization, etc.) or change of their surroundings (proteins, solvent, etc.) has great potential for technological applications [23, 32, 228]. Retinals are indeed the light-sensitive molecular units of natural opsin pigments and are tunable chromophores that can, for instance, achieve impressive color tuning if embedded in artificial proteins [93, 221] or modulate their excited state decay among different timescales in solution when synthetically functionalized [20, 196]. Fundamental understanding of retinal intrinsic photophysical and photochemical properties and elucidation of subtle environmental effects of the surroundings (proteins, solvent, organic/inorganic supports, etc.) are crucial for their potential applications in biomimetic electronic devices and optical memories [23, 32, 120, 228].

The synergy between advanced computational and experimental studies represents the main route to obtain essential insights into retinals photochemistry, allowing elucidation of both the intrinsic behavior of these chromophores and the role of environmental effects. Atomistic modeling of photoinduced processes in retinals, accompanied by simulations of spectroscopic experimental evidences, would allow the reconstruction of the photoisomerization molecular process with high temporal (femtosecond) resolution and with atomistic details. Moreover, assessment of the computational results by direct comparison with experimental data provides robust models that can be used to rationalize the photoinduced processes, and possibly, to predict the retinal responses to light in various molecular configurations and environmental conditions, which can open to *in silico* design of retinal-based photochromic devices.

In the last twenty years, our group has extensively employed computational methods to sort out the intrinsic spectral and photochemical properties of isolated retinal chromophores in vacuo [38, 71–73, 78, 172, 218]. These studies provided the basic background for a comprehensive understanding of retinal photochemistry in the natural pigments including the bovine Rhodopsin (Rh, see Fig. 2) and the proton pump bacteriorhodopsin (bR), as we have recently reviewed [178]. More recently, we have focused on the study of retinal photophysics and photochemistry associated with chemically modified retinals and to different environmental conditions, given the relevance of these studies for the exploitation of retinal-based chromophores for developing biomimetic optical devices. In particular, we have considered the photochemistry of all-trans protonated Schiff bases (PSB) in solution by comparing the photoinduced process of native PSB chromophore (i.e., with methylations in positions 9 and 13 as in Rh, see Fig. 2) and its 10-methylated derivative (10Me-PSB), and we started investigating the photophysics of the 11-cis unprotonated Schiff base (USB) embedded in UV-sensitive ancestral pigments as compared to that of the 11-cis PSB (PSB11) in Rh [28].

In this section, we review these studies and the associated simulations of spectroscopic data, since this will provide good examples illustrating the flexibility and the reliability of our QM/MM methodologies.

3.1 QM/MM Setup

Figure 2b shows the QM/MM scheme employed in the study of retinal photoisomerization in Rh [166], as implemented in COBRAMM [5, 6], with two main QM/MM

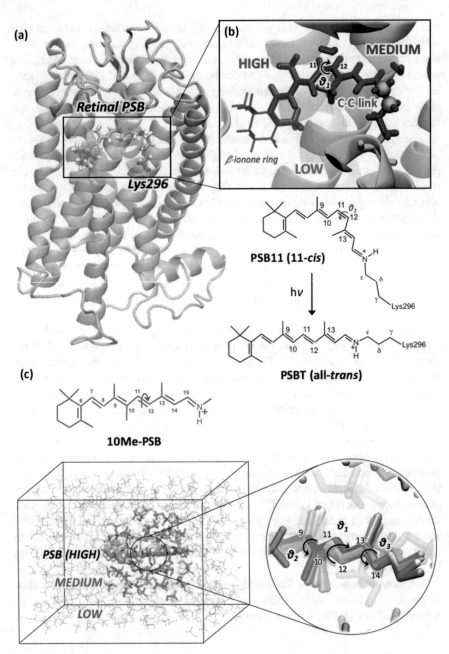

Fig. 2 (Continued)

◀ **Fig. 2** **a** X-ray structure of the bovine Rhodopsin (Rh, PDB: 1U19) with retinal chromophore covalently bound as protonated Schiff base (PSB) to the Lys296 amino acid residue and schematic representation of the 11-cis→all-trans photoisomerization. **b** QM/MM schemes adopted for the study of retinal photoisomerization in Rh, showing the three QM/MM (high, medium and low) layers, with a high layer that could involve the alkyl moiety of the β-ionone ring (otherwise in the medium layer), the atom-link position separating the high and medium layers, the latter involving two water molecules in the binding site and the atoms between the C_β of Lys 296 and the PSB nitrogen and the C_β of Lys 296. **c** QM/MM modeling of the 10-methylated derivative of the all-trans PSB in methanol solution, with high layer involving the chromophore (as in Rh) and the medium layer involving mobile (red sticks) solvent molecules, while the solvent molecules beyond 6 Å from the retinal heavy atoms are kept frozen in the low layer (gray sticks). The inset shows the photoisomerization process through an asynchronous bicycle pedal mechanism occurring in solution, in analogy to retinals bound to opsin proteins. Panel (c) has been reproduced with permission from Ref. [159] (Copyright 2017 American Chemical Society)

schemes employed that differ from each other for the size of their high layers, which could comprise the entire PSB unit or disregard the alkyl moiety of the β-ionone ring (that goes to the medium layer), depending on the type of QM computation performed. A Complete Active Space Self-Consistent Field (CASSCF) method with an active space of 12 electrons and 12 π orbitals—CAS(12,12)—is employed when the full-chromophore is included in the high layer for computing the MEP via the static approach. Instead, in order to minimize the computational cost without significant loss of accuracy, the model with reduced high layer was used in conjunction with a smaller CAS(10,10) active space for the molecular dynamics simulations of the excited state decay.

QM/MM geometry optimizations, frequency calculations and molecular dynamics have been performed with the CASSCF/6-31G*/AMBER method, involving the AMBER the ff99 force field [48] while the photoisomerization MEP has been refined with 2nd order Complete Active Space Perturbation Theory, namely with the CASPT2//CASSCF/6-31G*/AMBER approach. The medium layer (see Fig. 2b) included the region between the Schiff base nitrogen and the C_β of Lys 296, the two water molecules in the retinal binding site and, eventually, the alkyl moiety of the β-ionone ring. The rest of the system was set to the low layer and frozen in the positions of the high resolution (2.2 Å) X-ray data of bovine Rh [154]. It is worth noting that the PSB counterion (i.e., the E113 residue) was not included in the high layer to reduce the computational cost and in consideration of the *counterion quenching* effect due to the electrostatic field of the protein matrix, which we have demonstrated using the electrostatic embedding at the QM level [206].

Figure 2c shows, instead, the QM/MM scheme employed in the study of all-*trans* PSB and 10-Me derivative photoisomerization in methanol solution [58], as implemented in COBRAMM [222]. As in the case of the reduced high layer in Rh model, in order to minimize the computational cost, only the conjugated chains of the chromophores were included in the QM region with the boundary between the QM and MM regions set to place the alkyl moiety of the β-ionone ring in the medium layer. For the static computations the medium layer also included 9 movable solvent molecules, selected from the solvent molecules within 6 Å from the retinal heavy atoms (see Fig. 2c inset), according to a preliminary estimation of the molecules really moving along the photoisomerization MEP. Instead, in order to avoid any bias and to ensure full flexibility during the photoisomerization dynamics, all the solvent molecules within 6 Å from the retinal have been included in the 0 K overdamped classical MD simulations. All geometries were optimized at the CAS(12,12) level, using a state-average procedure including three states (SA3), given the importance of covalent S_2 state in the case of solvated PSB. Excited state dynamics simulations on S_1 have been carried out with the full active space CAS(12,12) and with first two roots included in the state-average procedure as implemented in Molpro-2010 [224].

As mentioned above, we have investigated the primary steps of photoisomerization of the 11-cis unprotonated Schiff base (USB) embedded in a UV-sensitive pigment, i.e., the Siberian hamster ultraviolet (SHUV) visual pigment. In this case, we were initially interested in the electronic structure at the excited state minimum and in assessing a QM/MM model that could reproduce the linear absorption spectrum and the appropriate protonation states of the USB and the E113 "counterion". Therefore, a QM/MM two-layers scheme was adopted, involving the E113 counterion besides the full retinal model as used for Rh (Fig. 2), and using the Amber99ffSB [90] force field. Relatively large state-averages (up to 10 excited states) were used due to the fact that the USB spectroscopic state was not found among the first five roots at the CASSCF levels and its root number is geometry dependent. To speed up the computations, the Cholesky decomposition approach [8, 9] was used as implemented in the Molcas 8.1 code [10]. In the employed CASPT2//CASSCF protocol, the CASSCF active space included the entire π system of the chromophore, i.e., CAS(12,12), except for the calculations of the $n\pi^*$ states involving a CAS(14,13) space. In this case, a comparison with TDDFT results were also reported, to highlight the importance of including double excitations for describing the photochemistry of the neutral USB chromophore. For all calculations, the 6-31G* basis set was used.

3.2 Initial Conditions

Rhodopsin X-ray structure has been resolved at 2.2 Å resolution [154], with a tertiary structure characterized by a bundle of seven transmembrane alpha helices (see Fig. 2), linked by protein loops (three in the extracellular side, E1-3, and three in the cyto-plasmic side, C1-3), providing the fundamental initial structure for our simulations. As shown in Fig. 3a, these X-ray data indicated the presence of two protonable amino acids with carboxylate groups nearby the PSB11 chromophore, i.e., the residue E181 on the extracellular *loop* 2 (E2) and the counterion E113. The protonation state of E181 in the dark-adapted state of Rh it has been a matter of discussion in the literature [24, 114, 118, 179, 204, 230, 231]. However, as shown in Fig. 3b, the protonation state of the E181 sidechain has negligible effects on the relative stabilities of the S_1 and S_2 excited states and on the photoisomerization path of PSB11.

To generate the initial conditions for the PSB11 dynamics in Rh, CAS(10,10)/6-31G*/AMBER QM/MM numerical frequencies have been calculated on the PSB11

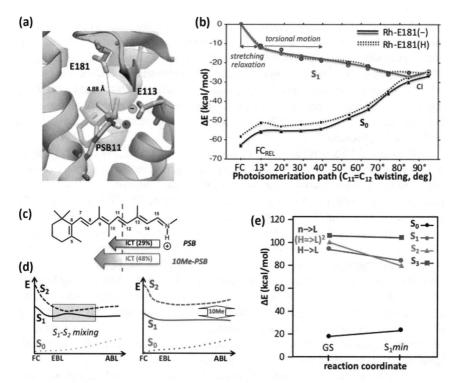

Fig. 3 **a** Protonable residues in the retinal binding pocket in Rh according to X-ray data, with E181 placed right above the central double bond and E113 counterion paired with PSB11. **b** PSB11 photoisomerization MEP in Rh, showing CASPT2 corrected S_0 and S_1 energy profiles along the QM(CASSCF)/MM relaxed scan on S_1, from the FC point to the twisted CI funnel, through a point (FCrel) where only skeletal bond stretchings have been relaxed. Two different MEPs with charged (full lines) and protonated (dotted lines) E181 are reported. **c** The percentage of (S_0) positive charge moving from the C_{12}–N to the C_5–C_{11} fragment in the S_1 state of solvated retinals, indicated as intramolecular charge transfer (ICT) by arrows. **d** Schematic representation of the effect of 10-methylation on the energy profiles of S_0, S_1, and S_2 states along the bond relaxation pathways on the S_1 PES of solvated retinals in methanol solution. **e** Evolution of energy levels of the singlet (S_N) excited states of the SHUV-USB along the S_1 relaxation pathway calculated at the CASPT2//CASSCF level, showing the roles of $(H \Rightarrow L)^2$ double excitation (i.e., the S_2 covalent state) and the n→L one-electron excitation (i.e., the S_3 $n\pi^\star$ state). Panels (a) and (b) have been reproduced from Ref. [178], with permission from the PCCP Owner Societies. Panels (c) and (d) have been reproduced with permission from Ref. [159] (Copyright 2017 American Chemical Society). Panel (e) is reprinted by permission from Springer Nature from Ref. [28] (Copyright 2016 Springer Nature)

ground state (GS) minimum, thermally sampling the vibrational modes at 300 K, including zero-point energy corrections and excluding high frequency C–H, N–H and O–H modes. The temperature distribution is considered only for the mobile part, while the protein fixed during thermal sampling and molecular dynamics is considered as a mean representation of the positions of the atoms, although at low (100 K) temperature. The extremely short timescale of the rhodopsin primary photochemical event investigated here (100 fs decay time) prevents protein thermalization and justifies this approximation.

In the case of PSB and the 10Me-PSB retinals in methanol, classical MD simulations were performed in order to obtain a representative starting geometry of each chromophore for subsequent computations of their photochemistry. The MD simulations were carried out using Amber-11 suite [37], with the ff99SB [219], and GAFF force fields [220] with RESP charges for the parametrization of the retinals. The overall system comprised ca. 3500 atoms with retinal chromophores solvated by a solvent box consisting of ca. 600 solvent molecules. Pre-equilibration with constant volume and temperature was performed, heating from 0 to 300 K, followed by 40-ns simulation runs at constant temperature (300 K) and pressure (1 atm) using Berendsen barostat as implemented in Amber-11. To extract a representative configuration of equilibrated solvent molecules surrounding the chromophores, the retinals coordinates were kept frozen during the dynamics, and the snapshots characterized by the lowest potential energies were selected as the initial structures for the subsequent QM/MM calculations of the two chromophores.

The X-ray structure of the SHUV pigment has not been resolved yet and our structural models were thus based on the uniquely available structure, a DFT-QM//MM optimized structure previously obtained by homology modeling based on the Rh crystal structure resolved at 2.2 Å, featuring 43% identity between the two primary sequences [191]. For the SHUV structure, the Rh's residue numbering is generally used to facilitate the comparison between the two photoreceptors, although the primary sequence alignment features a 5-residue shift in the N-terminus of the proteins.

3.3 PES Exploration

The retinal photoisomerization in Rh is the primary event in vision and it represents a paradigm for studying the molecular basis of vertebrate vision. The Rh absorption maximum lies in the visible at $\sim$500 nm and it induces the 11-*cis*$\rightarrow$all-*trans* isomerisation of the PSB11 (see Fig. 2a), triggering the protein conformational changes that initiate the signal transduction processes in vision [21].

The typical reaction coordinate on the S1 surface of retinals is characterized by two reactive modes, involving first the skeletal deformations (C=C bond relaxations) then a torsion about the reactive double bond. As indicated in Fig. 3b, static characterization of this reaction pathways for Rh, based on CASPT2/CASSCF/AMBER calculations, indicated that (in analogy with what found for isolated chromophores in vacuo) the ionic (bright) S_1 state and the covalent (dark) S_2 state do not interact along the photoisomerization MEP. This photoisomerization mechanism, known as the *two-state two-mode* (TSTM) model [78], is generally accompanied by a barrier-less relaxation of S_1 that ends at a CI with $\sim 80°$ twisted central double bond, where the S_1 and the S_0 surfaces efficiently cross. Such MEP topology is typical of a fast and efficient non-adiabatic *cis→trans* isomerization, as observed for PSB11 in Rh. In fact, the S_1 state of PSB11 in Rh is found (experimentally) to decay monoexpo-nentially with a lifetime of ~ 150 fs and to yield a unique all-*trans* photoproduct in ~ 200 fs and with a 67% quantum yield (QY) [55, 103, 184]. As shown in Fig. 3b, this MEP is hardly affected by the protonation state of E181 in the PSB11 binding site. Instead, by computing the MEP in a protein environment where all external charges have been switched off except for the counterion charge—which is one of the power-ful computational options features by QM/MM techniques—we have demonstrated how the counterion quenching by the protein environment is crucial to trigger a fast and efficient photoisomerization in Rh [206].

Analogously to Rh, but slightly less efficiently, the chromophore of the proton pump bacteriorhodopsin (bR) that converts light into chemical energy, i.e., a cova-lently bound all-*trans* PSB, features a 200 fs lifetime and a *trans→cis* isomerization reaction rate of ~ 500 fs along the C_{13}–C_{14} double bond, with a 65% QY for the 13-*cis* PSB photoproduct [55, 133]. Remarkably, in methanol solution the all-*trans→cis* photoisomerization yield the 11-*cis* isomer and it is significantly slower (~ 4 ps) [103] and less efficient (0.16 QY) [67]. However, the functionalization of the all-*trans* PSB with a methyl group at position C_{10} (10Me-PSB, Fig. 2c) recovers the protein-like excited state sub-ps dynamics, although with a relatively low QY (0.09) for the 11-cis isomer [196].

Our synergistic computational and experimental studies pointed toward a different interplay between the covalent (S_0 and S_2) states and the ionic (S_1) state as the reason for the various behaviors of native and chemically modified retinals embedded in different environmental conditions. If the covalent S_2 excited state gets involved in the photoisomerization process, a *three-state* model has to be invoked (see Fig. 3d), in contrast to the *two-state* model generally found in the natural proteins. Figure 3c–d shows the effect of 10-methylation on the energy profiles of S_0, S_1, and S_2 states along the bond relaxation pathways on the S_1 PES. The first reactive mode of retinal photoisomerization (i.e., the skeletal C=C bond relaxations) is associated with bond inversion of the alternate single/double bonds of the retinal polyene chain. The bond-length alternation (BLA) in retinals, defined as the difference between the average distances of single and double C-C bonds can be used to monitor the bond inversion during the photoisomerization, representing the main photoreaction coordinate in the early times of the excited state decay. Due to the different intramolecular charge transfer (ICT, calculated using atomic partial charges in the S_1 and S_0 states at the FC geometry) upon $S_0 \rightarrow S_1$ excitation observed for PSB and 10Me-PSB, covalent (S_0 and S_2) and ionic (S_1) states in retinals are subject to different relative stabilizations in the two retinals, this affecting their photochemistry (see Fig. 3c-d).

CASPT2//CASSCF/AMBER computations indicated the presence of an avoided crossing along the MEP of the native PSB, with stabilization of an excited state minimum with even bond-lengths (EBL), while a major effect of 10-methylation is to increase the S_1–S_2 gap along the bond-inversion coordinate (see Fig. 3d), overall favoring the formation of an excited state transient species with alternate bond-lengths (ABL). This ABL species represents a reactive configuration toward the subsequent torsion mode, thus explaining why only the 10-methylated chromophore can rapidly (<1 ps) populate the torsion modes and reach the CI seam of twisted structures, effectively recovering the protein-like behavior. These results have been corroborated by 0 K overdamped classical MD simulations of both the PSB and the 10Me-PSB solvated retinal chromophores, as depicted in Fig. 4a. In particular, the excited state dynamics of the 10Me-PSB was found to be associated with the typical space-saving bicycle pedal torsions along the ground state C-C double bonds (see Fig. 2c), yielding to a twisted CI geometry.

We have initiated the characterization of the excited state decay mechanism in the unprotonated retinal chromophore of the SHUV pigment by static computations [28]. Our CASPT2//CASSCF/AMBER computations of the MEP along the initial bond relaxation mode suggested a S_1/S_2 excited state surface crossing before reaching the S_1 excited state minimum, which features an EBL bond alternation. These results indicated a strong involvement of the covalent S_2 state in the SHUV-USB photochemistry, which is currently under more detailed investigation in our group.

3.4 Spectroscopy Simulations

One of the most widely used electronic spectroscopy to explore retinal photochemistry is certainly the time-resolved PP technique. In particular, SE signals out of the wave-packet dynamics during photoisomerization are generally used to monitor the decay of the bright (spectroscopic) S_1 state and to define its lifetime while the formation of photoproducts, after crossing the CI seam, can be evaluated by the appearance of photoinduced absorptions at time delays longer than the S_1 lifetime. In the presence of two different fluorescent states, two SE signals could be expected and associated with the two excited state transients along the S_1 surface, as observed in time-resolved PP experiments and computational studies of retinal systems [42, 93]. This also the case for the PSB and the 10Me-PSB retinals in methanol solution [58, 196]. As shown in Fig. 4b, our QM/MM estimations of SE values from the EBL and ABL configurations are found in good agreement with experimental data, with larger red-shift on the ABL emission than on the EBL one as an effect of 10-methylation being consistent with the experimental signals. Moreover, QM/MM computations predicted the presence of ESA signals in both PSB and the 10Me-PSB retinals in the spectral region between the two SEs, suggesting that these signals partially affect the overall appearance of the double emission bands detected experimentally [196]. Notably, these experimental spectra showed that the two SE signals rise at different delay times, further corroborating our proposal of two in time sequential fluorescent states, first EBL and then ABL, as shown in Fig. 3b.

Finally, the typical scenario with one SE signal red-shifting in time, followed by a rising photoinduced absorption associated with photoproduct formation is observed

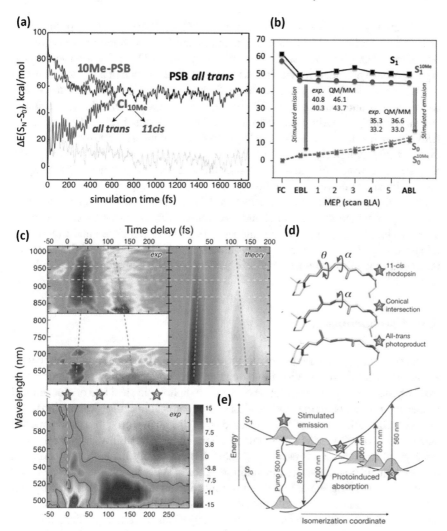

Fig. 4 a QM/MM energies from excited state dynamics of both PSB (black and gray lines for S_1 and S_0, respectively) and 10Me-PSB (red and blue lines, respectively) retinals in methanol solution, with latter showing fast photoisomerization. **b** Bond relaxation MEP on the S_1 PES as calculated by QM/MM optimized scan along a reaction coordinate defined by the BLA of the retinal backbone, for both PSB (black lines) and 10Me-PSB (red lines). Vertical emissions from the ABL and EBL geometries are indicated by arrows and compared with experimental stimulated emissions for both PSB (in black) and 10Me-PSB (in red), as reported in Ref. [196]. **c** Experimental and simulated differential transmission maps as a function of time delay and wavelength in the visible (bottom panel) and NIR (top panels) spectral regions. **d** Averaged structures of the chromophore at the initial 11-*cis* (blue star '1', $t = 50$ fs, $\alpha = -12.8°$), CI (blue star '2', $t = 110$ fs, $\alpha = -87.8°$) and final all-*trans* (blue star '3', $t = 200$ fs, $\alpha = -141.0°$) configurations. **e** Sketch of the S_0 and S_1 potential energy surfaces. Panels (a) and (b) have been reproduced with permission from Ref. [159] (Copyright 2017 American Chemical Society). Panels (c), (d) and (e) have been reproduced from Ref. [166]

experimentally in bovine Rh, as shown in Fig. 4c–e. These ultrafast spectroscopy experiments featured sub-20-fs time resolution and have been reported in conjunction with our QM/MM molecular dynamics simulations [166]. Such powerful synergic approach provided convincing evidences of the existence of a twisted CI that is reached within 80 fs after photoexcitation. Figure 4c, shows the comparison between the simulated and experimental PP spectra (reported as differential transmission maps), featuring at early probe delays the GSB (blue) signal (at $\sim$510 nm) and an ESA signal (namely a $S_1 \rightarrow S_n$ transition, in red) peaking at $\sim$500 nm that overlap in the visible region. By the first $\sim$80 fs the SE (blue) signal evolves following the S_1 relaxation, with a progressive red-shift the near-IR (NIR) region. The decay of the SE signal is followed (at probe delays > 80 fs) by the appearance of the photoinduced (red) signal at NIR wavelengths ($\sim$1000 nm), being associated with the wave packet approaching the CI funnel and then transferring from the S_1 surface to the photoproduct hot ground state. As shown in Fig. 4c, the simulated and experimental spectra agree almost quantitatively indicating that the analysis of the QM/MM molecular dynamics provides a realistic "movie" of retinal photochemistry in Rh.

3.5   Conclusions

In this section, we have reviewed our QM/MM studies on photoinduced processes in retinal systems and on the simulations of their corresponding electronic spectroscopy, accounting for modulation of the retinal photophysical and photochemical properties as function of the modification of the chemical formula (protonation states or functionalization) and of the surrounding environment (protein or solvent). We showed how our QM/MM methodology offers enough flexibility to properly treat the excited state decay of retinal PSB (or USB) covalently linked to a protein and also of retinal PSB derivatives in solution. These QM/MM schemes can be easily constructed by using the automatized routines included in the COBRAMM package and can be straightforwardly linked to classical MD simulations for considering suitable initial conditions to carry on the investigations on retinal photochemistry.

The COBRAMM package also offers the appropriate routines to explore the PES associated to the retinal S_1 photoisomerization pathway with multi-configurational/ multireference methodologies. These computations have shown the importance of accounting for various excited states relaxation pathways when studying retinal photochemistry, which demonstrated to be crucial for the characterization of the correct MEPs in both the solvated PSBs and the USB embedded in a UV-sensitive pigment. Finally, we showed how, by taking advantage of both static computations (i.e., MEP characterization) and mixed quantum-classical MD simulations, our QM/MM schemes also allow accurate simulations of retinal electronic spectroscopy.

The direct comparison of experimental and theoretical electronic spectra allowed, for instance, the characterization with atomistic resolution of two transient fluorescent species in solvated PSBs and of the conical intersection in the Rh protein. Thus, the retinal systems represent a good example that illustrates the flexibility of our

QM/MM methodologies and the reliability of the corresponding outcomes. In the next section, we will further illustrate the power of these methodologies with another class of fundamental biological light-sensitive molecules, i.e., the nucleobases of nucleic acids.

4 DNA Systems

Solar ultraviolet light is absorbed by DNA due to the intense absorption cross sections of its building blocks, the nucleobases. Electronic excitation of DNA can produce harmful photoproducts, even if the majority of excitation events do not lead to photoreactions as evidenced by the tiny photolesion quantum yield (much less than 1%). The importance of maintaining unaltered the genetic code had always created great interest toward the excited state's decay pathways along the reactive and nonreactive routes [50, 97, 139]. This background has stimulated a large amount of spectroscopy work, both experimental and theoretical, with the aim of better understanding the effect that UV light absorption has on DNA. As a consequence, nucleobases have become a benchmark system for developing nonlinear optical spectroscopy. Nowadays, isolated canonical nucleobases are well understood in terms of their fundamental photophysical properties: in this systems the ultrashort excited state lifetimes arise from an efficient radiationless mechanisms, consenting self-protection against UV damage. The focus is, therefore, progressively shifting toward increasing the complexity of the system, from single nucleobases in the presence of their native environment water [43, 44, 83, 109], the attached sugar (nucleosides) [129, 132, 159, 160] and phosphate (nucleotides) moieties [4] over dinucleotides [116, 149, 187] and small oligomers to DNA single and double strands [41, 45–47, 50, 76, 77, 97, 109, 128, 139]. Notably, the interaction with the environment, with the backbone of the DNA and π-stacking interaction radically alter the photophysical/photochemical behavior of DNA nucleobases compared with the isolated chromophore.

Hybrid QM/MM approaches constitute a practical tool for addressing the increasing complexity. In combination with the first-principles spectroscopy simulations, they provide the link between the molecular movie and the spectroscopic signals, thereby allowing to rationalize the experimental observations. On the one hand, linear spectroscopy techniques such as linear absorption (LA), circular vibrational dichroism (CVD), and third-order nonlinear techniques such as 2D IR/Raman and 2D ES show a progressively increasing sensitivity to the ground state conformational freedom of DNA oligomers and the reorganization dynamics of the immediate environment, thereby providing signatures of π-stacking arrangements [227], of DNA-chromophore intercalation [173], of the dynamics of hydrogen bond formation/breaking [64], etc. On the other hand, time-resolved transient techniques such as PP spectroscopy, third-order nonlinear 2D ES and fifth-order 2D IR/Raman are capable of resolving excited state deactivation mechanisms by resolving in frequency and time the buildup, decay and quantum beating of state specific spectroscopic signatures (SE and ESA in electronic spectroscopy, spectator modes in IR/Raman), as well as the passage through a conical intersection with a sub-fs temporal resolution [3, 111]. 2D techniques have become invaluable when dealing with multi-chromophoric systems due to their simultaneously high spectral and temporal resolution. [31, 192, 208] The ability to disentangle the evolution of signal contributions otherwise spectrally overlapping in conventional 1D experiments (LA, PP) makes them superior and opens the way for a more detailed interpretation of the photoinduced processes.

In the following, we will present spectroscopy simulations of DNA mononucleosides (5-methyl-(oxy/deoxy)-Cyd, (oxy/deoxy)-Cyd, Urd, Thd) and dinucleotides (ApA, ApU), as well as of thionated nucleobases (4-thio-U) in their native environment within the QM/MM framework.

- In Sects. 4.1 and 4.2, we will demonstrate how homogeneous and inhomogeneous broadening in LA and 2DES spectra can be accurately simulated from first principles by combining classical and Wigner conformational sampling, using only information from the ground state minumum [129, 187].
- In Sects. 4.3 and 4.4, we will show how excited state static information, obtained from PES exploration techniques in combination with multireference perturbation theory, allows to simulate transient spectral signatures (ESA, SE) in PP and 2DES [116, 132, 159, 160], thereby highlighting the ability of 2DES to resolve the dynamics of spectrally overlapping signals.
- In Sect. 4.5, we will finally show how an analytical model of the nonlinear spectral dynamics can be parametrized based on state-of-the-art electronic structure calculations and femtosecond mixed quantum-classical non-adiabatic molecular dynamics simulations. [30].

These spectroscopy simulations will be compared to experimental data, where available, and used to shed light into pending biochemical questions such as the preferential excited state deactivation pathway of cytosine and its 5-methyl derivatives, to distinguish the deactivation channels in single- and dinucleosides and to capture the intermediate dark state in the radiationless decay of thionucleotides.

4.1 Electronic Structure of DNA Nucleosides in Solution: Linear Absorption Spectroscopy of 5-Methylcytidine

Out of all DNA nucleobases cytidine exhibits the most complex electronic spectrum, characterized by three peaks above 200 nm arising due to four bright $\pi\pi^*$ transitions amidst four dark transitions of $n\pi^*$ type [15, 25, 26, 105, 136]. Here, we use 5-methylcytidine (5mCyd) to compare three protocols of progressing complexity for simulating LA spectra in solution:

(i) Static QM/PCM Approach: the chromophore, treated quantum-mechanically, is equilibrated in an implicit solvent (plus few explicit H-bonded water molecules in the aqueous solution case) described as a polarizable continuum (PCM) [135, 207].

(ii) Static QM/MM Approach: the chromophore, treated quantum-mechanically, is equilibrated in the presence of an explicit solvent described by a classical molecular force field in an electrostatic QM/MM scheme.

(iii) Dynamic QM/MM Approach: the conformational space of the solute-solvent is sampled via a classical molecular dynamics simulation and the Static QM/MM Approach is applied on top of each snapshot.

In particular, we simulate the spectra in three different solvents: water, tetrahydrofuran (THF), and acetonitrile (ACN) and water [129]. In the following we describe the details of the computational protocol.[1] The solvent effect evaluation is fundamental in photoactivated processes, because it could decide the interplay between the involved excited states, modulating the shape and the order of their potential energy surface (PES), including energy barriers leading to nonradiative ground state recovery [26, 95, 107, 119].

In general, PCM and QM/MM are among the most widely used models for describing solvation effects. PCM has the advantage of fully capturing the polarization effects of the solvation shell but lacks elements that are very important for biological systems. Three situations that are common in biosystems particularly require the use of an explicit solvent model: (1) when hydrogen bonds are formed between the chromophore and the environment, (2) when the environment is inhomogeneous and/or anisotropic, e.g., in a protein cavity, and (3) when we are considering a time-dependent process in which the dynamics of the solvent plays a significant role. To include electronic polarization effects within an explicit solvent model, several polarizable force fields methodologies are available and can be used in standard QM/MM schemes. Furthermore, a simpler alternative has been proposed in Ref. [99]. This methodology—called "Electronic Response of the Surroundings" (ERS)—combines the explicit solvent with a polarizable continuous that describes the instantaneous response of the environment.

4.1.1 Methodology

QM/MM setup. The setup was as follows:

(i) *Static QM/PCM Approach*: THF and ACN were described implicitly through PCM, whereas in the case of water, four solvent molecules from the first solvation shell were explicitly included in the QM layer (Fig. 5a), in order to account for hydrogen bonding [82, 182].

(ii) *Static QM/MM Approach and (iii) Dynamic QM/MM MD Approach*: the High Layer contained the chromophore 5mCyd, the Medium Layer encompassed the movable solvent molecules within a 5.5 Å distance, the Low Layer contained the rest of the solvent molecules.

---

[1]Note that the PES exploration step is omitted as we address solely the ground state conformational freedom of the system.

In all three models, the QM calculations were performed at the DFT level and with its time-dependent formulation (i.e., TDDFT) [127], utilizing the CAM-B3LYP functional [232], and the 6-31G* basis set with the software Gaussian 09 [69].

Boundary conditions. For all three methods an initial classical molecular dynamics simulation was run using Amber 11 package [37], and employing the f99 force field [219]. The nucleoside moiety was embedded in cubic solvent boxes of 12 Å, employing the TIP3P force filed [101] for water and the GAFF force field for THF and ACN. [220] The three systems were firstly heated for 1 ns at a pressure of 1 atm and with constant volume, starting from 0 to reach 300 K, and subsequently a 100 ns molecular dynamics was performed, recording snapshots every 200 fs (resulting in 500,000 snapshots).

(i) *Static QM/PCM Approach and (ii) Static QM/MM Approach*: A cluster analysis was applied over the recorded snapshots utilizing a root mean square (RMS) coordinate deviation . Two main conformation were populated, *syn* and *anti* (Fig. 5b), with the *syn* conformation being energetically more stable and hence more frequently visited due to a hydrogen bond between the base and the sugar moiety. Because we verified that the *syn/anti* computed spectra are qualitatively not depending on the choice of the conformer, therefore, *anti*-conformer (preferred in DNA structures) was selected and refined at the QM/PCM (i) / QM/MM (ii) levels of theory.

(iii) *Dynamic QM/MM Approach*: 200 snapshots were extracted in equidistant intervals of out of the recorded 500,000 snapshots and each one was refined at the QM/MM level of theory.

Spectroscopic modeling.

(i) *Static QM/PCM Approach and (ii) Static QM/MM Approach*: the LA spectra were modeled according to Eq. (3) at the refined geometry of the representative *anti*-conformer.

(iii) *Dynamic QM/MM Approach*: the LA spectra were modeled according to Eq. (4) by summing over all 200 refined geometries. In all the three protocols, the first ten excited states at the CAM-B3LYP/6-31G* QM level of theory were taken into consideration. A phenomenological Gaussian broadening with a half width at half maximum of 0.35 eV (i and ii) and of 0.1 eV (iii) was applied to each transition.

4.1.2   Discussion

Qualitatively speaking, the experimental spectra are fairly well reproduced by all three approaches, in particular, the appearance of two absorption bands and the red-shift of the ACN and THF spectra (red and blue in Fig. 5c), with respect to water (black in Fig. 5c). From a quantitative point of view, the *Static QM/PCM Approach* estimates are uniformly blue-shifted comparing with the experimental maxima by ca. 0.6 eV. To correctly predict the blue-shift for 5mCyd in water, we observed that was necessary to include the first solvation shell of solvent molecules in the QM part. While our calculations demonstrate that a hybrid explicit/implicit solvation model is capable of reproducing spectral shifts due to hydrogen bonding, at the same time this approach depends on an arbitrary number and position of the explicit solvent molecules. In a similar manner the *Static QM/MM Approach* suffers from the arbitrariness of the positions of the solvent molecules in the selected snapshot. Furthermore, it suffers from the lack of a mechanism for treating instantaneous polarization of the solvent upon excitation (as all QM/MM models based on electrostatic embedding), a feature present in PCM models. Therefore, the *Static QM/MM Approach* shows the largest deviations from the experiment (Fig. 5c). The *Dynamic QM/MM Approach* allows to alleviate the arbitrariness of the solvent coordinates and in part takes into account the broadening effect due to the solute fluctuations, outperforming the static approaches and correctly reproducing the spectral bandwidths (Fig. 5c).

4.2 Spectral Signatures of π-stacking in Solvated Dinucleosides: 2DES of Adenine-Adenine Monophosphate

As demonstrated in the previous section LA spectroscopy proves invaluable to discern the electronic structure of DNA/RNA mononucleosides and to study how the environment affects the energetics. Moving up the ladder of complexity, one is confronted quickly with a rapidly number of degrees of freedom. While in most mononucleosides there are two predominant conformers (*syn* and *anti*), with respect to the intramolecular hydrogen bond between the CH_2OH group of the sugar and the C=O group of the nucleobase, an ensemble of cooperating geometrical (inter-base distance and angle, shift, slide, stacking) [47] and electronic (exciton, excimer, charge transfer) factors are believed to be the reason why DNA/RNA polymers exhibit completely different photophysical properties [128, 185, 201]. It is, therefore, of importance to be able to distinguish the dynamics of multi-chromophoric DNA/RNA systems from the dynamics of their monomer build blocks. Naturally, dinucleosides present the most simple extension of complexity with respect to the mononucleosides, that allows to address the new photophysics. However, LA is no longer sufficient to reveal the electronic structure of the dimer due to the overlapping signals of different molecular motifs.

2DES with temporally coinciding pump and probe pulses[2] promises a way to overcome the shortcomings of LA and provides detailed knowledge about the Franck–Condon electronic structure of complex systems, as 2DES line shapes are particularly sensitive to inter-chromophore interactions. In particular, this emerging technique helps to disentangle inhomogeneous from homogeneous spectral broadening,

---

[2]Technically realized by keeping the delay between pump and probe short with respect to the molecular dynamics.

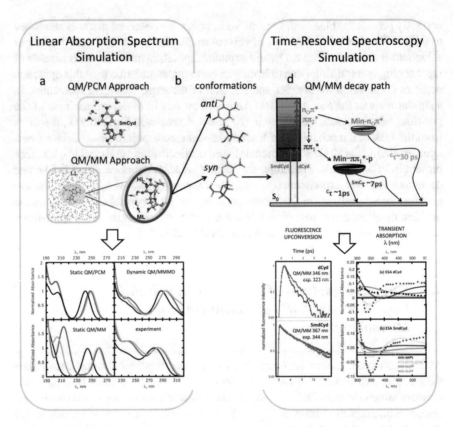

Fig. 5 **a** Representative schemes for the 5-methylcytidine (5mCyd) system used in the static QM/PCM approach (QM nucleoside + 4·H_2O) and, below, in the static and dynamic QM/MM calculations: water molecules in the low layer (LL) are in sticks, in the medium layer (ML) in tubes, and the 5-methylcytidine in the high layer (HL) is in ball and sticks representation. **b** Anti and syn conformers, the latter showing the base-sugar H-bond. **c** 5mCyd computed absorption spectra in different solvents: ACN (red), THF (blue), and water (black) adopting the static QM/PCM, static QM/MM, and dynamic QM/MM molecular dynamics approaches. Right, below panel shows the 5mCyd experimental normalized absorption spectra. **d** Comprehensive scheme for the deactivation mechanisms and experimental lifetimes of two deoxy-cytidine systems, dCyd and 5mdCyd in water solution. **e** Left panel: time-resolved fluorescence decays of dCyd at 330 nm (top panel) and 5mdCyd at 350 nm (bottom) in water (black) and ACN (red), for an excitation wavelength of 267 nm. Experimental and QM/MM fluorescence emission energies indicated inside. Right panel: calculated excited state absorption spectra calculated with CASPT2/MM (at the geometries optimized with MS-CASPT2 for each different stationary points) for dCyd (top) and 5mdCyd (bottom). Dots indicate the experimental TA (red and black for the short-living and long-living signals, respectively). Adapted with the permission from Refs. [129, 132]. Copyright 2016 and 2017 American Chemical Society

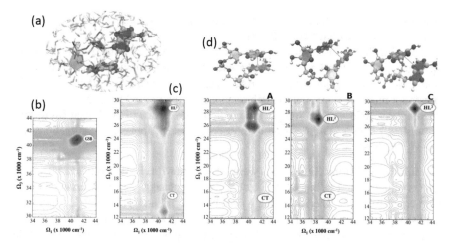

Fig. 6 **a** QM/MM partitioning of water-solvated ApA: both adenine nucleobases are included in the High Layer, the sugars, phosphate and water molecules are included in the Low Layer; **b, c** 2DES of ApA in water for the full conformational space, probed in the UV (**b**) and in the Visible (**c**); **d** 2DES spectra of representative geometrical conformations: π-stacking (A), T-stacking (B), and noninteracting (C) dimer; a red color denotes excited state absorption signals, a blue color denotes ground state bleach and stimulated emission, coinciding at waiting time $t_2 = 0$; Adapted from Ref. [187], with permission from The Royal Society of Chemistry

the former being related to the heterogeneity of the sample and informative of the timescales on which conformational interconversion occurs. Furthermore, as a nonlinear technique 2DES shows transient absorption signals spanning the far-IR to UV spectral window, some of which are characteristic "fingerprints" of interacting chromophores [147]. Their identification would allow to selectively track the photoinduced dynamics of stacked DNA/RNA oligomers and compare it to that of the monomers that constitute them.

In the following, we assess the applicability of 2DES to disentangle the conformational dynamics of ApA in water (Fig. 6a). We study the spectrum dependence on the stacking strength and typology, document the characteristic fingerprint arising due to inter-chromophore interaction, and give practical tips for their experimental detection.

4.2.1 Methodology

QM/MM setup. The dinucleoside was solvated in a water sphere, including a Na^+ as the counterion of the negatively charged phosphate group. As the scope of the study is to document characteristic spectral signatures of various types of stacking conformations we opted for a more rigid partitioning scheme, namely a HL/LL, having only the adenine nucleobases in the quantum mechanical layer while keeping the backbone and the surrounding waters frozen in order to reproduce as close as possible inter-base distance and orientation obtained via the classical configurational sampling (see **Boundary conditions**).

The CASSCF protocol was applied to the QM layer adopting a reduced active space size of 8 orbitals, i.e., 4 occupied π and 4 virtual π^*, equally distributed between both bases. This is a fair compromise between accuracy and computational cost for obtaining more accurate bond-length distances in the ground state than those given by the force field used to sample the configuration space of ApA (see **Boundary conditions**). The ANO-L basis set with valence double-ζ polarized contractions was used [225, 226].

Boundary conditions. Simulations were carried out with classical molecular dynamics, as implemented in the AMBER 11 software package [37], in a cubic box with side equal to 12 Å. The force fields parmbsc0 [161] and TIP3P [101] were used for the ApA moiety and water molecules, respectively. Configuration samples were extracted from the whole 100 ns trajectory, by taking one snapshot every 2.5 ps. RMS deviation cluster analysis with a threshold of 1.5 Å was carried out on the extracted 40,000 structures, resulting in a total of 89 different clusters. The 17 most populated clusters were identified and kept for further electronic structure calculations. These clusters account for the $\sim$75% of the whole conformational space sampled by ApA during the molecular dynamics trajectory. The geometry of the snapshots that are closest to the cluster centroids was refined adopting the QM/MM set up outlined above.

Spectroscopic modeling. 2D electronic spectra for the selected representative conformers were simulated in the "static picture" approximation utilizing a simplified version of Eq. (8), obtained in the limit of the pump and probe pulses temporally coinciding at the sample (i.e., waiting time $t_2 = 0$). This limit allows to compute the spectra of the individual conformers relying solely on the electronic structure of the Franck–Condon point (i.e., $\mathbb{G}_{e \to e'}(t_2)$ is omitted, $e' = e$). A complete spectrum, incorporating the inhomogeneous broadening of the signals was obtained by adding together the spectra of the single conformers, thereby weighting each spectrum according to the parent cluster population in the clustering analysis. A broadening of 1000 cm^{-1} was used throughout in order to reproduce the line shape of recently measured 2DUV experiments for adenine [171].

For the purposes of completeness and accuracy of the spectroscopy simulation the electronic structure of ApA was computed for over 100 states expanding the active space to 24 electrons in 18 orbitals, including the whole π occupied valence space of both adenine moieties and all the π^* virtual orbital of each moiety except the highest. To facilitate the calculations, the restricted active space extension of the CASSCF method, known as RASSCF, was utilized, in combination with second order perturbation theory RASPT2 in the single root flavor. Up to four holes/electrons were included in the RAS1/RAS3 subspaces, respectively, leading to the SS-RASPT2/SA-100-RASSCF(4, 12|0, 0|4, 6) protocol. Following the results of previously reported benchmarking [148], a nonstandard imaginary level shift [65] of 0.6 a.u. was used to obtain more reliable values for the transition energies.

4.2.2 Discussion

Figure 6b, c show the 2DES spectra of water-solvated ApA, excited in the near-UV and probed by UV or Visible (Vis) light, respectively. Adenine exhibits two energetically close lying bright states in the near-UV—the less intense L_b state and a more intense L_a— both of $\pi\pi^*$ character. Despite their exact order in water is still a matter of intense debate [104, 148, 183, 199], there is an agreement that the L_a state dominates the LA cross section. The interacting L_a states of the monomers in ApA are the main source of the observed trace along ω_3 for $\omega_1 = 41,000$ cm^{-1}, featuring a bleach signature (GSB) in the UV (blue in Fig. 6b) and multiple transient absorption (TA) signatures in the Vis (red in Fig. 6c). We note the broadening of the trace along ω_1, whose source is twofold: (a) the base-base interaction which is particularly sensitive to the relative orientation of the two chromophores, causing energy splitting of the equivalent L_a bands known as secondary splitting which can reach up to several hundred cm^{-1}; (b) the different local environment of each base due to the instantaneous water arrangement captured in every snapshot. The base-base interaction gives rise to the so called inhomogeneous broadening of the spectrum, clearly recognizable in the GSB (Fig. 6b), as the interconversion between different conformations is a slow process (compared to the duration of the experiment). The dynamics of the solvation shell is assumed to be fast (and thus Markovian) and induces a homogeneous broadening of the peaks.

The key fingerprint in the Vis probing window (Fig. 6c), is the doubly excited HOMO → LUMO transition (HL^2) which becomes accessible in two photon (i.e., pump-probe) spectroscopy as the outcome of two successive single HOMO → LUMO excitations. Nucleobase-specific HL^2 TA bands have been documented for all DNA/RNA monomers [188]. They allow to selectively follow the deactivation dynamics of the single nucleobases and their homo-multimers. On a side note, we point out that the doubly excited states are not trivially obtainable with single-reference methods and showcase the need to use multi-configurational techniques.

In the green-to-red window of the Vis spectrum ($\sim$15,000–20,000 cm^{-1} along ω_3) we note the appearance of several less intense bands, attributed mainly to CT between the two bases. It well known that the intensity of CT signals depends on the wavefunction overlap of the involved chromophores, therefore, these bands are intriguing as potential markers for quantifying the base-base interaction strength. Yet, we advise caution as our calculations reveal that presence of monomer excitations in the same spectral window, involving higher lying virtual orbitals LUMO+n (with $n \geq 1$) accessible in two photon spectroscopies as LUMO → LUMO+n transitions, which obscure the CT signals. A strategy for selectively detecting CT transitions has been demonstrated in spectroscopy simulations on oligopeptides by combining spectra recorded with cross-polarized pulse setups that suppress local signals [147].

Our computational strategy allows us to dissect the complete spectrum into contributions pertaining to the individual nuclear arrangements, thus enabling us to extract the specific base-base interaction behind a given peak. In order to do so, we have grouped the different clusters depending on the relative distance between the centers of mass. In Fig. 6d, panel A, we demonstrate how for short-distance (3–4 Å)

π-stacked arrangements, the aforementioned splitting of the L_a bands leads to a marked broadening of the spectral trace along ω_1. In Fig. 6d, panel B, we show that T-stacked conformations, do not exhibit a noticeable broadening of the trace due to the unfavorable orientation of the π-spaces of the two chromophores. However, they are characterized with relatively intense CT signals in the low-energy Vis window due to the proximity of the amino group of one of the base to the aromatic moiety on the opposing base. Finally, unstacked ApA conformations are shown to display very similar spectra to those of the monomers, lacking the broadening along ω_1 and CT signatures, as expected from these noninteracting species (Fig. 6d, panel C).

Overall, 2DES is shown to be a sensitive probe for inter-chromophore interactions. Its enhanced spectral resolution renders it potentially capable of separating contributions from different geometrical arrangements. However, the pronounced flexibility of small oligonucleosides results in broad unstructured spectra with strongly overlapping contributions, in particular in the region of the ground state bleach. The Visible window shows spectral fingerprints of base stacking but attention is advised due to the simultaneous detection of monomer signatures emerging in the same window.

4.3   Transient Signatures of the Ultrafast and Long-Lived UV-induced Decay Pathways in Solvated Pyrimidine Nucleosides

The intrinsic photostability of pyrimidine nucleobases/sides/tides is characterized by an efficient conversion of radiative to kinetic energy via nonharmful radiationless pathways. But, different time-resolved experiments and simulations revealed intricate excited state dynamics including multiple timescales, spanning from ultrafast sub-picosecond decay components to dozen/hundreds of ps until ns. [4, 17, 19, 75, 96, 97, 123, 137, 182]. Despite the fact that most of the available studies has associated the ultrafast dynamics to the HOMO to LUMO $\pi\pi^*$ transitions, there are relevant issues that are still subject to an intense scientific discussion, e.g., the involvement of additional bright $\pi\pi^*$ states in the photodynamics [36, 186]. Furthermore, the role of dark states, such as $n\pi^*$ and charge transfer states, in the excited state dynamics remains yet to be unambiguously demonstrated. Of particular interest regarding DNA/RNA photodamaging is their potential involvement in the excited state population trapping, responsible for observed long-lived decay components (up to the nanosecond timescale) in the spectral features [86, 87, 194] and thought as the prime cause for photolesions.

Among the nucleobases, cytosine derivatives (Fig. 5a), have the most debated mechanism for excited state decay. All quantum mechanical (QM) calculations show four close lying, potentially coupled states (previously discussed in Sect. 4.1), falling between $\sim$275 and 200 nm, in the region of the two lowest energy UV absorption bands (see Fig. 5c). All these different states could access to different conical intersections with the ground state, and the preferential nonradiative decay path is for this reason still unclear [18, 25, 79, 91, 98, 106, 122, 136]. Even small modifications of the chemical structure of the nucleobase can have a pronounced effect on the photophysics and photochemistry of cytosine. In this regard, the most frequent epigenetic modification in DNA is the 5-methylation, largely analyzed in molecular biology [29, 108, 112, 143, 153], because of its frequent presence in mutational DNA-sequences [141].

In several recent works [129, 132, 159, 160], we applied a unified scheme for modeling ultrafast and long-lived deactivation pathways and their transient spectroscopic features to unravel to photophysics of water-solvated pyrimidines—5-methyl-(oxy/deoxy)-Cyd, (oxy/deoxy)-Cyd, Urd, Thd—and used them to interpret fluorescence up-conversion (FU) and transient absorption (TA) experiments. In the following we outline the methodology used and the most important results.

4.3.1   Methodology

QM/MM setup. A three-layer approach was employed: the pyrimidine base was included in the High Layer, the Medium Layer consisted of the sugar and the water molecules within 5 Å from the base, the Low Layer contained the rest of the waters spanning a cubic box of 12 Å in diameter.

Concerning the QM part, the MS-CASPT2/SA-8-CASSCF protocol, in combination with the 6-31G* basis set for geometry optimizations and the ANO-L-VTZP for single-point calculations [225, 226], was used with an active space of 14 electrons in 10 orbitals, including the full valence π space and two lone pairs localized on the oxygen and nitrogen atoms. State averaging was performed over eight states.

Boundary conditions. It is well known that in mononucleosides two predominant conformers (*syn* and *anti*) could be populated (Fig. 5b). A cluster analysis performed along a 100 ns MM dynamics performed using the Amber-11 [37] suite and the ff99 force field [219]) gave as most probable geometry the *syn* conformation, because it is stabilized by an intramolecular hydrogen bond between the CH_2OH group of the sugar and the C=O group of the nucleobase (Fig. 5b). Therefore, this conformation was adopted for the subsequent QM/MM refinement on both the systems.

PES exploration. Critical points on the PES of the lowest singlet $\pi\pi^{*}$ and $n\pi^{*}$ states (minima, CI and intersystem crossings) were obtained through numerical CASPT2/MM gradients.

Spectroscopic modeling. Absorption cross sections were generated using according to Eq. 4 relying on a semi-classical (Wigner) sampling with 500 realizations. Thereby, high frequency normal modes of the nucleoside ($>2000\,cm^{-1}$) were kept frozen in the sampling procedure.

Emission and absorption features were modeled according to Eq. 7, at the critical points and compared to experimental data. For the purpose of characterizing the electronic structure of the monomer up to the far-UV as required for simulating TA features, the QM calculations at the critical points were performed averaging over

30 roots, i.e., at SS-CASPT2/SA-30-CASSCF(14,10) level of theory. The energies were convoluted with a Gaussian function of the half-width of 0.6 eV.

4.3.2 Discussion

Ultrafast decay paths. Figure 7c shows the computed ultrafast deactivation routes for all solvated pyrimidine nucleosides: Urd, Thd and Cyd. After photoexcitation to the bright S_1 ($\pi\pi^*$) state (HOMO $\rightarrow$ LUMO transition centered around ~4.65 eV), a steep path leads to a plateau comprising a pseudo-minimum, $min\pi\pi^* - plan$, characterized by a planar ring [19, 96, 137, 233] and a sightly more stable partially puckered minimum, $min\pi\pi^* - tw$ [82]. The existence of this plateau correlates with the experimentally observed broad emission [40, 158], resembling the calculated energy values close to the red arrow (experimental values in square brackets, Fig. 7c). Along the puckering coordinate an 'ethene-like' conical intersection ($CI\pi\pi^*/S_0$) can be accessed over a barrier which progressively increases in the sequence Urd, Thd, Cyd showing values of 0.09 eV, 0.12 eV, and 0.15 eV, respectively. Notably, an additional deactivation pathway along S_1 route exists (Fig. 7c), leading to the 'ring-opening' conical intersection ($CI\pi\sigma^*/S_0$), also reported in gas phase calculations. [19, 146, 177] The involved state is described by a transition from the HOMO to a σ^* orbital, localized on the N_1–C_2 bond. The $CI\pi\sigma^*/S_0$ is accessed over a slightly bigger barrier from the S_1 plateau following an identical trend of progressive increase in the sequence Urd, Thd, Cyd, characterized with barriers of 0.14 eV, 0.18 eV, and 0.24 eV, respectively.

Very remarkably, the computed trends agree with the trend followed by the ultrafast lifetimes registered with both fluorescence up-conversion (~100, 150 and 200 fs [155, 177]) and transient absorption (210, 540, and 720 fs [40, 156]) of oxy- and deoxy-pyrimidines. Moreover, all three systems show distinct contributions in the sub-ps spectra components which are reproduced by the simulations with a satisfactory agreement:

(a) a strong ESA in the near-UV between 300 and 450 nm (first blue vertical arrows in Fig. 7c), arising mainly from the planar $\pi\pi^*$ minima ($min\pi\pi^* - plan$), except for Urd where both the $min\pi\pi^* - plan$ and $min\pi\pi^* - tw$ contribute equally in this spectral region;
(b) a less intense ESA in the visible between 500 and 650 nm (second blue arrows in Fig. 7c), that can be assigned to the population of the twisted $\pi\pi^*$ conformations ($min\pi\pi^* - tw$).

These results confirm the hypothesis addressing the shortest ultrafast decay times (sub-ps) to the lowest $\pi\pi^*$ state, involving both, the 'ethene-like' ring puckering distortion path ($CI\pi\pi^*/S_0$) and the $\pi\sigma^*$ state (passing through the $min\pi\pi^* - plan$, Fig. 7c), leading to the ring-opening decay mechanism ($CI\pi\sigma^*/S_0$).

Deconstruction of the LA spectra of the solvated pyrimidine nucleosides into contributions from various electronic states (Fig. 7b) demonstrate that the second bright state $\pi_2\pi^*$ (HOMO-1 $\rightarrow$ LUMO transition, ~1 eV higher than the $\pi\pi^*$) gains

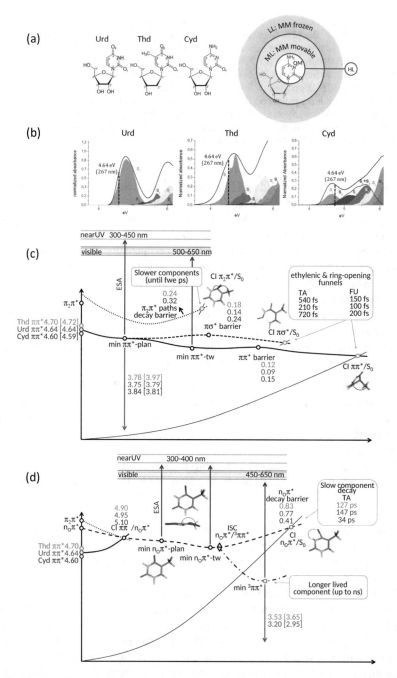

Fig. 7 (Continued)

◄ **Fig. 7** **a** Pyrimidine nucleosides (Urd, Thd, Cyd) and the QM/MM setup: the QM nucleobase in the High layer (HL), the MM sugar moiety and waters within 5 Å radius in the Medium Layer (ML), and the rest of waters in the Low Layer (LL). **b** MS-CASPT2/MM absorption cross sections, computed from 500 samples of a Wigner distribution. **c** and **d** panels show the evolution of the ground and lowest excited states for Urd, Thd, and Cyd along the computed ultrafast and longer-living relaxation paths, respectively. Solid and dotted lines are the $\pi\pi^\star$ and $\pi_2\pi^\star$ states. Dashed line is the $\pi\sigma^\star$ in the (**c**) panel and the $n_O\pi^\star$ in the (**d**) panel. Dashed-dotted line is the $^3\pi\pi^\star$ triplet state. Red and blue arrows are the fluorescence/phosphorescence and ESA energies, with the respective experimental values in brackets. Panels (**a**) and (**b**) are adapted with permission from Ref. [159]. Copyright 2017, American Chemical Society

in importance in Thy and Cyd and is predicted to account for roughly 5% and 20% of the excited population at 267 nm. The $\pi_2\pi^\star$ state decay path involves increasing energy barriers (0.24 and 0.32 eV for Thd and Cyd, respectively), comparing with $\pi\pi^\star$, to reach a crossing with S_0, suggesting a somewhat slower dynamics. This observation has been used to rationalize the experimentally observed two ultrafast decay times for these two nucleobases (the shortest in the sub-ps and the longer up to few ps), whereas only one lifetime has been reported for Urd (see different lifetimes associations in Fig. 7c).

It should be noted that ESAs is not well-suited to disentangle unambiguously the specific dynamical contributions coming from simultaneously populated channels, because the signals from ESA are characterized by broad, unstructured line shapes. Indeed, common TA signals characterize the $\pi\pi^\star$ and $\pi_2\pi^\star$ states.

Long-living decay paths. The involvement of the 'optically dark' states in the photoactivated processes of solvated nucleobases is not still clearly defined. Conical intersections between the firstly populated bright $\pi\pi^\star$ and the lower dark $n\pi^\star$ states, involving the lone pairs localized either on oxygen or nitrogen atoms, are here characterized (Fig. 7d).

A barrierless path, leading to a crossing with $n_O\pi^\star$ states, is predicted from the $\pi_2\pi^\star$ state which, as discussed above, is accessible in Thd and Cyd by pumping at 267 nm (Fig. 7b). [159] Furthermore, because the crossings between the lowest $\pi\pi^\star$ and the $n_O\pi^\star$ are found in the close to the Franck–Condon region, a non-adiabatic population transfer could also be possible, but just for Urd and Thd, because for Cyd the energy gap is too large (~0.5 eV, see $CI\pi\pi^\star/n_O\pi^\star$ energies in Fig. 7d). So, most of the initial wave packet could end, adiabatically or non-adiabatically, in the $n_O\pi^\star$ state.

On the PES of the $n\pi^\star$ state the three pyrimidine nucleobases show a swift decay leading toward a plateau with a planar pseudo-minimum ($minn_O\pi^\star - plan$) and a slightly more stable twisted minimum ($minn_O\pi^\star - tw$) characterized by the carbonyl out-of-plane motion. In contrast to the $\pi\pi^\star$ state PES, the path connecting the plateau and the CIs with S_0 ($CIn_O\pi^\star/S_0$) exhibit very large energy barriers, consistent with the tens/hundreds ps experimental lifetimes reported by pump–probe transient absorption measurements. Planar and twisted $n_O\pi^\star$ minima ($minn_O\pi^\star - plan$ and $minn_O\pi^\star - tw$) are, in fact, the main actors addressed to the 300–400 nm TA signals associated to the above lifetimes (first two blue arrows in Fig. 7d), instead not observed for $n_N \pi^*$ state [156].

Importantly, in the vicinity of the $n_O \pi^*$ twisted minima of all the three nucleosides, we characterized a singlet–triplet intersystem crossing region ($ISC^3 \pi \pi^* / n_O \pi^*$, Fig. 7d), with quite large spin-orbit coupling values, guaranteeing an efficient ISC rate. Furthermore, analyzing phosphorescence maxima of Thd and Urd in ethanol (3.65 eV and 2.95 eV, respectively [80]), they can be clearly addressed to the computet phosphorescent emission, obtained from the $^3 \pi \pi^*$ minima (placed at 3.53 and 3.20 eV, see red arrow in Fig. 7d), suggesting a direct relaxation along $^3 \pi \pi^*$ after the ISC processes. Additionally, the most pronounced TA band in the 450–650 nm region is assigned to a triplet $^3 \pi \pi^*$ vertical absorption (third blue arrow in Fig. 7d), in support of an experimental work on thymine [87] proposing the ISC involvement to justify the longer-lived component (up to ns). We predict that Urd and Cyd also are involving the same ISC processes due to the similarities between pyrimidine nucleosides, showing an analogous behavior.

A secondary possible decay path, through a CT state was tested, involving a sugar-to-base hydrogen/proton transfer process, evaluated in solution for the first time and showing a not-negligible barrier along its decay route, becoming eventually competitive just with other slow decay channels, in the order of ps and ns time ranges.

So, a unified deactivation scheme for the long-lived channels of pyrimidine nucleosides are delivered, where, mainly, the $^1 n_O \pi^*$ state is found to mediate the long-lived decay in the singlet manifold and act as the doorway for triplet population, thus accounting for the prominent recorded phosphorescence and, more generally, for the transient spectral signals registered up to the ns timescale [86, 87, 216].

dCyd versus 5mdCyd. As note above for all pyrimidines, the lowest $\pi \pi^*$ state of dCyd and 5mdCyd ($\pi \pi_1^*$, Fig. 5d) shows a minimum, separated by a barrier from an 'ethene-like' $S_0 / \pi \pi_1^*$ CI. The predicted values of the fluorescence maxima (345 nm and 367 nm for dCyd and 5mdCyd, respectively) are in good agreement with the experimentally observed (323 nm and 344 nm) [132] (Fig. 5e). The $\pi \pi^*$ PES of dCyd exhibits a small barrier (0.18 eV) toward the CI in agreement with the sub-picosecond lifetimes ($\sim$0.2 and $\sim$1 ps) detected by fluorescence up-conversion (FU) experiments (Fig. 5e) [132]. In the methylated system, instead, the CI is accessible over a larger energy barrier (0.3 eV), rationalizing the longer lifetime detected in the FU experiment ($\sim$7 ps). Correspondingly, 5mdCyd exhibits a larger fluorescence quantum yield [132]. TA spectroscopy corroborates the dominant role the $\pi \pi_1^*$ state in the short-time dynamics. As shown in Fig. 5e, the simulated TA features belonging to the $\pi \pi_1^*$ state (solid red and yellow lines) are consistent with the experiment characterized by a stimulated emission negative peak at 350 nm and two positive signals, one above 300 nm and another one between 400 and 500 nm (red dotted line, Fig. 5e).

As noted in the previous section, deconstruction of the LA spectra of Cyd (same holds for dCyd and 5mdCyd) into contributions from various electronic states demonstrate a significant overlap between the lowest two $\pi \pi^*$ bands at the wavelength of the excitation (267 nm). This suggests that the second bright $\pi \pi^*$ state ($\pi \pi_2^*$, Fig. 5d), is also partially populated upon excitation. Due to the energetic proximity between the $\pi\pi_2^*$ and the $n_O\pi^*$ and $n_N\pi^*$ states in the Frack-Condon region of dCyd bright state could potentially act as a doorway to the dark states. Exploration of the PES of the dark states suggest that the $n_O\pi^*$ state could trap the excited state population (whereas the $n_N\pi^*$ state decays barrierlessly to the lowest $\pi\pi^*$ state). This is further supported by TA simulations. Specifically, computed TA signatures of the $n_O\pi^*$ state (solid green line Fig. 5e upper panel) are consistent with the experimentally detected broad structureless positive signal spreading from the Visible to the UV (black dotted line, Fig. 5e upper panel), showing a dynamics on the time scale of several tens of picoseconds [121]. In 5mdCyd the dark states appear destabilized with respect to the bright states, which suggests their marginalized involvement in the deactivation. This interpretation is consistent with the fact that a lifetime in the order of tens of picoseconds has not been reported for 5mdCyd.

4.4 Disentangling Multiple Competing Deactivation Routes in Solvated Dinucleotides: 2DES of Adenine-Uracil Monophosphate

DNA multimers present notably different photophysics compared to the isolated nucleobases shaped by the interplay of the plethora of intra-chromophore deactivation pathways (see Sect. 4.3), and the inter-chromophore dynamics involving excitons, excimers, and charge transfer (CT) states, that are believed to prolong the excited state lifetime [46, 92, 130, 217, 234], causing the formation of photoproducts known as lesions [142, 202]. As a consequence, the number of the potentially competing deactivation pathways are considerable already in small oligomers, giving rise to congested 1D transient PP spectra unable to disentangle the origin of the numerous overlapping spectral signatures. To make things worse, the conformational freedom of DNA oligomers (see Sect. 4.2), adds to the complexity of the spectrum resulting in broad structureless bands that cover the entire probing window.

2DES has the potential to alleviate the issue of spectral congestion. By correlation the pump and probe wavelengths it allows not only to distinguish the individual conformers addressed by the pump pulse through their fingerprint spectral features (see Sect. 4.2), but also to follow the evolution of these features. More generally, in spectroscopic terms, 2DES provides a tool to correlate a spectral dynamics to the excitation wavelength, or speaking in chemical terms, to correlate a deactivation mechanism to an electronic transition falling under the envelope of the pump pulse.

In a proof of concept study, we demonstrated recently the superiority of 2DES with respect to conventional 1D techniques on the photophysics of adenine-uracil monophosphate (ApU) in aqueous solution. ApU shows a complex photodynamics consisting of three components of $\sim$2, 18, and 240 ps that have been measured with transient absorption spectroscopy after excitation at 267 nm. [201] In particular, we show how 2DES is capable of discerning and selectively following the dynamics of adenine and uracil local excited states spectral components from those components belonging to the A $\rightarrow$ U CT state.

4.4.1 Methodology

QM/MM setup. The dinucleoside was solvated in a water sphere consisting of 727 molecules, including a Na^+ as the counterion of the negatively charged phosphate group. The QM/MM setup consists of adenine and uracil nucleobases treated quantum-mechanically (i.e., in High Layer), the sugar and phosphate plus the counterion and a solvation shell encompassing 84 waters were included in the Medium Layer, the rest of the system was kept frozen in the Low Layer (Fig. 8a). Static PES exploration was performed treating the QM layer at the SA-CASSCF(8,8) level of theory taking four π orbitals from each base utilizing the 6-31+G* basis set with the ONIOM implementation of the QM/MM framework [198] as implemented in Gaussian 09 [69], whereas the electronic structure of the critical points was refined at the multi-configurational perturbation theory level, thereby enlarging the active space to 12 electrons and 12 orbitals (i.e., SS-CASPT2/SA-CASSCF(12,12)). The ANO-S basis set [165] was employed.

Boundary conditions. The solvated ApU could populate stacked and unstacked conformations, clearly distinguishable in NMR experiments [63]. For this reason, we assume that the time for interconversion between the different conformers exceeds the excited state lifetimes. Moreover, we expect that the photophysics of the unstacked systems resembles that of the single nucleotides [201]. Therefore, we focused only on the photophysics of the stacked conformer.

To sample the accessible conformational space, ten snapshots were extracted from a 1 ns ground-state classical molecular dynamics simulation at room temperature (300 K) and 1 atm. Subsequently, the electronic structure at the ten snapshots, computed with the protocol outlined in the *QM/MM setup* section, was used to select two representative conformations, one being the energetically most stable conformation in the ground state and one exhibiting the lowest CT energy (*FC-1* and *FC-2*, respectively, see Fig. 8b).

PES exploration. For both selected conformations we applied static PES exploration techniques (localizing minima and conical intersections) to explore the surfaces of the two lowest $\pi\pi^*$ states of adenine (labeled L_a and L_b) and the lowest $\pi\pi^*$ of uracil (labeled S(U)), as well as the lowest CT state of the nucleotide. These states were selected as the analysis of the ten snapshots along the classical MD simulation revealed that they absorb around 267 nm, that is the excitation wavelength for which the aforementioned lifetimes were measured in the transient absorption spectroscopy.

Spectroscopic modeling. The transient 2DES was simulated based on a few critical points along the photochemical MEP using Eq. 8. Specifically, the PES exploration helped establish a plausible mechanism and time frames for the depopulation of the involved states. For the purpose of characterizing the electronic structure of the dimer up to the far-UV, the QM calculations were performed averaging over 80 roots with an active space consisting of eight π orbitals per base, i.e., at the SS-CASPT2/SA-80-CASSCF(8,8) level of theory. 2D spectra were simulated for short (sub-100 fs), intermediate (sub-ps to few ps), and long (tens of ps) pump-probe delays utilizing

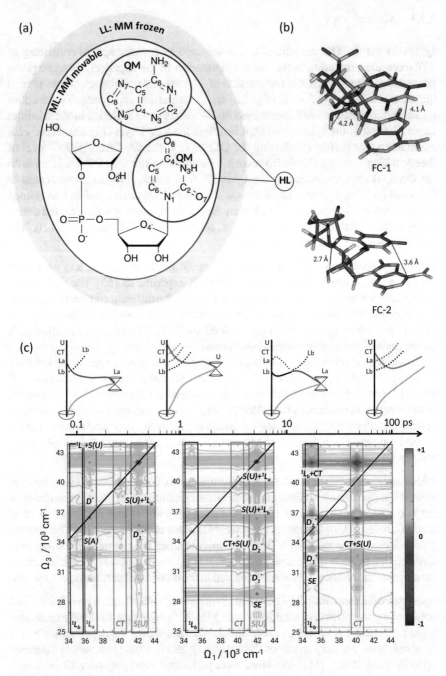

Fig. 8 (Continued)

◄ **Fig. 8 a** QM/MM setup: QM nucleobases are in the High Layer (HL), the MM sugar, phosphate, the counterion and a solvation shell encompassing 84 waters are in the Medium Layer (ML), the rest of the MM system was kept frozen in the Low Layer (LL, panel a). **b** Ground-state-optimized structures, FC-1 and FC-2 (including representative distances), one is the energetically most stable conformation and the other exhibits the lowest CT state energy. **c** Depiction of the time evolution proposed for the 2D nUV-pump nUV-probe spectra of stacked ApU. The arrow indicates the approximate timescale at which the spectra evolve, according to QM calculations. The bottom panels show the 2D spectra at the corresponding timescales (early, intermediate and late, from left to right). Reproduced from data reported in Ref. [116]

idealized broadband UV pump ($10,000 \, \text{cm}^{-1}$ bandwidth) and VIS -to-UV probe ($20,000 \, \text{cm}^{-1}$ bandwidth) pulses.

4.4.2 Discussion

The PES exploration revealed that, independent of the starting conformation, the L_a state of Ade relaxes to a CI (CIS_0/L_a) with the ground state through a barrierless path suggesting a sub-ps lifetime for this electronic state (Fig. 8c). Instead, the L_b of Ade and the S(U) of Ura relax to local minima, separated from the CI seam with the ground state by barriers, respectively. These states exhibit CI structures typical for the monomers, characterized by the typical ring distortion of the aromatic ring, characterized by an out-of-plane bending of the adenine C_2H and the uracil C_5H group, in CIS_0/L_a and $CIS_0/S(U)$ respectively (see Fig. 8c) [75]. Notably, a finite barrier of ~ 0.2 eV emerges on the PES of the S(U) state of Ura, not observed for the monomer, characterized by a sub-100 fs de-excitation lifetime [56, 201]. This suggests a considerably longer lifetime for the S(U) state in the few ps regimes. With a barrier close to 0.6 eV the L_b state of Ade suggests that this state survives even longer, possibly tens of ps.

The CT state shows dependence on the initial conformation: in one of the two selected geometries (*FC-1*) it exhibits a barrierless deactivation through a CI which restores the charge neutrality of the two bases, whereas in the second (*FC-2*) it is trapped in a minimum (stabilized by H-bonding between the two ribose moieties, Fig. 8c) with no access to the CI seam. Naturally, it can be speculated this minimum would contribute to the longest ES lifetime component (i.e., 240 ps). This hypothesis is supported by the calculated vertical emission energy from this minimum (2.14 eV), that is consistent with the emission of 2.7 eV associated with a decay component with a 72 ps time constant reported for an alternating d(AT)$_{10}$ oligomer, assigned to an AT excimer with A $\rightarrow$ T CT character [113]. Figure 8c presents transient 2D-UV simulations based on the outlined deactivation. We propose to focus on three pump-probe delay time windows

- early (i.e., sub-100 fs) times, resolving Franck–Condon dynamics before the decay of the L_a state kicks in (left panel);
- intermediate (i.e., sub-ps to few fs) times after the La decay is completed and before the S(U) decay kicks in (middle panel);

- later times, after the S(U) decay is completed and before the decay of the L_b state kicks in (right panel).

Within the femtosecond time window, intense signals arise from the L_a and S(U) manifolds, covering the much weaker bands of the L_b and CT transitions (left panel). On a picosecond timescale, the L_a state contributions disappear, thus allowing to resolve signals from population trapped on the L_b state in the near-UV spectral window as the other bright state S(U) has his spectral signatures in the far-UV (middle panel). Upon the decay of the S(U) state only contributions from the long-living L_b and CT states remain resolving signature ESA such a bright doubly exited states (labeled D*) and combination bands (e.g., CT+S(U)). It should be noted that, the signals line shapes are idealized. In fact, the trace associated with the CT state is expected to be significantly broader (and thus lower in intensity) because of the dependence of the absorption energy and oscillator strength of the CT state on the conformational dynamics.

Even if oversimplified, the presented 2DES simulations clearly demonstrate the superior spectral resolution of multidimensional optical techniques. The overlapping spectral signatures of the involved electronic states cannot be unambiguously discriminated by conventional 1D PP spectroscopy. In combination with state-of-the-art electronic structure calculations and non-adiabatic dynamics 2DUV could become a powerful tool for ab initio studies of the complex deactivation dynamics in polynucleotides.

4.5 Intersystem Crossing in Thionated Nucleobases Mediated by an Intermediate Dark State: Transient PP Spectroscopy on 4-Thiouracil

Thiobases [11, 13] are nucleobases where the canonical exocyclic carbonyl oxygen is exchanged with a sulfur atom. Due to their biological importance and the rising amount of phototerapeutic applications they recently attracted great attention. [169] Additionally, they represent perfect model systems to explain how just an atom substitution could influence the decay paths of the classical DNA or RNA bases: the absorption spectrum red-shift, induced by thionation, is the origin of other photochemical changings: [12, 51, 52, 131, 151, 167, 168, 174, 176, 203, 229]: while in the nucleobases the excited state deactivation happens on ultrafast timescale, decaying through an $S_1 \rightarrow S_0$ internal conversion [51, 51, 157], the thiobases are largely populating long-lived triplet states through intersystem crossing processes [168] arising on the sub-picosecond time range.

Recently, in a joint experimental and computational study [30] we tackled the decay mechanism of 4-thiouracil in water (Fig. 9a), and demonstrated that the intersystem crossing process mainly takes place through an intermediate dark state $n\pi^*$, populated within 100 fs, from the initially photoexcited bright $\pi\pi^*$ state. The mech-

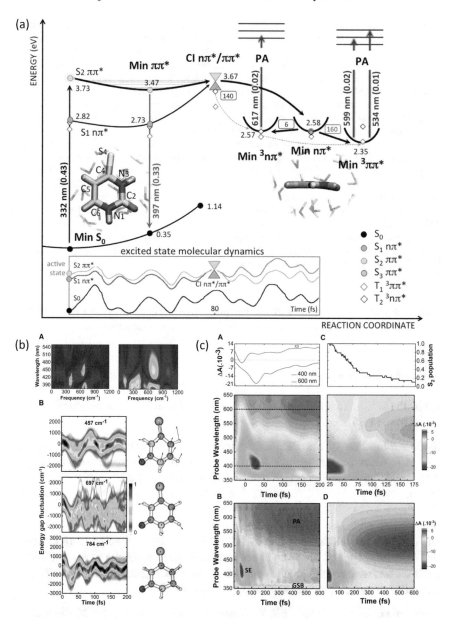

Fig. 9 (Continued)

◄ **Fig. 9 a** Water-solvated 4-thiouracil decay path from the $\pi\pi^\star$ bright state (S_2), calculated at
CASPT2/MM level. The vertical S_2 absorption wavelength is represented by the black arrow. The
stimulated emission from the Min $\pi\pi^\star$ bright state by the red arrow. The blue arrows represent
the photoinduced absorption from the triplet minima, originating the photoinduced absorption band
between 500 and 650 nm in the experimental transient absorption spectra. The dashed gray represents the secondary path $S_2 \rightarrow T_2 \rightarrow T_1$. Spin–orbit couplings are in text boxes. At the bottom, a
representative trajectory of non-adiabatic molecular dynamics [ground (S_0, black), $n\pi^\star$ (S_1, red),
and $\pi\pi^\star$ (S_2, green)]. The 'active' state, the bright $\pi\pi^\star$ before and the $n\pi^\star$ state after the conical
intersection, is shown in light blue. The day from the first to the second state takes place after ca.
80 fs. **b** (A) 2D Fourier transform of the residuals of the experimental (left) and theoretical (right)
ΔA dynamics. (B) Coherent energy gap fluctuations along selected vibrational modes at 457 cm^{-1}
(ring breathing), 697 cm^{-1} (H7 and H11 HOOP bending), and 784 cm^{-1} (H7, H11, and H12 out-
of-plane). The breathing mode originates the 400 cm^{-1} peak in the 2D Fourier transforms, instead
the two HOOP modes are responsible for the 680 cm^{-1} peaks. Intensities associated with trajectory
density. **c** ΔExperimental maps of 4-thiouracil in solution panels: (A) short scale from 30 fs until
200 fs and (B) long scale from 30 fs until 600 fs. Simulated panels: (C) short scale and (D) long
scale. At the top, panel (A) shows the experimental dynamics at 400 and 600 nm (dotted lines in
the map), panel (C) shows the calculated $S_2(\pi\pi^\star)$ population decay. Adapted with permission from
Ref. [30]. Copyright 2018 American Chemical Society

anism suggested on the basis of mixed quantum-classical non-adiabatic dynamics
was corroborated by fs time-resolved transient absorption simulations.

4.5.1 Methodology

QM/MM setup. Two different setups were used for the static and dynamic exploration of the PES.

Static exploration of the PES: The High Layer comprised the thio-nucleobase
(Fig. 9a). Water molecules in 3 Å distance were included in the Medium Layer.
The remaining water molecules constituting a spherical water droplet centered at
the chromophore with a radius of 12 Å were kept fixed in the Low Layer. The MS-
CASPT2/SA-CASSCF level of theory was applied to the QM layer. Thereby, the
active space consisted of the valence π-orbitals, the sulfur and oxygen lone pairs
and two bonding and antibonding carbon-sulfur σ orbitals, leading to 16 electrons
in 12 orbitals, applied on seven states state-averaged calculations. An ANO-RCC
basis set was used employing the contractions 6s5p3d2f1g on sulfur, 5s4p3d2f1g on
carbon/oxygen/nitrogen, and 4s3p2d1f on hydrogen [180].

Dynamic exploration of the PES: With respect to the static case, the Medium Layer
was extended to 5 Å to permit large distortions of the aromatic ring during the
excited state dynamics. The restricted version of the multireference perturbation
theory method, known as SS-RASPT2/SA-RASSCF was applied to the QM layer.
Thereby, the complete active space consisting of the valence π-orbitals, the sulfur and
the oxygen lone pairs was augmented by four extra-valence virtual orbitals allowing
for up to double excitations therein (i.e., RASSCF(12, 9|2, 4)). The state-averaging
was including the ground state and the lowest singlet $n\pi^\star$ and $\pi\pi^\star$ states.

Boundary conditions. Classical molecular dynamics simulation was carried out through the AMBER 12 suite [37] and the ff10 force field. Periodic boundary conditions were employed for a 40 Å x 39 Å x 35 Å box containing 1210 water molecules, characterized by the TIP3P force field[101]. A production run at 300 K and 1 atm was carried out for 500 ps, after an initial heating to room temperature in steps of 50K. The energetically most stable structure along the dynamics was taken for subsequent QM/MM refinement. Following the refinement (performed with the above described HL/ML/LL partitioning at the MP2 level) the critical nature of the relaxed structure was verified by a frequency calculation. The obtained normal modes and frequencies were subsequently utilized to sample initial conditions (geometries and velocities) for the excited state dynamics simulations. This was done via Wigner sampling at room temperature (300 K). From the sampling were excluded high frequency modes (i.e., C-H and N-H stretchings). Moreover, one normal mode showing a very low frequency (i.e., below $100\,cm^{-1}$)) was excluded from the sampling because we verified in preliminary samplings that excessive distortions along this mode were violating the harmonic approximation. Thirty structures, for which the lowest $\pi\pi^\star$ transition falls under the envelope of the experimentally utilized pump pulse, were selected.

PES exploration. *Static exploration of the PES:* Critical points on the PES of the lowest singlet and triplet $\pi\pi^\star$ and $n\pi^\star$ states (minima, conical intersections, and intersystem crossings) were obtained through numerical optimization. At every geometry the electronic structure of the singlet and triplet manifolds were verified by computing the lowest 7 state for each spin. Spin-orbit couplings between energetically close lying singlet and triplet states were computed in order to estimate the intersystem crossing propensity.

Dynamic exploration of the PES: Thirty snapshots were initiated in the bright $\pi\pi^\star$ state (S_2 in the Franck–Condon point) and propagated following Newton's equations of motion for the nuclei, fed with numerical gradients, for ca. 180 fs and a time step of 1.0 fs. Electron dynamics was treated with the Tully/Hammes-Shiffer modification [85], of the Tully's fewest switches surface hopping algorithm [209, 211], which relies on time-derivative couplings.

Spectroscopic modeling. In this study, we relied on the analytical model known as cumulant expansion of Gaussian fluctuations for the simulation of the third-order nonlinear response of the system recorded as transient pump-probe signal (see Sect. 2.4 for details). Utilizing the outcome of the electronic structure calculations, the nonadiabatic dynamics and the acquired experimental data, we elaborated a sequential model that reproduces the population decay following excitation of the bright $\pi\pi^\star$ state:

$$^1\pi\pi^\star \xrightarrow{k_{^1\pi\pi^\star}} {}^1n\pi^\star \xrightarrow{k_{^1n\pi^\star}} {}^3\pi\pi^\star \tag{9}$$

with $k_{^1\pi\pi^\star}$ set to ca. 67.5 fs and $k_{^1n\pi^\star}$ to 225 fs. The rate model was constructed by using lifetimes obtained from the dynamical simulations instead of experimental results. In this way, we were able to obtain all the parameters entering in the spectroscopic model from ab initio calculations. The mechanism was used to construct a rate model, whose solutions, i.e., the "waiting time"-dependent population in each state acts as weighting factors of the state specific signatures. The parameters of the composite line shape function used to describe the intramolecular vibrational dynamics (DHO model) and the inter action with the environment (OBO model) was obtained through normal mode analysis (DHO) and chosen so to reproduce the bandwidth in the linear absorption spectrum at room temperature (OBO).

4.5.2 Discussion

Figure 9c, panel B, shows the 4-thiouracil experimental two-dimensional maps of the differential absorption (ΔA) signal. Figure 9c, panel A, instead, shows the dynamics at both 400 and 600 nm (map below, dotted lines). The experiments visibly display a mismatch between the decay time of the stimulated emission (SE, blue signal) at 400 nm associated to the $\pi\pi^\star$ depopulation at $\sim$76 fs and the increasing of the photoinduced absorption (PA, red signal) at $\sim$600 nm, assigned to triplet-triplet absorption, furnishing a prove for a intersystem crossing (ISC) process at $\sim$225 fs: this is an experimental validation that ISC process largely takes place through an intermediate dark state, as previous theoretical works proposed [51, 124, 125, 131, 175, 181, 229, 235], and not directly from the photoexcited bright $\pi\pi^\star$ state. This hypothesis is corroborated by the static and dynamic exploration of the excited state PES of 4-thiouracil. The singlet $\pi\pi^\star$ population balistically reaches the singlet $n\pi^\star$ state through a sloped conical intersection (CI $n\pi^\star/\pi\pi^\star$, Fig. 9a) reached over a small barrier of $\sim$0.2 eV (a representative trajectory is shown in the inset of Fig. 9a). The $\pi\pi^\star$ population decay (Fig. 9c, panel C) was fitted with a time constant of 67.5 fs, resembling the 76 fs experimental time constant. While on the $\pi\pi^\star$ surface the system emits around 400 nm (see red arrow in Fig. 9a), perfectly matching the experimentally recorded SE. Instead, in the probed spectral window the singlet $n\pi^\star$ is completely dark (i.e., it possesses no SE or TA features). The calculations show that the minimum on the $^1n\pi^\star$ state (Min $n\pi^\star$, Fig. 9a) is isoenergetic with two triplet states, $^3n\pi^\star$ and $^3\pi\pi^\star$ respectively (black and red diamond labels in Fig. 9a). Due to a non-negligible spin-orbit coupling, the $n\pi^\star$ state becomes a doorway for the population of the triplet manifold leading to the buildup of its fingerprint PA between 500 and 650 nm, reproduceable in our calculations (triplet manifold TA signatures between 539 nm and 617 nm, Fig. 9a, blue arrow).

The static PES exploration indicates the presence of a secondary decay channel directly populating the triplet manifold out of the bright state, i.e., S_2 ($^1\pi\pi^\star$) $\rightarrow$ T_2 ($^3n\pi^\star$) $\rightarrow$ T_1 ($^3\pi\pi^\star$) (dashed gray arrows in Fig. 9a). Its possible involvement is supported by experimental photoinduced absorption (PA) signatures at 600–650 nm, appearing at the same decay time of $\pi\pi^\star$ state (76 fs), whose wavelength exactly matches the calculated T_2 photoinduced absorption.

Transient spectra constructed on the basis of the static and dynamics exploration of the PES (Fig. 9c, panel D), show a remarkable agreement with the experimental counterpart. Interestingly, the signal is modulated on the base of a complex oscillatory pattern associated with vibrational coherences impulsively excited by the sub-20 fs pump pulse. A 2D Fourier analysis of the experimental and simulated spectra oscillatory component has been executed to extract the frequencies of the observed coherent oscillations. Figure 9b, panel A, shows the obtained 2D maps revealing two dominant modes with frequencies of $680\,cm^{-1}$ and $400\,cm^{-1}$. A normal mode investigation permits to recognize the responsibles for the coherent beatings: a $457\,cm^{-1}$ breathing mode and two different hydrogen-out-of-plane bending modes with frequencies $697\,cm^{-1}$ and $784\,cm^{-1}$.

In summary, by incorporating the coupling of the electronic and intramolecular vibrational degrees of freedom in the simulation of the transient signals we reproduce spectral signatures such as quantum beatings that cannot be tackled with the "static" approaches presented in the previous examples. The remarkable agreement between experiment and simulation reinforces the validity of the underlying QM/MM and spectroscopic models.

5   Conclusions and Perspectives

In this chapter, we have described a simulation workflow tailored for the study of the photo-response of biological chromophores, that can lead from the modeling of the photoinduced dynamics to the prediction of both steady-state and transient spectroscopic experiments. This workflow is based on a subtractive QM/MM scheme and comprises a number of tools that have been implemented in or interfaced to COBRAMM, our software for QM/MM calculations.

The effectiveness of the proposed methodology has been illustrated by means of a few selected examples, in which important biological problems have been studied: the photoisomerization of the retinal and the excited state dynamics of DNA monomers and dimers. These cases show that the modeling of UV-Vis spectra of small biological chromophores embedded in complex condensed-phase environment is feasible, even with a computationally expensive QM description as MRPT. This choice of the electronic structure technique ensures a high level of accuracy of the results, that are then reliable and can be effectively used to assist the interpretation of experimental spectra.

However, the examples considered also illustrate the main limitation of the approach: the size of the chromophore that can be modeled is limited by the computational cost of the QM calculation. Moving in the directions of multi-chromophoric systems, e.g., light-harvesting systems involving several coupled pigment molecules or extended systems as DNA oligomers, the number of electronic and nuclear dimensions of QM part grows, preventing the use of standard high-level electronic structure methods [53, 190].

In this context, an alternative strategy needs to be pursued to maintain the necessary level of electronic structure accuracy, while alleviating the computational cost to treat larger systems. One very promising direction consists in taking advantage of the multi-chromophoric structure of the problem, exploiting the fact that in many cases electronic excitations are (semi)localized on a set of well separated chromophores.

Under these conditions, one can partition the full system in terms of the fragments on which the electronic excitation resides, each interacting with the neighbor fragments. The electronic states of the full system can then be represented on a basis of Frenkel configurations and the resulting exciton Hamiltonian comprises terms—the *site energies* and the *excitonic couplings*—that can readily be extracted from electronic structure calculations of each of the fragments, embedded in the electrostatic field of the rest of the system [53, 190]. This exciton model strategy then allows breaking the full problem in terms of a set of smaller electronic structure calculations, easily computed in parallel [195]. In addition to make large computations feasible, this goes in the direction of current technical advancements, by exploiting massively parallel platforms for high-performance computing.

Different groups have done some preliminary work in this direction, showing that the exciton model can be effectively integrated to the standard tools for dynamical and spectroscopic simulations. The exciton model can be coupled to FSSH dynamical scheme, by updating the parameters of the electronic Hamiltonian on-the-fly, as the nuclear configuration evolves in time [134, 195]. This requires to compute at each step the derivatives of the exciton model parameters with respect to the nuclear coordinates, and to map these and the non-adiabatic couplings from the Frenkel to the adiabatic basis [134].

The exciton model is also a convenient basis to simulate nonlinear spectroscopic experiments. In a recent work, Segatta et al. [189], parametrized a exciton Hamiltonian from accurate quantum chemical calculations for the Light-Harvesting 2 (LH2) complex of purple bacteria. Simulations of the 2DES spectra over the entire visible-near-infrared spectral region provided a comprehensive interpretation of the spectral signatures.

In the future, we will pursue this promising direction by extending the simulation workflow of COBRAMM to include exciton model based dynamical algorithms and spectroscopic simulation techniques. This will allow to consider extended multichromophoric systems without sacrificing the accuracy of the quantum chemical calculations, that is one of the key elements that enabled the success of the strategy described in this chapter.

Acknowledgements The authors acknowledge support from the European Union's Horizon 2020 research and innovation program under grant agreement No. 814492. M. G acknowledges the support from the European Research Council Advanced Grant STRATUS (ERC-2011-AdG No. 291198). I.R. acknowledges the use of HPC resources of the Pôle Scientifique de Modélisation Numérique (PSMN) at the École Normale Supérieure de Lyon, France. M.G. and I. R. acknowledge support by the French Agence National de la Recherche (FEMTO-2DNA, ANR-15-CE29-0010). Francesco Segatta and Vishal Kumar Jaiswal are gratefully acknowledged for useful discussions.

References

1. Abraham MJ, Murtola T, Schulz R, Páll S, Smith JC, Hess B, Lindahl E (2015) GROMACS: high performance molecular simulations through multi-level parallelism from laptops to supercomputers. SoftwareX 1–2:19–25. https://doi.org/10.1016/j.softx.2015.06.001
2. Abramavicius D, Valkunas L, Mukamel S (2007) Transport and correlated fluctuations in the nonlinear optical response of excitons. Europhys Lett 80(1):17005. https://doi.org/10.1209/0295-5075/80/17005
3. Abramavicius D, Palmieri B, Voronine DV, Šanda F, Mukamel S (2009) Coherent multidimensional optical spectroscopy of excitons in molecular aggregates. Quasiparticle versus supermolecule perspectives. Chem Rev 109(6):2350–2408. https://doi.org/10.1021/cr800268n
4. Altavilla SF, Segarra-Martí J, Nenov A, Conti I, Rivalta I, Garavelli M (2015) Deciphering the photochemical mechanisms describing the UV-induced processes occurring in solvated guanine monophosphate. Front Chem 3(APR). DOIurlhttps://doi.org/10.3389/fchem.2015.00029
5. Altoè P, Stenta M, Bottoni A, Garavelli M (2007a) A tunable QM/MM approach to chemical reactivity, structure and physico-chemical properties prediction. Theor Chem Acc 118(1):219–240. https://doi.org/10.1007/s00214-007-0275-9
6. Altoè P, Stenta M, Bottoni A, Garavelli M (2007b) COBRAMM: a tunable QM/MM approach to complex molecular architectures. Modelling the excited and ground state properties of sized molecular systems. AIP Conf Proc 963(1):491–505. https://doi.org/10.1063/1.2827033
7. Anda A, De Vico L, Hansen T, Abramavičius D (2016) Absorption and fluorescence lineshape theory for polynomial potentials. J Chem Theory Comput 12(12):5979–5989. https://doi.org/10.1021/acs.jctc.6b00997
8. Aquilante F, Lindh R, Bondo Pedersen T (2007) Unbiased auxiliary basis sets for accurate two-electron integral approximations. J Chem Phys 127(11):114107. https://doi.org/10.1063/1.2777146
9. Aquilante F, Malmqvist PÅ, Pedersen TB, Ghosh A, Roos BO (2008) Cholesky decomposition-based multiconfiguration second-order perturbation theory (CD-CASPT2): application to the spin-state energetics of Co III(diiminato)(NPh). J Chem Theory Comput 4(5):694–702. https://doi.org/10.1021/ct700263h
10. Aquilante F, Autschbach J, Carlson RK, Chibotaru LF, Delcey MG, De Vico L, Fdez Galván I, Ferré N, Frutos LM, Gagliardi L, Garavelli M, Giussani A, Hoyer CE, Li Manni G, Lischka H, Ma D, Malmqvist PÅ, Müller T, Nenov A, Olivucci M, Pedersen TB, Peng D, Plasser F, Pritchard B, Reiher M, Rivalta I, Schapiro I, Segarra-Martí J, Stenrup M, Truhlar DG, Ungur L, Valentini A, Vancoillie S, Veryazov V, Vysotskiy VP, Weingart O, Zapata F, Lindh R (2016) Molcas 8: new capabilities for multiconfigurational quantum chemical calculations across the periodic table. J Comput Chem 37(5):506–541. https://doi.org/10.1002/jcc.24221
11. Arslancan S, Martínez-Fernández L, Corral I (2017) Photophysics and photochemistry of canonical nucleobases' thioanalogs: from quantum mechanical studies to time resolved experiments. Molecules 22(6):998. https://doi.org/10.3390/molecules22060998
12. Ashwood B, Jockusch S, Crespo-Hernández CE (2017) Excited-state dynamics of the thiopurine prodrug 6-thioguanine: can N9-glycosylation affect its phototoxic activity? Molecules 22(3):379. https://doi.org/10.3390/molecules22030379
13. Ashwood B, Pollum M, Crespo-Hernández CE (2019) Photochemical and photodynamical properties of sulfur-substituted nucleic acid bases. Photochem Photobiol 95(1):33–58. https://doi.org/10.1111/php.12975
14. Atkins PW, Friedman RS (2011) Molecular quantum mechanics. Oxford University Press
15. Avila Ferrer FJ, Santoro F, Improta R (2014) The excited state behavior of cytosine in the gas phase: a TD-DFT study. Comput Theor Chem 1040–1041:186–194. https://doi.org/10.1016/j.comptc.2014.03.010
16. Barbatti M, Aquino AJ, Lischka H (2010a) The UV absorption of nucleobases: semi-classical ab initio spectra simulations. Phys Chem Chem Phys 12(19):4959–4967. https://doi.org/10.1039/b924956g

17. Barbatti M, Aquino AJ, Szymczak JJ, Nachtigallová D, Hobza P, Lischka H (2010b) Relaxation mechanisms of UV-photoexcited DNA and RNA nucleobases. Proc Natl Acad Sci USA 107(50):21453–21458. https://doi.org/10.1073/pnas.1014982107

18. Barbatti M, Aquino AJ, Szymczak JJ, Nachtigallová D, Lischka H (2011) Photodynamical simulations of cytosine: characterization of the ultrafast bi-exponential UV deactivation. Phys Chem Chem Phys 13(13):6145–6155. https://doi.org/10.1039/c0cp01327g

19. Barbatti M, Borin AC, Ullrich S (2015) Photoinduced processes in nucleic acids. Top Curr Chem 355:1–32. https://doi.org/10.1007/128_2014_569

20. Bassolino G, Sovdat T, Liebel M, Schnedermann C, Odell B, Claridge TD, Kukura P, Fletcher SP (2014) Synthetic control of retinal photochemistry and photophysics in solution. J Am Chem Soc 136(6):2650–2658. https://doi.org/10.1021/ja4121814

21. Baylor D (1996) How photons start vision. Proc Natl Acad Sci USA 93(2):560–565. https://doi.org/10.1073/pnas.93.2.560

22. Bearpark MJ, Robb MA, Bernhard Schlegel H (1994) A direct method for the location of the lowest energy point on a potential surface crossing. Chem Phys Lett 223(3):269–274. https://doi.org/10.1016/0009-2614(94)00433-1

23. Birge R (1990) Photophysics and molecular electronic applications of the rhodopsins. Annu Rev Phys Chem 41(1):683–733. https://doi.org/10.1146/annurev.physchem.41.1.683

24. Birge RR, Murray LP, Pierce BM, Akita H, Balogh-Nair V, Findsen LA, Nakanishi K (1985) Two-photon spectroscopy of locked-11-cis-rhodopsin: evidence for a protonated Schiff base in a neutral protein binding site. Proc Natl Acad Sci USA 82(12):4117–4121. https://doi.org/10.1073/pnas.82.12.4117

25. Blancafort L (2007) Energetics of cytosine singlet excited-state decay paths-a difficult case for CASSCF and CASPT2†. Photochem Photobiol 83(3):603–610. https://doi.org/10.1562/2006-05-29-ra-903

26. Blancafort L, Migani A (2007) Water effect on the excited-state decay paths of singlet excited cytosine. J Photochem Photobiol A Chem 190(2–3):283–289. https://doi.org/10.1016/j.jphotochem.2007.04.015

27. Bo C, Maseras F (2008) QM/MM methods in inorganic chemistry. Dalt Trans 22:2911. https://doi.org/10.1039/b718076d

28. Bonvicini A, Demoulin B, Altavilla SF, Nenov A, El-Tahawy MM, Segarra-Martí J, Giussani A, Batista VS, Garavelli M, Rivalta I (2016) Ultraviolet vision: photophysical properties of the unprotonated retinyl Schiff base in the Siberian hamster cone pigment. Theor Chem Acc 135(4):110. https://doi.org/10.1007/s00214-016-1869-x

29. Booth MJ, Branco MR, Ficz G, Oxley D, Krueger F, Reik W, Balasubramanian S (2012) Quantitative sequencing of 5-methylcytosine and 5-hydroxymethylcytosine at single-base resolution. Science 336(6083):934–937. https://doi.org/10.1126/science.1220671

30. Borrego-Varillas R, Teles-Ferreira DC, Nenov A, Conti I, Ganzer L, Manzoni C, Garavelli M, Maria De Paula A, Cerullo G (2018) Observation of the sub-100 femtosecond population of a dark state in a thiobase mediating intersystem crossing. J Am Chem Soc 140(47):16087–16093. https://doi.org/10.1021/jacs.8b07057

31. Borrego-Varillas R, Nenov A, Ganzer L, Oriana A, Manzoni C, Tolomelli A, Rivalta I, Mukamel S, Garavelli M, Cerullo G (2019) Two-dimensional UV spectroscopy: a new insight into the structure and dynamics of biomolecules. Chem Sci 10(43):9907–9921. https://doi.org/10.1039/c9sc03871j

32. Briand J, Bräm O, Réhault J, Léonard J, Cannizzo A, Chergui M, Zanirato V, Olivucci M, Helbing J, Haacke S (2010) Coherent ultrafast torsional motion and isomerization of a biomimetic dipolar photoswitch. Phys Chem Chem Phys 12(13):3178–3187. https://doi.org/10.1039/b918603d

33. Brixner T, Stiopkin IV, Fleming GR (2004) Tunable two-dimensional femtosecond spectroscopy. Opt Lett 29(8):884. https://doi.org/10.1364/ol.29.000884

34. Brixner T, Stenger J, Vaswani HM, Cho M, Blankenship RE, Fleming GR (2005) Two-dimensional spectroscopy of electronic couplings in photosynthesis. Nature 434(7033):625–628. https://doi.org/10.1038/nature03429

35. Brunk E, Rothlisberger U (2015) Mixed quantum mechanical/molecular mechanical molecular dynamics simulations of biological systems in ground and electronically excited states. Chem Rev 115(12):6217–6263. https://doi.org/10.1021/cr500628b

36. Buchner F, Nakayama A, Yamazaki S, Ritze HH, Lübcke A (2015) Excited-state relaxation of hydrated thymine and thymidine measured by liquid-jet photoelectron spectroscopy: experiment and simulation. J Am Chem Soc 137(8):2931–2938. https://doi.org/10.1021/ja511108u

37. Case DA, Ben-Shalom IY, Brozell SR, Cerutti DS, Cheatham TE III, Cruzeiro VWD, Darden TA, Duke RE, Ghoreishi D, Gilson MK (2018) Amber 2018:2018

38. Cembran A, González-Luque R, Altoè P, Merchán M, Bernardi F, Olivucci M, Garavelli M (2005) Structure, spectroscopy, and spectral tuning of the gas-phase retinal chromophore: the β-Ionone "Handle" and alkyl group effect. J Phys Chem A 109(29):6597–6605. https://doi.org/10.1021/jp052068c

39. Cho M (2008) Coherent two-dimensional optical spectroscopy. Chem Rev 108(4):1331–1418. https://doi.org/10.1021/cr078377b

40. Cohen B, Crespo-Hernández CE, Kohler B (2004) Strickler-Berg analysis of excited singlet state dynamics in DNA and RNA nucleosides. Faraday Discuss 127:137–147. https://doi.org/10.1039/b316939a

41. Conti I, Garavelli M (2018) Evolution of the excitonic state of DNA stacked thymines: intra-base $\pi\pi^*$ $\longrightarrow$ S 0 decay paths account for ultrafast (subpicosecond) and longer (>100 ps) deactivations. J Phys Chem Lett 9(9):2373–2379. https://doi.org/10.1021/acs.jpclett.8b00698

42. Conti I, Bernardi F, Orlandi G, Garavelli M (2006) Substituent controlled spectroscopy and excited state topography of retinal chromophore models: fluorinated and methoxy-substituted protonated Schiff bases. Mol Phys 104(5–7):915–924. https://doi.org/10.1080/00268970500417911

43. Conti I, Di Donato E, Negri F, Orlandi G (2009a) Revealing Excited State Interactions by Quantum-Chemical Modeling of Vibronic Activities: The R2PI Spectrum of Adenine†. J Phys Chem A 113(52):15265–15275. https://doi.org/10.1021/jp905795n

44. Conti I, Garavelli M, Orlandi G (2009b) Deciphering low energy deactivation channels in adenine. J Am Chem Soc 131(44):16108–16118. https://doi.org/10.1021/ja902311y

45. Conti I, Altoè P, Stenta M, Garavelli M, Orlandi G (2010) Adenine deactivation in DNA resolved at the CASPT2//CASSCF/AMBER level. Phys Chem Chem Phys 12(19):5016. https://doi.org/10.1039/b926608a

46. Conti I, Nenov A, Höfinger S, Flavio Altavilla S, Rivalta I, Dumont E, Orlandi G, Garavelli M (2015) Excited state evolution of DNA stacked adenines resolved at the CASPT2//CASSCF/Amber level: from the bright to the excimer state and back. Phys Chem Chem Phys 17(11):7291–7302. https://doi.org/10.1039/C4CP05546B

47. Conti I, Martínez-Fernández L, Esposito L, Hofinger S, Nenov A, Garavelli M, Improta R (2017) Multiple electronic and structural factors control cyclobutane pyrimidine dimer and 6–4 thymine-thymine photodimerization in a DNA duplex. Chem - A Eur J 23(60):15177–15188. https://doi.org/10.1002/chem.201703237

48. Cornell WD, Cieplak P, Bayly CI, Gould IR, Merz KM, Ferguson DM, Spellmeyer DC, Fox T, Caldwell JW, Kollman PA (1995) A second generation force field for the simulation of proteins, nucleic acids, and organic molecules. J Am Chem Soc 117(19):5179–5197. https://doi.org/10.1021/ja00124a002

49. Cowan ML, Ogilvie JP, Miller RJ (2004) Two-dimensional spectroscopy using diffractive optics based phased-locked photon echoes. Chem Phys Lett 386(1–3):184–189. https://doi.org/10.1016/j.cplett.2004.01.027

50. Crespo-Hernández CE, Cohen B, Hare PM, Kohler B (2004) Ultrafast excited-state dynamics in nucleic acids. Chem Rev 104(4):1977–2019. https://doi.org/10.1021/cr0206770

51. Cui G, Fang WH (2013) State-specific heavy-atom effect on intersystem crossing processes in 2-thiothymine: a potential photodynamic therapy photosensitizer. J Chem Phys 138(4):044315. https://doi.org/10.1063/1.4776261

52. Cui G, Thiel W (2014) Intersystem crossing enables 4-thiothymidine to act as a photosensitizer in photodynamic therapy: an ab initio QM/MM study. J Phys Chem Lett 5(15):2682–2687. https://doi.org/10.1021/jz501159j

53. Curutchet C, Mennucci B (2017) Quantum chemical studies of light harvesting. Chem Rev 117(2):294–343. https://doi.org/10.1021/acs.chemrev.5b00700
54. Dapprich S, Komáromi I, Byun K, Morokuma K, Frisch MJ (1999) A new ONIOM implementation in Gaussian98. Part I. The calculation of energies, gradients, vibrational frequencies and electric field derivatives. J Mol Struct THEOCHEM 461–462:1–21. https://doi.org/10.1016/S0166-1280(98)00475-8
55. Dartnall HJ (1967) The visual pigment of the green rods. Vision Res 7(1–2):1–16. https://doi.org/10.1016/0042-6989(67)90022-3
56. de La Harpe K, Crespo-Hernández CE, Kohler B (2009) The excited-state lifetimes in a G-C DNA duplex are nearly independent of helix conformation and base-pairing motif. Chem Phys Chem 10(9–10):1421–1425. https://doi.org/10.1002/cphc.200900004
57. Demoulin B, El-Tahawy MMT, Nenov A, Garavelli M, Le Bahers T (2016) Intramolecular photo-induced charge transfer in visual retinal chromophore mimics: electron density-based indices at the TD-DFT and post-HF levels. Theor Chem Acc 135(4):96. https://doi.org/10.1007/s00214-016-1815-y
58. Demoulin B, Altavilla SF, Rivalta I, Garavelli M (2017) Fine tuning of retinal photoinduced decay in solution. J Phys Chem Lett 8(18):4407–4412. https://doi.org/10.1021/acs.jpclett.7b01780
59. Dresselhaus T, Neugebauer J, Knecht S, Keller S, Ma Y, Reiher M (2015) Self-consistent embedding of density-matrix renormalization group wavefunctions in a density functional environment. J Chem Phys 142(4):044111. https://doi.org/10.1063/1.4906152
60. Dreuw A, Wormit M (2015) The algebraic diagrammatic construction scheme for the polarization propagator for the calculation of excited states. Wiley Interdiscip Rev Comput Mol Sci 5(1):82–95. https://doi.org/10.1002/wcms.1206
61. Du L, Lan Z (2015) An on-the-fly surface-hopping program jade for nonadiabatic molecular dynamics of polyatomic systems: implementation and applications. J Chem Theory Comput 11(4):1360–1374. https://doi.org/10.1021/ct501106d
62. Duster AW, Garza CM, Aydintug BO, Negussie MB, Lin H (2019) Adaptive partitioning QM/MM for molecular dynamics simulations: 6. Proton transport through a biological channel. J Chem Theory Comput 15(2):892–905. https://doi.org/10.1021/acs.jctc.8b01128
63. Ezra FS, Lee CH, Kondo NS, Danyluk SS, Sarma RH (1977) Conformational properties of purine-pyrimidine and pyrimidine-purine dinucleoside monophosphates. Biochemistry 16(9):1977–1987. https://doi.org/10.1021/bi00628a035
64. Fingerhut BP, Elsaesser T (2019) Noncovalent interactions of hydrated DNA and RNA mapped by 2D-IR spectroscopy. Springer Ser Opt Sci 226:171–195. https://doi.org/10.1007/978-981-13-9753-0
65. Forsberg N, Malmqvist PÅ (1997) Multiconfiguration perturbation theory with imaginary level shift. Chem Phys Lett 274(1–3):196–204. https://doi.org/10.1016/S0009-2614(97)00669-6
66. Fosso-Tande J, Nguyen TS, Gidofalvi G, DePrince AE (2016) Large-scale variational two-electron reduced-density-matrix-driven complete active space self-consistent field methods. J Chem Theory Comput 12(5):2260–2271. https://doi.org/10.1021/acs.jctc.6b00190
67. Freedman KA, Becker RS (1986) Comparative investigation of the photoisomerization of the protonated and unprotonated n-butylamine schiff bases of 9-cis-, 11-cis-, 13-cis-, and all-trans-retinals. J Am Chem Soc 108(6):1245–1251. https://doi.org/10.1021/ja00266a020
68. Friesner RA, Guallar V (2005) Ab initio quantum chemical and mixed quantum mechanics/molecular mechanics (QM/MM) methods for studying enzymatic catalysis. Annu Rev Phys Chem 56(1):389–427. https://doi.org/10.1146/annurev.physchem.55.091602.094410
69. Frisch MJ, Trucks GW, Schlegel HB, Scuseria GE, Robb MA, Cheeseman JR, Scalmani G, Barone V, Petersson GA, Nakatsuji H, Li X, Caricato M, Marenich AV, Bloino J, Janesko BG, Gomperts R, Mennucci B, Hratchian HP, Ortiz JV, Izmaylov AF, Sonnenberg JL, Williams-Young D, Ding F, Lipparini F, Egidi F, Goings J, Peng B, Petrone A, Henderson T, Ranasinghe D, Zakrzewski VG, Gao J, Rega N, Zheng G, Liang W, Hada M, Ehara M, Toyota K, Fukuda R, Hasegawa J, Ishida M, Nakajima T, Honda Y, Kitao O, Nakai H, Vreven T, Throssell K,

Montgomery Jr JA, Peralta JE, Ogliaro F, Bearpark MJ, Heyd JJ, Brothers EN, Kudin KN, Staroverov VN, Keith TA, Kobayashi R, Normand J, Raghavachari K, Rendell AP, Burant JC, Iyengar SS, Tomasi J, Cossi M, Millam JM, Klene M, Adamo C, Cammi R, Ochterski JW, Martin RL, Morokuma K, Farkas O, Foresman JB, Fox DJ (2016) Gaussian~16 Revision C.01

70. Froese RDJ, Musaev DG, Morokuma K (1998) Theoretical study of substituent effects in the diimine-M(II) catalyzed ethylene polymerization reaction using the IMOMM method. J Am Chem Soc 120(7):1581–1587. https://doi.org/10.1021/ja9728334

71. Garavelli M, Celani P, Bernardi F, Robb MA, Olivucci M (1997) The C5H6NH2+ protonated Shiff base: An ab initio minimal model for retinal photoisomerization. J Am Chem Soc 119(29):6891–6901. https://doi.org/10.1021/ja9610895

72. Garavelli M, Vreven T, Celani P, Bernardi F, Robb MA, Olivucci M (1998) Photoisomerization path for a realistic retinal chromophore model: the nonatetraeniminium cation. J Am Chem Soc 120(6):1285–1288. https://doi.org/10.1021/ja972695i

73. Garavelli M, Negri F, Olivucci M (1999) Initial excited-state relaxation of the Isolated 11-cis protonated schiff base of retinal: evidence for in-plane motion from ab initio quantum chemical simulation of the resonance raman spectrum. J Am Chem Soc 121(5):1023–1029. https://doi.org/10.1021/ja981719y

74. Ghosh A, Ostrander JS, Zanni MT (2017) Watching proteins wiggle: mapping structures with two-dimensional infrared spectroscopy. Chem Rev 117(16):10726–10759. https://doi.org/10.1021/acs.chemrev.6b00582

75. Giussani A, Segarra-Martí J, Roca-Sanjuán D, Merchán M (2015) Excitation of nucleobases from a computational perspective I: reaction paths. Top Curr Chem 355:57–98. https://doi.org/10.1007/128_2013_501

76. Giussani A, Segarra-Martí J, Nenov A, Rivalta I, Tolomelli A, Mukamel S, Garavelli M (2016) Spectroscopic fingerprints of DNA/RNA pyrimidine nucleobases in third-order nonlinear electronic spectra. Theor Chem Acc 135(5):121. https://doi.org/10.1007/s00214-016-1867-z

77. Giussani A, Conti I, Nenov A, Garavelli M (2018) Photoinduced formation mechanism of the thymine-thymine (6–4) adduct in DNA; a QM(CASPT2//CASSCF):MM(AMBER) study. Faraday Discuss 207:375–387. https://doi.org/10.1039/C7FD00202E

78. Gonzalez-Luque R, Garavelli M, Bernardi F, Merchan M, Robb MA, Olivucci M, González-Luque R, Garavelli M, Bernardi F, Merchán M, Robb MA, Olivucci M (2000) Computational evidence in favor of a two-state, two-mode model of the retinal chromophore photoisomerization. Proc Natl Acad Sci 97(17):9379–9384. https://doi.org/10.1073/pnas.97.17.9379

79. Gonzalez-vazquez J, Gonzalez L (2010) A time-dependent picture of the ultrafast deactivation of keto-cytosine including three-state conical intersections. Chem Phys Chem 11(17):3617–3624. https://doi.org/10.1002/cphc.201000557

80. Görner H (1990) Phosphorescence of nucleic acids and DNA components at 77 K. J Photochem Photobiol B Biol 5(3–4):359–377. https://doi.org/10.1016/1011-1344(90)85051-W

81. Granucci G, Persico M (2007) Critical appraisal of the fewest switches algorithm for surface hopping. J Chem Phys 126(13):134114. https://doi.org/10.1063/1.2715585

82. Gustavsson T, Bányász À, Lazzarotto E, Markovitsi D, Scalmani G, Frisch MJ, Barone V, Improta R, (2006) Singlet excited-state behavior of uracil and thymine in aqueous solution: A combined experimental and computational study of 11 uracil derivatives. J Am Chem Soc 128(2):607–619. https://doi.org/10.1021/ja056181s

83. Gustavsson T, Improta R, Markovitsi D (2010) DNA/RNA: building blocks of life under UV irradiation. J Phys Chem Lett 1(13):2025–2030. https://doi.org/10.1021/jz1004973

84. Hamm P, Lim M, Hochstrasser RM (1998) Structure of the amide I band of peptides measured by femtosecond nonlinear-infrared spectroscopy. J Phys Chem B 102(31):6123–6138. https://doi.org/10.1021/jp9813286

85. Hammes-Schiffer S, Tully JC (1994) Proton transfer in solution: molecular dynamics with quantum transitions. J Chem Phys 101(6):4657–4667. https://doi.org/10.1063/1.467455

86. Hare PM, Crespo-Hernández CE, Kohler B (2007) Internal conversion to the electronic ground state occurs via two distinct pathways for pyrimidine bases in aqueous solution. Proc Natl Acad Sci USA 104(2):435–440. https://doi.org/10.1073/pnas.0608055104

87. Hare PM, Middleton CT, Mertel KI, Herbert JM, Kohler B (2008) Time-resolved infrared spectroscopy of the lowest triplet state of thymine and thymidine. Chem Phys 347(1–3):383–392. https://doi.org/10.1016/j.chemphys.2007.10.035

88. Heyden A, Lin H, Truhlar DG (2007) Adaptive partitioning in combined quantum mechanical and molecular mechanical calculations of potential energy functions for multiscale simulations. J Phys Chem B 111(9):2231–2241. https://doi.org/10.1021/jp0673617

89. Hochstrasser RM (2007) Two-dimensional spectroscopy at infrared and optical frequencies. Proc Natl Acad Sci USA 104(36):14190–14196. https://doi.org/10.1073/pnas.0704079104

90. Hornak V, Abel R, Okur A, Strockbine B, Roitberg A, Simmerling C (2006) Comparison of multiple amber force fields and development of improved protein backbone parameters. Proteins Struct Funct Genet 65(3):712–725. https://doi.org/10.1002/prot.21123

91. Hudock HR, Martínez TJ (2008) Excited-state dynamics of cytosine reveal multiple intrinsic subpicosecond pathways. Chem Phys Chem 9(17):2486–2490. https://doi.org/10.1002/cphc.200800649

92. Huix-Rotllant M, Brazard J, Improta R, Burghardt I, Markovitsi D (2015) Stabilization of mixed frenkel-charge transfer excitons extended across both strands of guanine-cytosine DNA duplexes. J Phys Chem Lett 6(12):2247–2251. https://doi.org/10.1021/acs.jpclett.5b00813

93. Huntress MM, Gozem S, Malley KR, Jailaubekov AE, Vasileiou C, Vengris M, Geiger JH, Borhan B, Schapiro I, Larsen DS, Olivucci M (2013) Toward an understanding of the retinal chromophore in rhodopsin mimics. J Phys Chem B 117(35):10053–10070. https://doi.org/10.1021/jp305935t

94. Hybl JD, Ferro AA, Jonas DM (2001) Two-dimensional Fourier transform electronic spectroscopy. J Chem Phys 115(14):6606–6622. https://doi.org/10.1063/1.1398579

95. Improta R, Barone V (2004) Absorption and fluorescence spectra of uracil in the gas phase and in aqueous solution: a TD-DFT quantum mechanical study. J Am Chem Soc 126(44):14320–14321. https://doi.org/10.1021/ja0460561

96. Improta R, Barone V (2008) The excited states of adenine and thymine nucleoside and nucleotide in aqueous solution: a comparative study by time-dependent DFT calculations. Theor Chem Acc 120(4–6):491–497. https://doi.org/10.1007/s00214-007-0404-5

97. Improta R, Santoro F, Blancafort L (2016) Quantum mechanical studies on the photophysics and the photochemistry of nucleic acids and nucleobases. Chem Rev 116(6):3540–3593. https://doi.org/10.1021/acs.chemrev.5b00444

98. Ismail N, Blancafort L, Olivucci M, Kohler B, Robb MA (2002) Ultrafast decay of electronically excited singlet cytosine via a $\pi,\pi*$ to $n0,\pi*$ state switch. J Am Chem Soc 124(24):6818–6819. https://doi.org/10.1021/ja0258273

99. Jacquemin D, Perpète EA, Laurent AD, Assfeld X, Adamo C (2009) Spectral properties of self-assembled squaraine-tetralactam: a theoretical assessment. Phys Chem Chem Phys 11(8):1258. https://doi.org/10.1039/b817720a

100. Jonas DM (2003) Two-dimensional femtosecond spectroscopy. Annu Rev Phys Chem 54(1):425–463. https://doi.org/10.1146/annurev.physchem.54.011002.103907

101. Jorgensen WL, Chandrasekhar J, Madura JD, Impey RW, Klein ML (1983) Comparison of simple potential functions for simulating liquid water. J Chem Phys 79(2):926–935. https://doi.org/10.1063/1.445869

102. van der Kamp MW, Mulholland AJ (2013) Combined quantum mechanics/molecular mechanics (QM/MM) methods in computational enzymology. Biochemistry 52(16):2708–2728. https://doi.org/10.1021/bi400215w

103. Kandori H, Sasabe H, Nakanishi K, Yoshizawa T, Mizukami T, Shichida Y (1996) Real-time detection of 60-fs isomerization in a rhodopsin analog containing eight-membered-ring retinal. J Am Chem Soc 118(5):1002–1005. https://doi.org/10.1021/ja951665h

104. Khani SK, Faber R, Santoro F, Hättig C, Coriani S (2019) UV absorption and magnetic circular dichroism spectra of purine, adenine, and guanine: a coupled cluster study in vacuo

and in aqueous solution. J Chem Theory Comput 15(2):1242–1254. https://doi.org/10.1021/acs.jctc.8b00930

105. Kistler KA, Matsika S (2007) Radiationless decay mechanism of cytosine: an ab initio study with comparisons to the fluorescent analogue 5-methyl-2-pyrimidinone. J Phys Chem A 111(14):2650–2661. https://doi.org/10.1021/jp0663661

106. Kistler KA, Matsika S (2008) Three-state conical intersections in cytosine and pyrimidinone bases. J Chem Phys 128(21):215102. https://doi.org/10.1063/1.2932102

107. Kistler KA, Matsika S (2009) Solvatochromic shifts of uracil and cytosine using a combined multireference configuration interaction/molecular dynamics approach and the fragment molecular orbital method. J Phys Chem A 113(45):12396–12403. https://doi.org/10.1021/jp901601u

108. Ko M, Huang Y, Jankowska AM, Pape UJ, Tahiliani M, Bandukwala HS, An J, Lamperti ED, Koh KP, Ganetzky R, Liu XS, Aravind L, Agarwal S, Maciejewski JP, Rao A (2010) Impaired hydroxylation of 5-methylcytosine in myeloid cancers with mutant TET2. Nature 468(7325):839–843. https://doi.org/10.1038/nature09586

109. Kohler B (2010) Nonradiative decay mechanisms in DNA model systems. J Phys Chem Lett 1(13):2047–2053. https://doi.org/10.1021/jz100491x

110. Kolano C, Helbing J, Kozinski M, Sander W, Hamm P (2006) Watching hydrogen-bond dynamics in a β-turn by transient two-dimensional infrared spectroscopy. Nature 444(7118):469–472. https://doi.org/10.1038/nature05352

111. Kowalewski M, Fingerhut BP, Dorfman KE, Bennett K, Mukamel S (2017) Simulating coherent multidimensional spectroscopy of nonadiabatic molecular processes: from the infrared to the X-ray regime. Chem Rev 117(19):12165–12226. https://doi.org/10.1021/acs.chemrev.7b00081

112. Krebs JE, Goldstein ES, Kilpatrick ST (2011) Lewin's Genes X. Jones \& Barlett Publishers, BOPTurlington, MA

113. Kwok WM, Ma C, Phillips DL (2009) "Bright" and "Dark" excited states of an alternating at oligomer characterized by femtosecond broadband spectroscopy. J Phys Chem B 113(33):11527–11534. https://doi.org/10.1021/jp906265c

114. Lewis JW, Szundi I, Kazmi MA, Sakmar TP, Kliger DS (2004) Time-resolved photointermediate changes in rhodopsin glutamic acid 181 mutants. Biochemistry 43(39):12614–12621. https://doi.org/10.1021/bi0495811

115. Li B, Johnson AE, Mukamel S, Myers AB (1994) The Brownian oscillator model for solvation effects in spontaneous light emission and their relationship to electron transfer. J Am Chem Soc 116(24):11039–11047. https://doi.org/10.1021/ja00103a020

116. Li Q, Giussani A, Segarra-Martí J, Nenov A, Rivalta I, Voityuk AA, Mukamel S, Roca-Sanjuán D, Garavelli M, Blancafort L (2016) Multiple decay mechanisms and 2D-UV spectroscopic fingerprints of singlet excited solvated adenine-uracil monophosphate. Chem - A Eur J 22(22):7497–7507. https://doi.org/10.1002/chem.201505086

117. Lin H, Truhlar DG (2007) QM/MM: what have we learned, where are we, and where do we go from here? Theor Chem Acc 117(2):185–199. https://doi.org/10.1007/s00214-006-0143-z

118. Lüdeke S, Beck M, Yan EC, Sakmar TP, Siebert F, Vogel R (2005) The role of Glu181 in the photoactivation of rhodopsin. J Mol Biol 353(2):345–356. https://doi.org/10.1016/j.jmb.2005.08.039

119. Ludwig V, do Amaral MS, da Costa ZM, Borin AC, Canuto S, Serrano-Andrés L (2008) 2-Aminopurine non-radiative decay and emission in aqueous solution: a theoretical study. Chem Phys Lett 463(1–3):201–205. https://doi.org/10.1016/j.cplett.2008.08.031

120. Lumento F, Zanirato V, Fusi S, Busi E, Latterini L, Elisei F, Sinicropi A, Andruniów T, Ferré N, Basosi R, Olivucci M (2007) Quantum chemical modeling and preparation of a biomimetic photochemical switch. Angew Chemie 119(3):418–424. https://doi.org/10.1002/ange.200602915

121. Ma C, Cheng CCW, Chan CTL, Chan RCT, Kwok WM (2015) Remarkable effects of solvent and substitution on the photo-dynamics of cytosine: a femtosecond broadband time-resolved

fluorescence and transient absorption study. Phys Chem Chem Phys 17(29):19045–19057. https://doi.org/10.1039/c5cp02624e

122. Mai S, Marquetand P, Richter M, González-Vázquez J, González L (2013) Singlet and triplet excited-state dynamics study of the keto and enol tautomers of cytosine. ChemPhysChem 14(13):2920–2931. https://doi.org/10.1002/cphc.201300370

123. Mai S, Richter M, Marquetand P, González L (2014) Excitation of nucleobases from a computational perspective II: dynamics. Top Curr Chem 355:99–153. https://doi.org/10.1007/128_2014_549

124. Mai S, Marquetand P, González L (2016a) Intersystem crossing pathways in the noncanonical nucleobase 2-thiouracil: a time-dependent picture. J Phys Chem Lett 7(11):1978–1983. https://doi.org/10.1021/acs.jpclett.6b00616

125. Mai S, Pollum M, Martínez-Fernández L, Dunn N, Marquetand P, Corral I, Crespo-Hernández CE, González L (2016b) The origin of efficient triplet state population in sulfur-substituted nucleobases. Nat Commun 7(1):13077. https://doi.org/10.1038/ncomms13077

126. Mai S, Gattuso H, Monari A, González L (2018) Novel Molecular-Dynamics-Based Protocols for Phase Space Sampling in Complex Systems. Front Chem 6(OCT):495. https://doi.org/10.3389/fchem.2018.00495

127. Maitra NT (2016) Perspective: fundamental aspects of time-dependent density functional theory. J Chem Phys 144(22):220901. https://doi.org/10.1063/1.4953039

128. Markovitsi D, Gustavsson T, Vayá I (2010) Fluorescence of DNA duplexes: from model helices to natural DNA. J Phys Chem Lett 1(22):3271–3276. https://doi.org/10.1021/jz101122t

129. Martínez-Fernández L, Pepino AJ, Segarra-Martí J, Banyasz A, Garavelli M, Improta R (2016) Computing the absorption and emission spectra of 5-methylcytidine in different solvents: a test-case for different solvation models. J Chem Theory Comput 12(9):4430–4439. https://doi.org/10.1021/acs.jctc.6b00518

130. Martinez-Fernandez L, Zhang Y, de La Harpe K, Beckstead AA, Kohler B, Improta R (2016) Photoinduced long-lived charge transfer excited states in AT-DNA strands. Phys Chem Chem Phys 18(31):21241–21245. https://doi.org/10.1039/C6CP04550B

131. Martínez-Fernández L, Granucci G, Pollum M, Crespo-Hernández CE, Persico M, Corral I (2017a) Decoding the molecular basis for the population mechanism of the triplet phototoxic precursors in UVA light-activated pyrimidine anticancer drugs. Chem - A Eur J 23(11):2619–2627. https://doi.org/10.1002/chem.201604543

132. Martínez-Fernández L, Pepino AJ, Segarra-Martí J, Jovaišaitei J, Vaya I, Nenov A, Markovitsi D, Gustavsson T, Banyasz A, Garavelli M, Improta R (2017b) Photophysics of deoxycytidine and 5-methyldeoxycytidine in solution: a comprehensive picture by quantum mechanical calculations and femtosecond fluorescence spectroscopy. J Am Chem Soc 139(23):7780–7791. https://doi.org/10.1021/jacs.7b01145

133. Mathies RA, Brito Cruz CH, Pollard WT, Shank CV (1988) Direct observation of the femtosecond excited-state cis-trans isomerization in bacteriorhodopsin. Science 240(4853):777–779. https://doi.org/10.1126/science.3363359

134. Menger MF, Plasser F, Mennucci B, González L (2018) Surface Hopping within an Exciton Picture. An Electrostatic Embedding Scheme. J Chem Theory Comput 14(12):6139–6148. https://doi.org/10.1021/acs.jctc.8b00763

135. Mennucci B, Tomasi J, Cammi R, Cheeseman JR, Frisch MJ, Devlin FJ, Gabriel S, Stephens PJ (2002) Polarizable continuum model (PCM) calculations of solvent effects on optical rotations of chiral molecules. J Phys Chem A 106(25):6102–6113. https://doi.org/10.1021/jp020124t

136. Merchán M, Serrano-Andrés L (2003) Ultrafast internal conversion of excited cytosine via the lowest $\pi\pi$ electronic singlet state. J Am Chem Soc 125(27):8108–8109. https://doi.org/10.1021/ja0351600

137. Mercier Y, Santoro F, Reguero M, Improta R (2008) The decay from the dark $n\pi^*$ excited state in uracil: An integrated CASPT2/CASSCF and PCM/TD-DFT study in the gas phase and in water. J Phys Chem B 112(35):10769–10772. https://doi.org/10.1021/jp804785p

138. Metz S, Kästner J, Sokol AA, Keal TW, Sherwood P (2014) Chem Shell—a modular software package for QM/MM simulations. Wiley Interdiscip Rev Comput Mol Sci 4(2):101–110. https://doi.org/10.1002/wcms.1163

139. Middleton CT, de La Harpe K, Su C, Law YK, Crespo-Hernández CE, Kohler B (2009) DNA excited-state dynamics: from single bases to the double helix. Annu Rev Phys Chem 60(1):217–239. https://doi.org/10.1146/annurev.physchem.59.032607.093719

140. Milota F, Sperling J, Nemeth A, Mančal T, Kauffmann HF (2009) Two-dimensional electronic spectroscopy of molecular excitons. Acc Chem Res 42(9):1364–1374. https://doi.org/10.1021/ar800282e

141. Momparler RL, Bovenzi V (2000) DNA methylation and cancer. J Cell Physiol 183(2):145–154. https://doi.org/10.1002/(SICI)1097-4652(200005)183:2<145::AID-JCP1>3.0.CO;2-V

142. Morrison H (1990) Bioorganic Photochemistry: Photochemistry and the nucleic acids. No. v. 1 in Wiley-Interscience publication, Wiley

143. Motorin Y, Lyko F, Helm M (2010) 5-methylcytosine in RNA: detection, enzymatic formation and biological functions. Nucleic Acids Res 38(5):1415–1430. https://doi.org/10.1093/nar/gkp1117

144. Mukamel S (1995) Principles of nonlinear optical spectroscopy. Oxford University Press, New York

145. Mukamel S (2000) Multidimensional femtosecond correlation spectroscopies of electronic and vibrational excitations. Annu Rev Phys Chem 51(1):691–729. https://doi.org/10.1146/annurev.physchem.51.1.691

146. Nachtigallovaí D, Aquino AJ, Szymczak JJ, Barbatti M, Hobza P, Lischka H (2011) Nonadiabatic dynamics of uracil: population split among different decay mechanisms. J Phys Chem A 115(21):5247–5255. https://doi.org/10.1021/jp201327w

147. Nenov A, Rivalta I, Cerullo G, Mukamel S, Garavelli M (2014) Disentangling peptide configurations via two-dimensional electronic spectroscopy: ab initio simulations beyond the frenkel exciton hamiltonian. J Phys Chem Lett 5(4):767–771. https://doi.org/10.1021/jz5002314

148. Nenov A, Giussani A, Segarra-Martí J, Jaiswal VK, Rivalta I, Cerullo G, Mukamel S, Garavelli M (2015a) Modeling the high-energy electronic state manifold of adenine: calibration for nonlinear electronic spectroscopy. J Chem Phys 142(21):212443. https://doi.org/10.1063/1.4921016

149. Nenov A, Segarra-Martí J, Giussani A, Conti I, Rivalta I, Dumont E, Jaiswal VK, Altavilla SF, Mukamel S, Garavelli M (2015b) Probing deactivation pathways of DNA nucleobases by two-dimensional electronic spectroscopy: first principles simulations. Faraday Discuss 177:345–362. https://doi.org/10.1039/C4FD00175C

150. Nenov A, Borrego-Varillas R, Oriana A, Ganzer L, Segatta F, Conti I, Segarra-Marti J, Omachi J, Dapor M, Taioli S, Manzoni C, Mukamel S, Cerullo G, Garavelli M (2018a) UV-light-induced vibrational coherences: the key to understand kasha rule violation in trans-azobenzene. J Phys Chem Lett 9(7):1534–1541. https://doi.org/10.1021/acs.jpclett.8b00152

151. Nenov A, Conti I, Borrego-Varillas R, Cerullo G, Garavelli M (2018b) Linear absorption spectra of solvated thiouracils resolved at the hybrid RASPT2/MM level. Chem Phys 515:643–653. https://doi.org/10.1016/j.chemphys.2018.07.025

152. Norrish RG, Porter G (1949) Chemical reactions produced by very high light intensities [1]. Nature 164(4172):658. https://doi.org/10.1038/164658a0

153. Ogino M, Taya Y, Fujimoto K (2008) Highly selective detection of 5-methylcytosine using photochemical ligation. Chem Commun 45:5996. https://doi.org/10.1039/b813677g

154. Okada T, Sugihara M, Bondar AN, Elstner M, Entel P, Buss V (2004) The retinal conformation and its environment in rhodopsin in light of a new 2.2 Å crystal structure. J Mol Biol 342(2):571–583. https://doi.org/10.1016/j.jmb.2004.07.044

155. Onidas D, Markovitsi D, Marguet S, Sharonov A, Gustavsson T (2002) Fluorescence properties of DNA nucleosides and nucleotides: a refined steady-state and femtosecond investigation. J Phys Chem B 106(43):11367–11374. https://doi.org/10.1021/jp026063g

156. Pecourt JM, Peon J, Kohler B (2001) DNA excited-state dynamics: ultrafast internal conversion and vibrational cooling in a series of nucleosides. J Am Chem Soc 123(42):10370–10378. https://doi.org/10.1021/ja0161453

157. Pecourt JML, Peon J, Kohler B (2000) Ultrafast internal conversion of electronically excited RNA and DNA nucleosides in water. J Am Chem Soc 122(38):9348–9349. https://doi.org/10.1021/ja0021520

158. Peon J, Zewail AH (2001) DNA/RNA nucleotides and nucleosides: Direct measurement of excited-state lifetimes by femtosecond fluorescence up-conversion. Chem Phys Lett 348(3–4):255–262. https://doi.org/10.1016/S0009-2614(01)01128-9

159. Pepino AJ, Segarra-Martí J, Nenov A, Improta R, Garavelli M (2017) Resolving ultrafast photoinduced deactivations in water-solvated pyrimidine nucleosides. J Phys Chem Lett 8(8):1777–1783. https://doi.org/10.1021/acs.jpclett.7b00316

160. Pepino AJ, Segarra-Martí J, Nenov A, Rivalta I, Improta R, Garavelli M (2018) UV-induced long-lived decays in solvated pyrimidine nucleosides resolved at the MS-CASPT2/MM level. Phys Chem Chem Phys 20(10):6877–6890. https://doi.org/10.1039/c7cp08235e

161. Pérez A, Marchán I, Svozil D, Sponer J, Cheatham TE, Laughton CA, Orozco M (2007) Refinement of the AMBER force field for nucleic acids: improving the description of α/γ conformers. Biophys J 92(11):3817–3829. https://doi.org/10.1529/biophysj.106.097782

162. Perlík V, Šanda F (2017) Vibrational relaxation beyond the linear damping limit in two-dimensional optical spectra of molecular aggregates. J Chem Phys 147(8):084104. https://doi.org/10.1063/1.4999680

163. Pezeshki S, Lin H (2011) Adaptive-partitioning redistributed charge and dipole schemes for QM/MM dynamics simulations: on-the-fly relocation of boundaries that pass through covalent bonds. J Chem Theory Comput 7(11):3625–3634. https://doi.org/10.1021/ct2005209

164. Picchiotti A, Nenov A, Giussani A, Prokhorenko VI, Miller RJD, Mukamel S, Garavelli M (2019) Pyrene, a test case for deep-ultraviolet molecular photophysics. J Phys Chem Lett 10(12):3481–3487. https://doi.org/10.1021/acs.jpclett.9b01325

165. Pierloot K, Dumez B, Widmark PO, Roos BO (1995) Density matrix averaged atomic natural orbital (ANO) basis sets for correlated molecular wave functions. Theor Chim Acta 90(2–3):87–114. https://doi.org/10.1007/BF01113842

166. Polli D, Altoè P, Weingart O, Spillane KM, Manzoni C, Brida D, Tomasello G, Orlandi G, Kukura P, Mathies RA, Garavelli M, Cerullo G (2010) Conical intersection dynamics of the primary photoisomerization event in vision. Nature 467(7314):440–443. https://doi.org/10.1038/nature09346

167. Pollum M, Crespo-Hernández CE (2014) Communication: the dark singlet state as a doorway state in the ultrafast and efficient intersystem crossing dynamics in 2-thiothymine and 2-thiouracil. J Chem Phys 140(7):071101. https://doi.org/10.1063/1.4866447

168. Pollum M, Jockusch S, Crespo-Hernández CE (2015a) Increase in the photoreactivity of uracil derivatives by doubling thionation. Phys Chem Chem Phys 17(41):27851–27861. https://doi.org/10.1039/c5cp04822b

169. Pollum M, Martínez-Fernández L, Crespo-Hernández CE (2015b) Photochemistry of nucleic acid bases and their thio- and aza-analogues in solution. Top Curr Chem 355:245–355. https://doi.org/10.1007/128_2014_554

170. Porter G (1950) Flash photolysis and spectroscopy. A new method for the study of free radical reactions. Proc R Soc London Ser A Math Phys Sci 200(1061):284–300. https://doi.org/10.1098/rspa.1950.0018

171. Prokhorenko VI, Picchiotti A, Pola M, Dijkstra AG, Miller RJD (2016) New insights into the photophysics of DNA nucleobases. J Phys Chem Lett 7(22):4445–4450. https://doi.org/10.1021/acs.jpclett.6b02085

172. Rajput J, Rahbek DB, Andersen LH, Hirshfeld A, Sheves M, Altoè P, Orlandi G, Garavelli M (2010) Probing and modeling the absorption of retinal protein chromophores in vacuo. Angew Chemie 122(10):1834–1837. https://doi.org/10.1002/ange.200905061

173. Ramakers LAI, Hithell G, May JJ, Greetham GM, Donaldson PM, Towrie M, Parker AW, BOPTurley GA, Hunt NT (2017) 2D-IR spectroscopy shows that optimized DNA minor

groove binding of Hoechst33258 follows an induced fit model. J Phys Chem B 121(6):1295–1303. https://doi.org/10.1021/acs.jpcb.7b00345

174. Reichardt C, Crespo-Hernaíndez CE (2010) Room-temperature phosphorescence of the DNA monomer analogue 4-thiothymidine in aqueous solutions after UVA excitation. J Phys Chem Lett 1(15):2239–2243. https://doi.org/10.1021/jz100729w

175. Reichardt C, Crespo-Hernández CE (2010) Ultrafast spin crossover in 4-thiothymidine in an ionic liquid. Chem Commun 46(32):5963–5965. https://doi.org/10.1039/c0cc01181a

176. Reichardt C, Guo C, Crespo-Hernández CE (2011) Excited-state dynamics in 6-thioguanosine from the femtosecond to microsecond time scale. J Phys Chem B 115(12):3263–3270. https://doi.org/10.1021/jp112018u

177. Richter M, Mai S, Marquetand P, González L (2014) Ultrafast intersystem crossing dynamics in uracil unravelled by ab initio molecular dynamics. Phys Chem Chem Phys 16(44):24423–24436. https://doi.org/10.1039/c4cp04158e

178. Rivalta I, Nenov A, Garavelli M (2014) Modelling retinal chromophores photoisomerization: From minimal models in vacuo to ultimate bidimensional spectroscopy in rhodopsins. Phys Chem Chem Phys 16(32):16865–16879. https://doi.org/10.1039/c3cp55211j

179. Röhrig UF, Guidoni L, Rothlisberger U (2002) Early steps of the intramolecular signal transduction in rhodopsin explored by molecular dynamics simulations. Biochemistry 41(35):10799–10809. https://doi.org/10.1021/bi026011h

180. Roos BO, Lindh R, Malmqvist PÅ, Veryazov V, Widmark PO (2004) Main group atoms and dimers studied with a new relativistic ANO basis Set. J Phys Chem A 108(15):2851–2858. https://doi.org/10.1021/jp031064+

181. Ruckenbauer M, Mai S, Marquetand P, González L (2016) Photoelectron spectra of 2-thiouracil, 4-thiouracil, and 2,4-dithiouracil. J Chem Phys 144(7):074303. https://doi.org/10.1063/1.4941948, https://doi.org/10.1063/1512.02905

182. Santoro F, Barone V, Gustavsson T, Improta R (2006) Solvent effect on the singlet excited-state lifetimes of nucleic acid bases: a computational study of 5-fluorouracil and uracil in acetonitrile and water. J Am Chem Soc 128(50):16312–16322. https://doi.org/10.1021/ja0657861

183. Santoro F, Improta R, Fahleson T, Kauczor J, Norman P, Coriani S (2014) Relative stability of the L a and L b excited states in adenine and guanine: direct evidence from TD-DFT calculations of MCD spectra. J Phys Chem Lett 5(11):1806–1811. https://doi.org/10.1021/jz500633t

184. Schoenlein RW, Peteanu LA, Mathies RA, Shank CV (1991) The first step in vision: femtosecond isomerization of rhodopsin. Science 254(5030):412–415. https://doi.org/10.1126/science.1925597

185. Schreier WJ, Gilch P, Zinth W (2015) Early events of DNA photodamage. Annu Rev Phys Chem 66(1):497–519. https://doi.org/10.1146/annurev-physchem-040214-121821

186. Segarra-Martí J, Francés-Monerris A, Roca-Sanjuán D, Merchán M (2016) Assessment of the potential energy hypersurfaces in thymine within multiconfigurational theory: CASSCF vs. CASPT2. Molecules 21(12):1666. https://doi.org/10.3390/molecules21121666

187. Segarra-Martí J, Jaiswal VK, Pepino AJ, Giussani A, Nenov A, Mukamel S, Garavelli M, Rivalta I (2018a) Two-dimensional electronic spectroscopy as a tool for tracking molecular conformations in DNA/RNA aggregates. Faraday Discuss 207:233–250. https://doi.org/10.1039/C7FD00201G

188. Segarra-Martí J, Mukamel S, Garavelli M, Nenov A, Rivalta I (2018b) Towards accurate simulation of two-dimensional electronic spectroscopy. Top Curr Chem 376(3):24. https://doi.org/10.1007/s41061-018-0201-8

189. Segatta F, Cupellini L, Jurinovich S, Mukamel S, Dapor M, Taioli S, Garavelli M, Mennucci B (2017) A quantum chemical interpretation of two-dimensional electronic spectroscopy of light-harvesting complexes. J Am Chem Soc 139(22):7558–7567. https://doi.org/10.1021/jacs.7b02130

190. Segatta F, Cupellini L, Garavelli M, Mennucci B (2019) Quantum chemical modeling of the photoinduced activity of multichromophoric biosystems. Chem Rev 119(16):9361–9380. https://doi.org/10.1021/acs.chemrev.9b00135

191. Sekharan S, Mooney VL, Rivalta I, Kazmi MA, Neitz M, Neitz J, Sakmar TP, Yan EC, Batista VS (2013) Spectral tuning of ultraviolet cone pigments: an interhelical lock mechanism. J Am Chem Soc 135(51):19064–19067. https://doi.org/10.1021/ja409896y

192. Selig U, Schleussner CF, Foerster M, Langhojer F, Nuernberger P, Brixner T (2010) Coherent two-dimensional ultraviolet spectroscopy in fully noncollinear geometry. Opt Lett 35(24):4178. https://doi.org/10.1364/ol.35.004178

193. Senn HM, Thiel W (2006) QM/MM Methods for Biological Systems. In: Reiher M (ed) At. Approaches Mod. Biol., Springer Berlin Heidelberg, pp 173–290. https://doi.org/10.1007/128_2006_084

194. Serrano-Andrés L, Merchán M (2009) Are the five natural DNA/RNA base monomers a good choice from natural selection?. A photochemical perspective. J Photochem Photobiol C Photochem Rev 10(1):21–32. https://doi.org/10.1016/j.jphotochemrev.2008.12.001

195. Sisto A, Stross C, Van Der Kamp MW, O'Connor M, McIntosh-Smith S, Johnson GT, Hohenstein EG, Manby FR, Glowacki DR, Martinez TJ (2017) Atomistic non-adiabatic dynamics of the LH2 complex with a GPU-accelerated: ab initio exciton model. Phys Chem Chem Phys 19(23):14924–14936. https://doi.org/10.1039/c7cp00492c

196. Sovdat T, Bassolino G, Liebel M, Schnedermann C, Fletcher SP, Kukura P (2012) Backbone modification of retinal induces protein-like excited state dynamics in solution. J Am Chem Soc 134(20):8318–8320. https://doi.org/10.1021/ja3007929

197. Stone KW, Gundogdu K, Turner DB, Li X, Cundiff ST, Nelson KA (2009) Two-quantum 2D FT electronic spectroscopy of biexcitons in GaAs quantum wells. Science 324(5931):1169–1173. https://doi.org/10.1126/science.1170274

198. Svensson M, Humbel S, Froese RDJ, Matsubara T, Sieber S, Morokuma K (1996) ONIOM: A Multilayered Integrated MO + MM Method for Geometry Optimizations and Single Point Energy Predictions. A Test for Diels-Alder Reactions and Pt(P(t -Bu) 3) 2 + H 2 Oxidative Addition. J Phys Chem 100(50):19357–19363. https://doi.org/10.1021/jp962071j

199. Szalay PG, Watson T, Perera A, Lotrich VF, Bartlett RJ (2012) Benchmark Studies on the Building Blocks of DNA. 1. Superiority of Coupled Cluster Methods in Describing the Excited States of Nucleobases in the Franck–Condon Region. J Phys Chem A 116(25):6702–6710. https://doi.org/10.1021/jp300977a

200. Tahawy MMTE (2017) Modelling Spectral Tunability and Photoisomerization Mechanisms in Natural and Artificial Retinal Systems. PhD thesis, alma. https://doi.org/10.6092/unibo/amsdottorato/7829

201. Takaya T, Su C, de La Harpe K, Crespo-Hernandez CE, Kohler B (2008) UV excitation of single DNA and RNA strands produces high yields of exciplex states between two stacked bases. Proc Natl Acad Sci 105(30):10285–10290. https://doi.org/10.1073/pnas.0802079105

202. Taylor JS (1994) Unraveling the molecular pathway from sunlight to skin cancer. Acc Chem Res 27(3):76–82. https://doi.org/10.1021/ar00039a003

203. Teles-ferreira DC, Conti I, Borrego-Varillas R, Nenov A, van Stokkum IHM, Ganzer L, Manzoni C, Paula AMD, Cerullo G, Garavelli M (2020) A Unified Experimental/Theoretical Description of the Ultrafast Photophysics of Single and Double Thionated Uracils. Chem - A Eur J 26(1):336–343

204. Terakita A, Yamashita T, Shichida Y (2000) Highly conserved glutamic acid in the extracellular IV-V loop in rhodopsins acts as the counterion in retinochrome, a member of the rhodopsin family. Proc Natl Acad Sci USA 97(26):14263–14267. https://doi.org/10.1073/pnas.260349597

205. Thyrhaug E, Tempelaar R, Alcocer MJ, Žídek K, Bína D, Knoester J, Jansen TL, Zigmantas D (2018) Identification and characterization of diverse coherences in the Fenna-Matthews-Olson complex. Nat Chem 10(7):780–786. https://doi.org/10.1038/s41557-018-0060-5

206. Tomasello G, Gloria OG, Altoè P, Stenta M, Luis SA, Merchán M, Orlandi G, Bottoni A, Garavelli M (2009) Electrostatic control of the photoisomerization efficiency and optical properties in visual pigments: On the role of counterion quenching. J Am Chem Soc 131(14):5172–5186. https://doi.org/10.1021/ja808424b

207. Tomasi J, Mennucci B, Cammi R (2005) Quantum mechanical continuum solvation models. Chem Rev 105(8):2999–3093. https://doi.org/10.1021/cr9904009
208. Tseng Ch, Matsika S, Weinacht TC (2009) Two-dimensional ultrafast fourier transform spectroscopy in the deep ultraviolet. Opt Express 17(21):18788. https://doi.org/10.1364/oe.17.018788
209. Tully JC (1990) Molecular dynamics with electronic transitions. J Chem Phys 93(2):1061–1071. https://doi.org/10.1063/1.459170
210. Tully JC (1998) Mixed quantum-classical dynamics. Faraday Discuss 110:407–419. https://doi.org/10.1039/a801824c
211. Tully JC, Pkeston RK (1971) Trajectory surface hopping approach to nonadiabatic molecular collisions: The reaction of H+ with D2. J Chem Phys 55(2):562–572. https://doi.org/10.1063/1.1675788
212. Tuna D, Lu Y, Koslowski A, Thiel W (2016) Semiempirical quantum-chemical orthogonalization-corrected methods: benchmarks of electronically excited states. J Chem Theory Comput 12(9):4400–4422. https://doi.org/10.1021/acs.jctc.6b00403
213. Turner DB, Nelson KA (2010) Coherent measurements of high-order electronic correlations in quantum wells. Nature 466(7310):1089–1091. https://doi.org/10.1038/nature09286
214. Tuttle T, Thiel W (2008) OMx-D: Semiempirical methods with orthogonalization and dispersion corrections. Implementation and biochemical application. Phys Chem Chem Phys 10(16):2159–2166. https://doi.org/10.1039/b718795e
215. Valiev M, Bylaska E, Govind N, Kowalski K, Straatsma T, Van Dam H, Wang D, Nieplocha J, Apra E, Windus T, de Jong W (2010) NWChem: a comprehensive and scalable open-source solution for large scale molecular simulations. Comput Phys Commun 181(9):1477–1489. https://doi.org/10.1016/j.cpc.2010.04.018
216. Vayá I, Gustavsson T, Miannay FA, Douki T, Markovitsi D (2010) Fluorescence of natural DNA: from the femtosecond to the nanosecond time scales. J Am Chem Soc 132(34):11834–11835. https://doi.org/10.1021/ja102800r
217. Vayá I, Brazard J, Huix-Rotllant M, Thazhathveetil AK, Lewis FD, Gustavsson T, Burghardt I, Improta R, Markovitsi D (2016) High-energy long-lived mixed frenkel-charge-transfer excitons: from double stranded (AT) n to natural DNA. Chem - A Eur J 22(14):4904–4914. https://doi.org/10.1002/chem.201504007
218. Vreven T, Bernardi F, Garavelli M, Olivucci M, Robb MA, Schlegel HB (1997) Ab Initio photoisomerization dynamics of a simple retinal chromophore model. J Am Chem Soc 119(51):12687–12688. https://doi.org/10.1021/ja9725763
219. Wang J, Cieplak P, Kollman PA (2000) How well does a restrained electrostatic potential (RESP) model perform in calculating conformational energies of organic and biological molecules? J Comput Chem 21(12):1049–1074. https://doi.org/10.1002/1096-987X(200009)21:12<1049::AID-JCC3>3.0.CO;2-F
220. Wang J, Wolf RM, Caldwell JW, Kollman PA, Case DA (2004) Development and testing of a general Amber force field. J Comput Chem 25(9):1157–1174. https://doi.org/10.1002/jcc.20035
221. Wang W, Nossoni Z, Berbasova T, Watson CT, Yapici I, Lee KSS, Vasileiou C, Geiger JH, Borhan B (2012) Tuning the electronic absorption of protein-embedded all-trans-retinal. Science 338(6112):1340–1343. https://doi.org/10.1126/science.1226135
222. Weingart O, Nenov A, Altoè P, Rivalta I, Segarra-Martí J, Dokukina I, Garavelli M (2018) COBRAMM 2.0—a software interface for tailoring molecular electronic structure calculations and running nanoscale (QM/MM) simulations. J Mol Model 24(9):271. https://doi.org/10.1007/s00894-018-3769-6
223. Werner HJ, Knowles PJ, Knizia G, Manby FR (2012a) MOLPRO, A Package of Ab Initio Programs
224. Werner HJ, Knowles PJ, Knizia G, Manby FR, Schütz M (2012b) Molpro: a general-purpose quantum chemistry program package. Wiley Interdiscip Rev Comput Mol Sci 2(2):242–253. https://doi.org/10.1002/wcms.82

225. Widmark PO, Malmqvist PÅ, Roos BO (1990) Density matrix averaged atomic natural orbital (ANO) basis sets for correlated molecular wave functions - I. First row atoms. Theor Chim Acta 77(5):291–306. https://doi.org/10.1007/BF01120130

226. Widmark PO, Persson BJ, Roos BO (1991) Density matrix averaged atomic natural orbital (ANO) basis sets for correlated molecular wave functions - II. Second row atoms. Theor Chim Acta 79(6):419–432. https://doi.org/10.1007/BF01112569

227. Widom JR, Johnson NP, von Hippel PH, Marcus AH (2013) Solution conformation of 2-aminopurine dinucleotide determined by ultraviolet two-dimensional fluorescence spectroscopy. New J Phys 15(2):025028. https://doi.org/10.1088/1367-2630/15/2/025028

228. Wise KJ, Gillespie NB, Stuart JA, Krebs MP, Birge RR (2002) Optimization of bacteriorhodopsin for bioelectronic devices. Trends Biotechnol 20(9):387–394. https://doi.org/10.1016/S0167-7799(02)02023-1

229. Xie BB, Wang Q, Guo WW, Cui G (2017) The excited-state decay mechanism of 2,4-dithiothymine in the gas phase, microsolvated surroundings, and aqueous solution. Phys Chem Chem Phys 19(11):7689–7698. https://doi.org/10.1039/c7cp00478h

230. Yan EC, Kazmi MA, De S, Chang BS, Seibert C, Marin EP, Mathies RA, Sakmar TP (2002) Function of extracellular loop 2 in rhodopsin: Glutamic acid 181 modulates stability and absorption wavelength of metarhodopsin II. Biochemistry 41(11):3620–3627. https://doi.org/10.1021/bi0160011

231. Yan ECY, Kazmi MA, Ganim Z, Hou JM, Pan D, Chang BSW, Sakmar TP, Mathies RA (2003) Retinal counterion switch in the photoactivation of the G protein-coupled receptor rhodopsin. Proc Natl Acad Sci 100(16):9262–9267. https://doi.org/10.1073/pnas.1531970100

232. Yanai T, Tew DP, Handy NC (2004) A new hybrid exchange-correlation functional using the Coulomb-attenuating method (CAM-B3LYP). Chem Phys Lett 393(1–3):51–57. https://doi.org/10.1016/j.cplett.2004.06.011

233. Zhang X, Herbert JM (2014) Excited-state deactivation pathways in uracil versus hydrated uracil: Solvatochromatic shift in the $1n\pi^*$ state is the key. J Phys Chem B 118(28):7806–7817. https://doi.org/10.1021/jp412092f

234. Zhang Y, de La Harpe K, Beckstead AA, Improta R, Kohler B (2015) UV-induced proton transfer between DNA strands. J Am Chem Soc 137(22):7059–7062. https://doi.org/10.1021/jacs.5b03914

235. Zou X, Dai X, Liu K, Zhao H, Song D, Su H (2014) Photophysical and photochemical properties of 4-thiouracil: time-resolved ir spectroscopy and DFT studies. J Phys Chem B 118(22):5864–5872. https://doi.org/10.1021/jp501658a

Polarizable Embedding as a Tool to Address Light-Responsive Biological Systems

Peter Hartmann, Peter Reinholdt, and Jacob Kongsted

Abstract Quantum mechanical response theory represents a convenient way of addressing calculations of excited states and their properties in molecular systems. Today, quantum chemical response theory is a well-established tool within computational quantum chemistry for isolated molecules. In this chapter, we discuss how to extend the capabilities of quantum mechanical response theory to molecules subjected to an environment—this being either the case of a simple solvent or the more challenging case of a biological matrix such as a protein. Specifically, we will be concerned with a detailed and coherent presentation of the theoretical background and derivation of the polarizable embedding model. For this model, we will derive and discuss the physical and mathematical expressions for response functions and equations based on the physical picture of a chromophore embedded into an environment. Following the derivation of the polarizable embedding scheme, we discuss how to connect response functions and their residues to the calculation of selected molecular properties such as one- and two-photon absorption strengths. Finally, we discuss a few results for specific molecular systems showing the generally broad applicability of the method developed.

Keywords Polarizable embedding · Response theory · Local fields · Two-photon absorption · Beta-lactoglobulin

1 Introduction

Within the field of computational chemistry, the inclusion of environmental effects is of considerable interest. Since most chemistry takes place in either a condensed phase or a biological matrix such as, e.g., a protein, environmental effects are important for a broad range of applications. For example, in organic chemistry, the effect of a solvent

P. Hartmann · P. Reinholdt · J. Kongsted (✉)
Department of Physics, Chemistry and Pharmacy, University of Southern Denmark, Campusvej 55, 5230 Odense, Denmark
e-mail: kongsted@sdu.dk

© Springer Nature Switzerland AG 2021 143
T. Andruniów and M. Olivucci (eds.), *QM/MM Studies of Light-responsive Biological Systems*, Challenges and Advances in Computational Chemistry and Physics 31, https://doi.org/10.1007/978-3-030-57721-6_3

can be a major factor governing the outcome of a reaction. The UV/vis absorption properties of chromophores often change drastically depending on the environment, giving rise to solvatochromism [41]. Furthermore, biochemical applications that use a dye as an optical probe by exploiting the change in the one- or two-photon spectra between the solvated state and the state where it is bound to a protein or membrane are important in many practical applications [43].

The inclusion of environmental effects by including the environment directly in the quantum mechanical (QM) calculation is straightforward, but quickly reaches its limits regarding system size. Also, due to the long-range electrostatic interactions, a fairly large part of the environment has to be included, which limits the accuracy of the methods that can be employed. In recent years the use of embedding methods for the inclusion of environmental effects has become increasingly popular due to their efficient yet still relatively accurate nature compared to a full quantum mechanical treatment. Embedding methods split a system into the core region and the environment, where the former is the main focus of interest and is, therefore, treated with a method exhibiting high accuracy, while the latter is treated at a lower level of theory [18, 19, 28, 81].

Of the embedding approaches, especially the class of hybrid quantum-classical approaches enable calculations on large systems of sizes impossible to treat with purely quantum mechanical means. The simplest of such hybrid approaches are the dielectric continuum models like the polarizable continuum model (PCM) [71] and the conductor-like screening model (COSMO) [4, 32, 33, 67, 72], which are by now implemented in many of the available electronic structure programs. Due to their continuum nature, these methods implicitly include the sampling of the conformations of the environment, and as a result, only a single calculation is needed to model the environment, sparing the need for lengthy molecular dynamics simulations or other configurational sampling methods. This, and the fact that all necessary parameters like atomic radii and the dielectric constant of the environment are predefined, makes these methods very easy to use. However, despite all the advantages of the continuum models, their simplicity comes with the major drawback of being unable to model the directionality of specific intermolecular interactions such as hydrogen bonding or π-π stacking, resulting in the loss of the possibility to model environment anisotropies as found in, e.g., the binding site of a protein. In cases where a more realistic description of the environment is required, discrete embedding models come into play. Those methods describe the core region with quantum mechanics while at the same time keeping the atomistic detail of the environment intact by employing molecular mechanics (QM/MM) [13, 63, 64, 77]. However, realistically modeling the environment comes at the price of an increase in both computational cost and complexity. This increase in computational cost is not only linked to the atomistic description of the environment as such but also that in discrete approaches, the sampling of environmental conformations has to be included explicitly. This means performing a molecular dynamics (MD) simulation or the use of some comparable sampling methodology prior to any actual production calculations. From these calculations, a number of snapshots are extracted and used for the following QM/MM calculations, replacing the single calculation needed when employing continuum models with a number of calculations employing the discrete embedding approach, leading to the above-mentioned increase in computational expenses.

The increased complexity is a consequence of the fact that in discrete models for each atom of the environment, a set of parameters is needed instead of just describing everything through the dielectric constant. Those parameters are, depending on the level of sophistication, charges, dipoles, and possibly higher-order multipoles, isotropic or anisotropic polarizabilities, and parameters for non-electrostatic interactions as dispersion or exchange-repulsion. Thus, the setup of a discrete embedding calculation is inherently more complicated than the one for a calculation employing a continuum model. Contrary to the latter, it may be necessary to calculate these parameters before one can perform the actual calculation. On the other hand, the increased complexity makes the discrete models much more flexible than the continuum models and they can, therefore, be expected to be applicable to a broader range of systems with comparable accuracy [68].

In the following, we focus on one discrete embedding model, in particular, namely the polarizable embedding (PE) model [38, 52, 53]. As discussed in the following, the PE model is especially suited for performing calculations relating both to the ground and excited states, as well as spectroscopic parameters. Within the class of discrete embedding methods designed for describing excited states and spectroscopic properties building on the principles of quantum mechanical (linear) response theory, polarization is usually described based on induced dipoles [10, 29, 53, 84] or by fluctuating charges (FQ) [36]. In this chapter, we provide a detailed discussion of the theoretical background of the PE model, starting with a review on standard intermolecular Rayleigh–Schrödinger perturbation theory applied to the division of a supersystem. This discussion will form the basis for the derivation of the PE model. Hereafter we will derive the working equations for the PE model within a self-consistent-field (SCF) parametrization. Following this, we derive the PE model within a quantum mechanical response picture [39]. We continue by discussing how response functions can be linked to experimental properties [47]. Finally, we end this contribution by showing a few examples of the use of the PE model to realistic chemical systems.

2 Theoretical Background of the PE Model

The polarizable embedding (PE) model [38, 52, 53] is a fragment-based QM/MM method belonging to the class of discrete environment embedding approaches, meaning that the environmental parameters are obtained by splitting the environment into smaller fragments and then performing quantum mechanical calculations on each fragment [20, 68]. By using this approach, the model will yield results that are in close agreement with full QM calculations, but at a much reduced computational cost [44]. The permanent charge distribution of the environment is modeled by a distributed multipole expansion, with the expansion points typically placed at the position of the atoms. To account for polarization of the environment, each expansion point is also assigned a polarizability.

The PE model has been combined with a number of electronic-structure methods including Hartree-Fock (HF) and Kohn-Sham density functional theory (DFT) [52, 53], but also correlated wave function-based methods such as coupled cluster (CC) [62, 65], multiconfigurational self-consistent-field theory (MCSCF) [22], multiconfigurational short-range DFT (MC-srDFT) [23], the second-order polarization propagator approach (SOPPA) [12], and also in a relativistic framework using DFT [24]. The user can thus choose among a wide range of methods the one that performs in a reasonable way regarding both accuracy and cost-efficiency.

2.1 Inter-Subsystem Perturbation Theory

The theoretical basis for the polarizable embedding model has been outlined in Ref. [50], but will for consistency briefly be discussed below. We begin our discussion by providing a theoretical basis of the polarizable embedding model rooted in inter-system perturbation theory. In order to treat different regions of a composite system at different levels of theory, a subdivision of the system into subsystems is required. This could be a simple division into two subsystems in case of, e.g., a fluorophore in a protein, but the equations for the PE model can be derived more generally for a subdivision into N subsystems. One of the subsystems will later be specified as the QM-region, but this is not necessary at this stage of the derivation. The Hamiltonian for the complete system can be written as a sum of subsystem Hamiltonians and interaction operators between those subsystems,

$$\hat{H}^{\text{tot}} = \sum_{a=1}^{N} \hat{H}^a + \sum_{a>b}^{N} \hat{V}^{ab} \tag{1}$$

where the sum over the indices $a > b$ of the interaction operators avoids double counting of the interactions [69]. In the following derivations, it will be assumed that the subsystems are nonoverlapping, and in doing so, each subsystem is implicitly assigned a number of electrons. This approximation becomes more and more valid in the long-range limit, but the condition is not enforced. Therefore, the inter-subsystem antisymmetrization and thus exchange-repulsion is neglected, potentially leading to deviations from a corresponding supermolecular model.

Since the wave functions of the subsystems are assumed to be nonoverlapping, the wave function of the total system $|0\rangle$ can be written as a Hartree product. Within the second quantization formalism, this can be written as [50]

$$|0\rangle = \prod_{a=1}^{N} |a\rangle = \left(\prod_{a=1}^{N} \hat{\psi}^a \right) |vac\rangle \tag{2}$$

with $\hat{\psi}^a$ being the wave function operator for system a, consisting of a string of creation operators. The derivation of the main contributions to the interaction energy of the subsystems can be performed by employing standard Rayleigh–Schrödinger perturbation theory, expanding the total energy of the system as

$$E^{\text{tot}} = E^{(0)} + E^{(1)} + E^{(2)} + \dots \tag{3}$$

and treating the interaction operator as the perturbation. It is convenient to combine the individual subsystem Hamiltonians in Eq. (1), into a total, unperturbed Hamiltonian $\hat{H}$, and summarize the interaction operators into a perturbation operator $\hat{V}$

$$\hat{H}^{\text{tot}} = \sum_{a=1}^{N} \hat{H}^a + \sum_{a>b}^{N} \hat{V}^{ab} = \hat{H} + \hat{V} \tag{4}$$

The Schrödinger equation for the unperturbed system is then

$$\hat{H} |0\rangle = E^{(0)} |0\rangle \tag{5}$$

where $E^{(0)}$ is the sum of unperturbed subsystem energies. The energy terms up to second order consist of a first-order electrostatic term

$$E^{(1)} = \langle 0| \hat{V} |0\rangle = \sum_{a>b}^{N} \langle ab| \hat{V}^{ab} |ab\rangle \tag{6}$$

and a second order term

$$E^{(2)} = -\sum_{i}' \frac{\langle 0|\hat{V}|i\rangle \langle i|\hat{V}|0\rangle}{E_i^{(0)} - E_0^{(0)}} = -\sum_{a>b}^{N} \sum_{mn}' \frac{\langle ab|\hat{V}^{ab}|a^m b^n\rangle \langle a^m b^n|\hat{V}^{ab}|ab\rangle}{E_{a,m}^{(0)} + E_{b,n}^{(0)} - E_{a,0}^{(0)} - E_{b,0}^{(0)}}. \tag{7}$$

Here the assumption of nonoverlapping subsystems has the implication that $E_i^{(0)}$ can be rewritten as $E_{a,m}^{(0)} + E_{b,n}^{(0)}$ because excitations can only occur within a specific subsystem. A prime in a single sum indicates that zero is excluded from the index summed over, while a prime in a double sum indicates that not both indices summed over can be zero. The indices are changed upon switching from $|i\rangle = |a^m b^n\rangle$ in order to emphasize that there is no inherent relation between them. The significance of $E^{(2)}$ becomes visible when it is split into two terms according to

$$E^{(2)} = -\sum_{a=1}^{N} \sum_{a\neq b} \sum_{m}' \frac{\langle ab|\hat{V}^{ab}|a^m b\rangle \langle a^m b|\hat{V}^{ab}|ab\rangle}{E_{a,m}^{(0)} - E_{a,0}^{(0)}}$$
$$- \sum_{a>b}^{N} \sum_{mn}' \frac{\langle ab|\hat{V}^{ab}|a^m b^n\rangle \langle a^m b^n|\hat{V}^{ab}|ab\rangle}{E_{a,m}^{(0)} + E_{b,n}^{(0)} - E_{a,0}^{(0)} - E_{b,0}^{(0)}} \tag{8}$$

which can be recognized as

$$E^{(2)} = \sum_{a=1}^{N} E_a^{\text{ind}} + \sum_{a>b}^{N} E_{ab}^{\text{disp}}$$ (9)

where E_a^{ind} is the (linear) polarization energy of fragment a and E_{ab}^{disp} is the (linear) dispersion energy of fragments a and b. The linear dispersion energy (and higher-order polarization and dispersion energies) is neglected in the polarizable embedding model. We now proceed to describe how the first and second order corrections to the energy included in PE can be evaluated.

2.2 Electrostatic Interaction Energy

The inter-fragment interaction part of the total Hamiltonian can, in second quantiza-tion, be written as [38]

$$\hat{V}^{ab} = \sum_{n\in b}^{M_b} Z_n \sum_{pq\in a} v_{pq}(\mathbf{R}_n)\hat{E}_{pq} + \sum_{m\in a}^{M_a} Z_m \sum_{rs\in b} v_{rs}(\mathbf{R}_m)\hat{E}_{rs}$$

$$+ \sum_{pq\in a}\sum_{rs\in b} v_{pq,rs}\hat{E}_{pq}\hat{E}_{rs} + \sum_{m\in a}^{M_a}\sum_{n\in b}^{M_b} \frac{Z_m Z_n}{|\mathbf{R}_n - \mathbf{R}_m|}$$ (10)

where in the first two terms, which describe the electron-nucleus interactions between the different fragments, the one-electron integrals are

$$v_{pq}(\mathbf{R}) = -\int \frac{\phi_p^*(\mathbf{r})\phi_q(\mathbf{r})}{|\mathbf{R} - \mathbf{r}|}d\mathbf{r}$$ (11)

and in the third term, which describes the electron-electron repulsion between the different fragments, the two-electron integrals are

$$v_{pq,rs} = \int \frac{\phi_p^*(\mathbf{r})\phi_q(\mathbf{r})\phi_r^*(\mathbf{r}')\phi_s(\mathbf{r}')}{|\mathbf{r} - \mathbf{r}'|}d\mathbf{r}d\mathbf{r}'.$$ (12)

Since the electrons in the two-electron integrals belong to different fragments, the cor-responding two-electron operator is essentially reduced to a product of one-electron operators. This greatly simplifies its evaluation as there is no exchange, and only the classical electrostatic interactions remain. The fourth term is the nuclear repul-sion energy between the two fragments. The expressions above still rely on a wave function-based description of all the electrons in the total system. For realizing only a partial QM treatment of the total system, it is desirable to find an expression for the electrostatic potential (ESP) of a subsystem. This can be done in terms of a multipole expansion [69]. The ESP of subsystem a can be written as

$$V_a^{\text{esp}}(\mathbf{R}) = \sum_{pq \in a} v_{pq}(\mathbf{R}) D_{pq}^a + \sum_{m \in a}^{M_a} \frac{Z_m}{|\mathbf{R} - \mathbf{R}_m|} \tag{13}$$

where D_{pq}^a is an element of the density matrix of subsystem a. We now proceed by expanding the ESP in a Taylor expansion around an origin $\mathbf{R}_a$ inside fragment a. For this, it is convenient to first introduce a multi-index notation [50]. Let $k = (k_x, k_y, k_z)$ be a 3-tuple, where $k_x, k_y, k_z \in \mathbb{N}$, with each member of the tuple associated to the indicated component of a cartesian coordinate. The norm of a multi-index be defined as $|k| = k_x + k_y + k_z$, the factorial $k! = k_x! k_y! k_z!$ and the multi-index power of a vector $\mathbf{r}^k = x^{k_x} y^{k_y} z^{k_z}$. The resulting Taylor expansion may very conveniently be written in terms of the interaction tensors as

$$V_a^{\text{esp}}(\mathbf{R}_b)$$

$$= -\sum_{pq \in a} D_{pq}^a \int \frac{\phi_p^*(\mathbf{r}) \phi_q(\mathbf{r})}{|\mathbf{R}_b - \mathbf{r}|} d\mathbf{r} + \sum_{m \in a}^{M_a} \frac{Z_m}{|\mathbf{R}_b - \mathbf{R}_m|}$$

$$= \sum_{|k|=0}^{\infty} \frac{(-1)^{|k|}}{k!} T_{ab}^{(k)} \left[-\sum_{pq \in a} D_{pq}^a \int \phi_p^*(\mathbf{r}) \phi_q(\mathbf{r})(\mathbf{r} - \mathbf{R}_a)^k d\mathbf{r} + \sum_{m \in a}^{M_a} Z_m (\mathbf{R}_m - \mathbf{R}_a)^k \right]$$

$$= \sum_{|k|=0}^{\infty} \frac{(-1)^{|k|}}{k!} T_{ab}^{(k)} M_a^{(k)} \tag{14}$$

In the last step, we have introduced the multipole moments of fragment a

$$M_a^{(k)} = -\sum_{pq \in a} D_{pq}^a \int \phi_p^*(\mathbf{r}) \phi_q(\mathbf{r})(\mathbf{r} - \mathbf{R}_a)^k d\mathbf{r} + \sum_{m \in a}^{M_a} Z_m (\mathbf{R}_m - \mathbf{R}_a)^k \tag{15}$$

which is a charge for $|k| = 0$, a dipole for $|k| = 1$ and so on. We can now obtain an expression for the expectation value of the electrostatic interaction energy in terms of an approximate interaction operator

$$E_{ab}^{\text{es}} \approx \langle ab | \hat{V}_{ab}^{\text{esp}} | ab \rangle \tag{16}$$

where the operator is expressed in terms of the ESP. In second quantization the corresponding operator can be written as

$$\hat{V}_{ab}^{\text{esp}} = \sum_{|k|=0}^{K_a} \frac{(-1)^{|k|}}{k!} \hat{T}_{ab}^{(k)} \hat{M}_a^{(k)} \tag{17}$$

where we have truncated the expansion at order K_a. The Operator $\hat{T}_{ab}^{(k)}$ is defined by

$$\hat{T}_{ab}^{(k)} = -\sum_{rs\in b} \int \phi_r^*(\mathbf{r}')\nabla_{\mathbf{r}'}^k \left(\frac{1}{|\mathbf{r}' - \mathbf{R}_a|}\right) \phi_s(\mathbf{r}')\mathrm{d}\mathbf{r}'\,\hat{E}_{rs} + \sum_{n\in b}^{M_b} T_{an}^{(k)} Z_n, \qquad (18)$$

where $T_{an}^{(k)} = \nabla_n^k \frac{1}{|\mathbf{R}_n - \mathbf{R}_a|}$ is a k'th order interaction tensor [50, 69]. Introducing the field integrals

$$t_{rs}^{(k)}(\mathbf{R}_a) = -\int \phi_r^*(\mathbf{r}')\nabla_{\mathbf{r}'}^k \left(\frac{1}{|\mathbf{r}' - \mathbf{R}_a|}\right) \phi_s(\mathbf{r}')\mathrm{d}\mathbf{r}', \qquad (19)$$

allows, together with Eq. 15, to express the approximate electrostatic interaction energy compactly as

$$E_{ab}^{\mathrm{es}} \approx \sum_{|k|=0}^{K_a} \frac{(-1)^{|k|}}{k!} M_a^{(k)} \sum_{rs\in b} t_{rs}^{(k)}(\mathbf{R}_a) D_{rs}^b + \sum_{|k|=0}^{K_a} \frac{(-1)^{|k|}}{k!} M_a^{(k)} \sum_{n\in b}^{M_b} T_{an}^{(k)} Z_n \qquad (20)$$

which can be thought of as consisting of the interactions between the electrons of fragment B with the multipoles of fragment A (first term) and the interactions between the nuclei of fragment B with the multipoles of fragment A (second term). The problem of a multipole expansion in the above form is that it can exhibit slow convergence or even diverge, especially at shorter distances. This can be improved upon by switching from the single-site expansion above, where all the multipoles are placed at $\mathbf{R}_a$ in fragment a, to a distributed multipole expansion, where we distribute the moments at expansion sites S in fragment a [69]. This introduces an additional sum over the expansion sites

$$V_a^{\mathrm{esp}}(\mathbf{R}_b) \approx \sum_{s\in a}^{S_a} \sum_{|k|=0}^{K_s} \frac{(-1)^{|k|}}{k!} T_{ab}^{(k)} M_a^{(k)} \qquad (21)$$

which essentially leads to the electrostatic interaction energy for the distributed multipole expansion

$$E_{ab}^{\mathrm{es}} \approx \sum_{s\in a}^{S_a} \sum_{|k|=0}^{K_s} \frac{(-1)^{|k|}}{k!} M_s^{(k)} \sum_{rs\in b} t_{rs}^{(k)}(\mathbf{R}_s) D_{rs}^b + \sum_{s\in a}^{S_a} \sum_{|k|=0}^{K_s} \frac{(-1)^{|k|}}{k!} M_s^{(k)} \sum_{n\in b}^{M_b} T_{sn}^{(k)} Z_n$$

$$(22)$$

2.3 Induction Energy

Before we derive the equations for the induction energy from a quantum mechanical point of view, we shall briefly discuss the underlying classical physics. We will only treat induced dipoles in the classical discussion. In a linear approximation the induced dipole moment of fragment a is equal to the (linear) polarizability of fragment a times the total electric field $\mathbf{F}^{tot}$ at $\mathbf{R}_a$. A good model to show this is the Drude model, in which an induced dipole is modeled by two charges $\pm q_D$ connected by a harmonic potential. Due to the definition of the electric field, the force exerted on the charges equals $q_D \mathbf{F}^{tot}$ and we get for the separation $\mathbf{d}$ of the two charges $\pm q_D$

$$q_D \mathbf{F}^{tot} = k\mathbf{d} \tag{23}$$

The induced dipole moment is easily seen to be

$$\boldsymbol{\mu}^{ind} = q_D \mathbf{d} = \frac{q_D^2 \mathbf{F}^{tot}}{k} = \alpha \mathbf{F}^{tot} \tag{24}$$

where we defined the polarizability α. From the potential energy of a harmonic oscillator we also find that the self-energy necessary to create the induced dipole is

$$E_{drude}^{self} = \frac{1}{2}k\mathbf{d}^2 = \frac{1}{2}\frac{q_D^2}{\alpha}\left(\frac{q_D \mathbf{F}^{tot}}{k}\right)^2$$
$$= \frac{1}{2}\boldsymbol{\mu}^{ind}\alpha^{-1}\boldsymbol{\mu}^{ind} \tag{25}$$

This could in a more general way, which does not rely on the Drude model, also be obtained from [2]

$$E^{self} = \int_0^{\mu_a^{ind}} \mathbf{F}^{tot}(\mathbf{R}_a)d\boldsymbol{\mu}_a^{ind} = \int_0^{\mu_a^{ind}} \boldsymbol{\mu}_a^{ind}\alpha_a^{-1}d\boldsymbol{\mu}_a^{ind}$$
$$= \frac{1}{2}\boldsymbol{\mu}_a^{ind}\alpha_a^{-1}\boldsymbol{\mu}_a^{ind} \tag{26}$$

The final aim is to describe an ensemble of multipoles with assigned dipole-dipole polarizabilities and find the optimum induced dipoles, which correspond to the minimum energy of the system. In such an ensemble the total field consists of the fields of the static $[\mathbf{F}^m(\mathbf{R}_a)]$ and induced $[(\mathbf{F}^{ind}(\mathbf{R}_a)]$ charge distributions of fragments b, as well as any externally applied field $[\mathbf{F}^{ext}]$. The total electrostatic energy of the induced dipoles is

$$E_{es}^{tot} = -\sum_a \boldsymbol{\mu}_a \mathbf{F}^{tot}(\mathbf{R}_a)$$

$$= -\sum_a \boldsymbol{\mu}_a \left[\frac{1}{2}\mathbf{F}^{ind}(\mathbf{R}_a) + \mathbf{F}^m(\mathbf{R}_a) + \mathbf{F}^{ext}(\mathbf{R}_a)\right] \quad (27)$$

$$= -\frac{1}{2}\sum_a \boldsymbol{\mu}_a \sum_{a \neq b} \mathbf{T}_{ba}^{(2)} \boldsymbol{\mu}_b - \sum_a \boldsymbol{\mu}_a \mathbf{F}^m(\mathbf{R}_a) - \sum_a \boldsymbol{\mu}_a \mathbf{F}^{ext}(\mathbf{R}_a)$$

where the factor $\frac{1}{2}$ was introduced to avoid double counting of induced-dipole inter-action energy. In order to obtain the total induction energy of the system, the total electrostatic energy of all the induced dipoles needs to be corrected for the self-energy that is necessary to form these induced dipoles. From Eq. 26, the total self-energy is

$$E_{self}^{tot} = \frac{1}{2}\sum_a \boldsymbol{\mu}_a^{ind} \boldsymbol{\alpha}_a^{-1} \boldsymbol{\mu}_a^{ind}, \quad (28)$$

and the total induction energy is then

$$E_{ind}^{tot} = \frac{1}{2}\sum_a \boldsymbol{\mu}_a^{ind} \boldsymbol{\alpha}_a^{-1} \boldsymbol{\mu}_a^{ind} - \frac{1}{2}\sum_a \boldsymbol{\mu}_a^{ind} \sum_{a \neq b} \mathbf{T}_{ba}^{(2)} \boldsymbol{\mu}_b^{ind} - \sum_a \boldsymbol{\mu}_a^{ind} \left[\mathbf{F}^m(\mathbf{R}_a) + \mathbf{F}^{ext}(\mathbf{R}_a)\right]$$

$$(29)$$

By introducing the matrix $\mathbf{B}$, which is of dimensions $3N \times 3N$

$$\mathbf{B} = \begin{pmatrix} \boldsymbol{\alpha}_1^{-1} & -\mathbf{T}_{12}^{(2)} & -\mathbf{T}_{13}^{(2)} & \cdots & -\mathbf{T}_{1N}^{(2)} \\ -\mathbf{T}_{21}^{(2)} & \boldsymbol{\alpha}_2^{-1} & -\mathbf{T}_{23}^{(2)} & \cdots & -\mathbf{T}_{2N}^{(2)} \\ -\mathbf{T}_{31}^{(2)} & -\mathbf{T}_{32}^{(2)} & \boldsymbol{\alpha}_3^{-1} & \cdots & -\mathbf{T}_{3N}^{(2)} \\ \vdots & \vdots & \vdots & \ddots & \vdots \\ -\mathbf{T}_{N1}^{(2)} & -\mathbf{T}_{N2}^{(2)} & -\mathbf{T}_{N3}^{(2)} & \cdots & \boldsymbol{\alpha}_N^{-1} \end{pmatrix}, \quad (30)$$

the field vector $\mathbf{F} = \mathbf{F}^m + \mathbf{F}^{ext}$, and the dipole vector $\boldsymbol{\mu}^{ind}$

$$\mathbf{F} = \begin{pmatrix} \mathbf{F}(\mathbf{R}_1) \\ \mathbf{F}(\mathbf{R}_2) \\ \vdots \\ \mathbf{F}(\mathbf{R}_N) \end{pmatrix} ; \quad \boldsymbol{\mu}^{ind} = \begin{pmatrix} \boldsymbol{\mu}_1^{ind} \\ \boldsymbol{\mu}_2^{ind} \\ \vdots \\ \boldsymbol{\mu}_N^{ind} \end{pmatrix} \quad (31)$$

Eq. 29 can be rewritten as

$$E_{ind}^{tot} = \frac{1}{2}\left(\boldsymbol{\mu}^{ind}\right)^T \mathbf{B}\boldsymbol{\mu}^{ind} - \left(\boldsymbol{\mu}^{ind}\right)^T \mathbf{F} \quad (32)$$

In order to obtain an optimizing condition for the induction energy we minimize Eq. 32, with respect to the induced dipoles

$$\frac{\partial E_{\text{ind}}^{\text{tot}}}{\partial \boldsymbol{\mu}^{\text{ind}}} = \mathbf{B} \boldsymbol{\mu}^{\text{ind}} - \mathbf{F} = 0 \tag{33}$$

We thus obtain the induced dipoles corresponding to the minimum energy from

$$\mathbf{B} \boldsymbol{\mu}^{\text{ind}} = \mathbf{F}. \tag{34}$$

By writing out Eq. 34, we obtain that the optimum dipoles are given by

$$\begin{aligned}
\boldsymbol{\mu}_a^{\text{ind}} &= \alpha_a \left[\sum_{a \neq b} \mathbf{F}_b^{(k)}(\mathbf{R}_a) + \sum_{a \neq b} \mathbf{T}_{ba}^{(2)} \boldsymbol{\mu}_b^{\text{ind}} + \mathbf{F}^{\text{ext}}(\mathbf{R}_a) \right] \\
&= \alpha_a \left[\mathbf{F}_b^m(\mathbf{R}_a) + \mathbf{F}_b^{\text{ind}}(\mathbf{R}_a) + \mathbf{F}^{\text{ext}}(\mathbf{R}_a) \right] \\
&= \alpha_a \mathbf{F}^{\text{tot}}(\mathbf{R}_a)
\end{aligned} \tag{35}$$

which is the intuitive result. Using this expression for the optimum induced dipoles we finally arrive at

$$E_{\text{ind}}^{\text{tot}} = -\frac{1}{2} \sum_a \boldsymbol{\mu}_a^{\text{ind}} \left[\mathbf{F}^m(\mathbf{R}_a) + \mathbf{F}^{\text{ext}}(\mathbf{R}_a) \right] \tag{36}$$

After discussing the classical physics, we now turn to the derivation of the quantum mechanical analog. The definition of the induction energy E_a^{ind}, which arises from the electronic charge distributions of the fragments polarizing each other, is given again in Eq. 37, where the polarization of the fragments can be found in the contributions from the excited states.

$$E_a^{\text{ind}} = -\sum_{a \neq b} \sum_m{}' \frac{\langle ab | \hat{V}^{ab} | a^m b \rangle \langle a^m b | \hat{V}^{ab} | ab \rangle}{E_{a,m}^{(0)} - E_{a,0}^{(0)}} \tag{37}$$

In terms of the approximate interaction operator this becomes

$$E_a^{\text{ind}} \approx \tilde{E}_a^{\text{ind}} = -\sum_{a \neq b} \sum_m{}' \frac{\langle ab | \hat{V}_{ab}^{\text{esp}} | a^m b \rangle \langle a^m b | \hat{V}_{ab}^{\text{esp}} | ab \rangle}{E_{a,m}^{(0)} - E_{a,0}^{(0)}} \tag{38}$$

We now substitute in Eq. 17, for the approximate interaction operator, which gives

$$
\begin{aligned}
\tilde{E}_a^{ind} &= -\sum_{a\neq b}\sum_{|k|=0}^{K_a}\sum_{|l|=0}^{K_a}{\sum_m}' \frac{(-1)^{|k+l|}}{k!l!} \frac{\langle ab|\hat{T}_{ab}^{(k)}\hat{M}_a^{(k)}|a^m b\rangle\langle a^m b|\hat{T}_{ab}^{(l)}\hat{M}_a^{(l)}|ab\rangle}{E_{a,m}^{(0)} - E_{a,0}^{(0)}} \\
&= -\sum_{a\neq b}\sum_{|k|=0}^{K_a}\sum_{|l|=0}^{K_a} \frac{(-1)^{|k+l|}}{k!l!}\langle b|\hat{T}_{ab}^{(k)}|b\rangle {\sum_m}' \frac{\langle a|\hat{M}_a^{(k)}|a^m\rangle\langle a^m|\hat{M}_a^{(l)}|a\rangle}{E_{a,m}^{(0)} - E_{a,0}^{(0)}}\langle b|\hat{T}_{ab}^{(l)}|b\rangle \\
&= -\sum_{a\neq b}\sum_{|k|=1}^{K_a}\sum_{|l|=1}^{K_a} \frac{(-1)^{|k+l|}}{k!l!}\langle b|\hat{T}_{ab}^{(k)}|b\rangle {\sum_m}' \frac{\langle a|\hat{M}_a^{(k)}|a^m\rangle\langle a^m|\hat{M}_a^{(l)}|a\rangle}{E_{a,m}^{(0)} - E_{a,0}^{(0)}}\langle b|\hat{T}_{ab}^{(l)}|b\rangle
\end{aligned}
$$

$$(39)$$

Truncating this expression at first order, and recognizing the expression for the electric fields and polarizability, this equation can be written as

$$
\tilde{E}_a^{ind} = -\frac{1}{2}\mathbf{F}(\mathbf{R}_a)\boldsymbol{\mu}_a^{ind} \tag{40}
$$

which is Eq. 36. Equation 40 is essentially one summand in Eq. 36, leading to the final expression for the polarization energy

$$
\tilde{E}_{ind}^{tot} = -\frac{1}{2}\sum_a \mathbf{F}(\mathbf{R}_a)\boldsymbol{\mu}_a^{ind} \tag{41}
$$

The induced dipole moments of fragment a are obtained from

$$
\boldsymbol{\mu}_a^{ind} = \boldsymbol{\alpha}_a\left(\mathbf{F}(\mathbf{R}_a) + \sum_{a\neq b}\mathbf{T}_{ba}^{(2)}\boldsymbol{\mu}_b^{ind}\right). \tag{42}
$$

These are the optimum induced dipoles derived at the beginning of this section (Eq. 35). For completeness, we mention that going from a single-site multipole expansion to a distributed one introduces an additional summation over the expansion sites

$$
\begin{aligned}
\tilde{E}_a^{ind} &= -\frac{1}{2}\sum_{s\in a}^{S_a}\mathbf{F}(\mathbf{R}_s)\boldsymbol{\mu}_s^{ind} \\
\tilde{E}_{ind}^{tot} &= -\frac{1}{2}\sum_a\sum_{s\in a}^{S_a}\mathbf{F}(\mathbf{R}_s)\boldsymbol{\mu}_s^{ind}
\end{aligned}
$$

$$(43)$$

Equation 42 can be setup for each fragment of the supersystem, where it matters not if we use a single-site multipole expansion or a distributed one. By using the definitions of the **B**-matrix and the field- and induced dipole-vectors in Eqs. 30 and 31, the obtained set of equations can then again be reformulated as a matrix-vector equation to obtain the induced dipole moments.

$$
\boldsymbol{\mu}^{ind} = \mathbf{B}^{-1}\mathbf{F} \tag{44}
$$

Using this notation, the total induction energy can be written as the sum of the individual induction energies as

$$\tilde{E}^{\text{tot}}_{\text{ind}} = -\frac{1}{2}\mathbf{F}\boldsymbol{\mu}^{\text{ind}} \tag{45}$$

It is important to note that the induction energy is doubly dependent on the electric field—it enters once directly in the equation above and indirectly in the determination of the induced dipoles.

2.4 Formulation in the SCF Framework

So far, we have derived the method using exact state theory, where the influence of other fragments b on the gas-phase wave function of a specific fragment a can be calculated by perturbation theory and the corresponding wave function corrections. However, the wave functions of molecules usually have to be calculated using approximate methods, such as Hartree-Fock or DFT. Thus, there is a need to include the effect of the other fragments b during the optimization of the wave function. This is also necessary for the subsequent application of response theory. In HF and DFT, we need to obtain a modified version of the Fock- or KS-operator, which can account for the effects of the other fragments b. We will now proceed to show how such an operator can be derived.

2.4.1 General Derivation of the Effective Operator

When deriving expressions of the effective embedding operator in an SCF framework, the procedure is similar to the one used in the derivation of the canonical HF equations, due to the choice of the form of the wave function, but with some important differences, e.g., in the summation. The energy of the Hartree product is

$$
\begin{aligned}
E^{\text{tot}} &= \langle abcd \ldots n| \, \hat{H}^{\text{tot}} \, |abcd \ldots n\rangle \\
&= \langle abcd \ldots n| \left(\sum_{a=1}^{n} \hat{H}^a + \sum_{a<b}^{n} \hat{V}^{ab} \right) |abcd \ldots n\rangle \\
&= \sum_{a=1}^{n} \langle a| \, \hat{H}^a \, |a\rangle + \sum_{a=1}^{n} \sum_{b>a}^{n} \langle ab| \, \hat{V}^{ab} \, |ab\rangle \\
&= \sum_{a=1}^{n} \langle a| \, \hat{H}^a \, |a\rangle + \frac{1}{2} \sum_{a=1}^{n} \sum_{b\neq a}^{n} \langle ab| \, \hat{V}^{ab} \, |ab\rangle
\end{aligned}
\tag{46}
$$

Since the fragments are nonoverlapping, the fragment wave functions are orthonormal

$$\langle a \,|\, b \rangle = \delta_{ab} \tag{47}$$

We now construct the functional L as

$$
\begin{aligned}
L &= E^{\text{tot}} - \sum_{a=1}^{n} \sum_{b=1}^{n} \epsilon_{ab} \left(\langle a \,|\, b \rangle - \delta_{ab} \right) \\
&= \sum_{a=1}^{n} \langle a | \, \hat{H}^a \, | a \rangle + \frac{1}{2} \sum_{a=1}^{n} \sum_{b \neq a}^{n} \langle ab | \, \hat{V}^{ab} \, | ab \rangle - \sum_{a=1}^{n} \sum_{b=1}^{n} \epsilon_{ab} \left(\langle a \,|\, b \rangle - \delta_{ab} \right)
\end{aligned}
\tag{48}
$$

where the Lagrange-multipliers ϵ_{ab} are the elements of a hermitian matrix (thus $\epsilon_{ab} = \epsilon_{ba}^*$), since L is real and $\langle a \,|\, b \rangle = \langle b \,|\, a \rangle^*$. We now minimize L, which is equivalent to minimizing the total energy, with respect to the variation of the wave functions of the fragments in the form

$$a \to a + \delta a \tag{49}$$

Thus, we set the variation of L equal to zero

$$\delta L = \delta E^{\text{tot}} - \sum_{a=1}^{n} \sum_{b=1}^{n} \epsilon_{ab} \delta \langle a \,|\, b \rangle = 0 \tag{50}$$

This can be expanded to

$$
\begin{aligned}
0 = \delta L &= \sum_{a=1}^{n} \delta \langle a | \, \hat{H}^a \, | a \rangle + \frac{1}{2} \sum_{a=1}^{n} \sum_{b \neq a}^{n} \delta \langle ab | \, \hat{V}^{ab} \, | ab \rangle - \sum_{a=1}^{n} \sum_{b=1}^{n} \epsilon_{ab} \delta \langle a \,|\, b \rangle \\
&= \sum_{a=1}^{n} \langle \delta a | \, \hat{H}^a \, | a \rangle + \frac{1}{2} \sum_{a=1}^{n} \sum_{b \neq a}^{n} \left(\langle \delta ab | \, \hat{V}^{ab} \, | ab \rangle + \langle a \delta b | \, \hat{V}^{ab} \, | ab \rangle \right) \\
&\quad - \sum_{a=1}^{n} \sum_{b=1}^{n} \epsilon_{ab} \langle \delta a \,|\, b \rangle + c.c. \\
&= \sum_{a=1}^{n} \langle \delta a | \, \hat{H}^a \, | a \rangle + \sum_{a=1}^{n} \sum_{b \neq a}^{n} \langle \delta ab | \, \hat{V}^{ab} \, | ab \rangle - \sum_{a=1}^{n} \sum_{b=1}^{n} \epsilon_{ab} \langle \delta a \,|\, b \rangle + c.c.
\end{aligned}
\tag{51}
$$

where the fact that a and b are only dummy indices was used and where "c.c." stands for "complex conjugate". We now rewrite this result in the form

$$0 = \delta L = \sum_{a=1}^{n} \langle \delta a | \left[\hat{H}^a \, | a \rangle + \sum_{b \neq a}^{n} \langle b | \, \hat{V}^{ab} \, | b \rangle \, | a \rangle - \sum_{b=1}^{n} \epsilon_{ab} \, | b \rangle \right] + c.c. \tag{52}$$

Since δa is arbitrary, the term in square brackets has to be zero. This gives

$$\left[\hat{H}^a + \sum_{b \neq a}^{n} \langle b| \hat{V}^{ab} |b\rangle\right] |a\rangle = \sum_{b=1}^{n} \epsilon_{ab} |b\rangle \qquad a = 1, 2, 3, \ldots, n \qquad (53)$$

which can, by a unitary transformation which leaves the operators in Eq. 53, unchanged, be transformed to

$$\left[\hat{H}^a + \sum_{b \neq a}^{n} \langle b'| \hat{V}^{ab} |b'\rangle\right] |a'\rangle = \epsilon_a' |a'\rangle \qquad a = 1, 2, 3, \ldots, n \qquad (54)$$

where the prime indicates the transformed wave function. We will omit the prime in the following to avoid overloading the notation. The left-hand side of Eqs. 53 and 54, is equivalent to an effective Fock- or KS-operator, where the usual form of these operators is extended by the interactions with the other fragments.

So far, we have not distinguished between the fragments. To achieve a differentiated treatment, we select one fragment (a) as the core region, and assign the rest of the fragments (b) to the environment. The environment is assumed to be only linearly responsive, so only terms up to second order in the interaction in the electrostatic potential of the environment is retained leading to [39]

$$\left[\hat{H}^a + \sum_{b \neq a}^{n} \langle b| \hat{V}^{ab} |b\rangle^{(1)} + \sum_{b \neq a}^{n} \langle b| \hat{V}^{ab} |b\rangle^{(2)}\right] |a\rangle = \epsilon_a |a\rangle , \qquad (55)$$

It is important to note that the effective Hamiltonian in the square brackets in Eq. 55, used for the optimization of the wave function of a is dependent on the permanent and induced charge distributions of fragments b and, due to the latter, is thereby dependent on the density of a. Thus, the effective Hamiltonian is nonlinear and Eq. 55, requires an iterative solution. To be more precise this already entered in Eq. 54. Explicit expressions for the two terms in the order expansion in Eq. 55, will be detailed in Eqs. 60 and 61.

In the last step, we eliminate the wave function of the fragments b of the environment in Eq. 55, by representing the permanent and linear induced charge distributions of fragments b by multipole expansions, i.e., assigning permanent multipole moments and anisotropic polarizabilities to expansion sites in the environment. We thus arrive at a description of the environmental electrostatic effects through an effective embedding operator $\hat{v}^{\text{PE}}$

$$\left[\hat{H}^a + \hat{v}^{\text{PE}}\right] |a\rangle = \epsilon_a |a\rangle \qquad (56)$$

where the polarization of the QM-region is accounted for by the inclusion of the environment in the wave function optimization process via the embedding operator and the polarization of the environment is partly due to the field originating from the QM-region, as we shall show below.

2.4.2 The PE Energy Functional and the Embedding Operator

From Eqs. 55 and 56, the energy of subsystem a can be decomposed as

$$
\begin{aligned}
\epsilon_a^{\text{tot}} &= \langle a| \left[\hat{H}^a + \sum_{b \neq a}^{n} \langle b| \hat{V}^{ab} |b\rangle^{(1)} + \sum_{b \neq a}^{n} \langle b| \hat{V}^{ab} |b\rangle^{(2)} \right] |a\rangle \\
&= \langle a| \hat{H}^a |a\rangle + \sum_{b \neq a}^{n} \langle ab| \hat{V}^{ab} |ab\rangle^{(1)} + \sum_{b \neq a}^{n} \langle ab| \hat{V}^{ab} |ab\rangle^{(2)} \\
&= \epsilon_a^{\text{QM}} + \epsilon^{\text{es}} + \epsilon^{\text{ind}}
\end{aligned}
\tag{57}
$$

where ϵ_a^{QM} is the energy of the quantum region. Since we include the effect of the environment in the optimization of the wave function, this term already includes the (wave function) polarization energy due to the environment. ϵ^{es} is the electrostatic interaction energy of the quantum region with the environment and ϵ^{ind} is the polarization energy of the environment. In this picture, the energy functional of the whole composite system reads

$$
E^{\text{tot}} = \epsilon_a^{\text{QM}} + \epsilon_a^{\text{es}} + \epsilon^{\text{ind}} + \epsilon^{\text{env}}
\tag{58}
$$

where ϵ^{env} is the internal *electrostatic* energy of the environment, excluding the energy associated with the induced dipoles, which is already included in ϵ^{ind}. This is due to the way ϵ^{ind} is defined (see below Eq. 61), as in this equation both $\mathbf{F}(\mathbf{R}_s)$ and $\boldsymbol{\mu}_s^{\text{ind}}$ are determined dependent on both the quantum region, as well as the environment.

We now proceed to find an expression for the embedding operator $\hat{v}^{\text{PE}}$. Due to the form the energy expression in Eq. 57, takes, we take the embedding operator $\hat{v}^{\text{PE}}$ to be of a similar form

$$
\hat{v}^{\text{PE}} = \hat{v}^{\text{es}} + \hat{v}^{\text{ind}}
\tag{59}
$$

By introducing a multipole expansion in Eq. 57, we can rewrite the electrostatic and induction energy terms according to

$$\epsilon^{es} = \sum_{b \neq a} \sum_{s \in b} \sum_{|k|=0}^{K_s} \frac{(-1)^{|k|}}{k!} M_s^{(k)} \sum_{pq \in a} t_{pq}^{(k)}(\mathbf{R}_s) D_{pq}^a$$

$$+ \sum_{b \neq a} \sum_{s \in b} \sum_{|k|=0}^{K_s} \frac{(-1)^{|k|}}{k!} M_s^{(k)} \sum_{n \in a}^{M_a} T_{sn}^{(k)} Z_n \qquad (60)$$

and

$$\epsilon^{ind} = -\frac{1}{2} \sum_{b \neq a} \sum_{s \in b}^{S_b} \mathbf{F}(\mathbf{R}_s) \boldsymbol{\mu}_s^{ind} \qquad (61)$$

where we used the expressions obtained in the previous section. The fields by which the induced dipoles of the environment are multiplied in Eq. 61, is the sum of the fields from the permanent multipoles of the environment $\mathbf{F}^m(\mathbf{R}_s)$ plus any externally applied field to this environment. The key point is that we shall view the field that the quantum region exerts onto the environment as an "external" field. The polarization of the environment is, in this picture, partly due to the "external" field originating from the QM-region, while the polarization of the QM-region is accounted for by the inclusion of the environment in the wave function optimization process via the embedding operator. The field the QM-region produces consists of the fields from the nuclei of the core $\mathbf{F}^n(\mathbf{R}_s)$ and the fields from the electronic charge distribution $\langle a | \hat{\mathbf{F}}^e(\mathbf{R}_s) | a \rangle$, where a component of the electric-field operator is

$$\hat{F}_\alpha^e(\mathbf{R}_s) = -\sum_{pq \in a} \int \phi_p^*(\mathbf{r}') \nabla_{r_{\alpha'}} \left(\frac{1}{|\mathbf{R}' - \mathbf{R}_s|} \right) \phi_q(\mathbf{r}') d\mathbf{r}' \hat{E}_{pq}$$

$$= -\sum_{pq \in a} \int \phi_p^*(\mathbf{r}') \left(\frac{R_{\alpha,s} - r_\alpha}{|\mathbf{R}_s - \mathbf{R}'|^3} \right) \phi_q(\mathbf{r}') d\mathbf{r}' \hat{E}_{pq} \qquad (62)$$

$$= \sum_{pq \in a} t_{\alpha,pq}(\mathbf{R}_s) \hat{E}_{pq}^a$$

Therefore, the field in Eq. 61, takes the form

$$\mathbf{F}(\mathbf{R}_s) = \mathbf{F}^m(\mathbf{R}_s) + \mathbf{F}^n(\mathbf{R}_s) + \langle a | \hat{\mathbf{F}}^e(\mathbf{R}_s) | a \rangle \qquad (63)$$

The induced dipoles are again obtained from

$$\boldsymbol{\mu}^{ind} = \mathbf{B}^{-1} \mathbf{F} \qquad (64)$$

and are thus the optimum dipoles of the polarizable sites in the environment exposed to the fields of the induced and permanent multipoles from the environment and the "external" field from the QM-region.

Using this we can rewrite Eq. 61, as

$$\epsilon^{\text{ind}} = -\frac{1}{2} \sum_{b \neq a} \sum_{s \in b}^{S_b} \mathbf{F}^m(\mathbf{R}_s) \boldsymbol{\mu}_s^{\text{ind}} - \frac{1}{2} \sum_{b \neq a} \sum_{s \in b}^{S_b} \mathbf{F}^n(\mathbf{R}_s) \boldsymbol{\mu}_s^{\text{ind}}$$
$$- \frac{1}{2} \sum_{b \neq a} \sum_{s \in b}^{S_b} \langle a | \hat{\mathbf{F}}^e(\mathbf{R}_s) | a \rangle \, \boldsymbol{\mu}_s^{\text{ind}} \tag{65}$$

From Eqs. 60 and 61, we can directly extract the two operators $\hat{v}^{\text{es}}$ and $\hat{v}^{\text{ind}}$

$$\hat{v}^{\text{es}} = \sum_{b \neq a} \sum_{s \in b}^{S_b} \sum_{|k|=0}^{K_s} \frac{(-1)^{|k|}}{k!} M_s^{(k)} \sum_{pq \in a} t_{pq}^{(k)}(\mathbf{R}_s) \hat{E}_{pq}^a \tag{66}$$

$$\hat{v}^{\text{ind}} = -\sum_{b \neq a} \sum_{s \in b}^{S_b} \sum_{\alpha=x,y,z} \mu_{\alpha,s}^{\text{ind}}(\mathbf{F}) \hat{F}_\alpha^e(\mathbf{R}_s) \tag{67}$$

where in the last step, we used the electric field integrals defined in the previous section. Using the definitions of these operators electrostatic interaction and induction energies become

$$\epsilon^{\text{es}} = \langle a | \hat{v}^{\text{es}} | a \rangle + \sum_{b \neq a} \sum_{s \in b}^{S_b} \sum_{|k|=0}^{K_s} \frac{(-1)^{|k|}}{k!} M_s^{(k)} \sum_{n \in a}^{M_a} T_{sn}^{(k)} Z_n \tag{68}$$

$$\epsilon^{\text{ind}} = -\frac{1}{2} \sum_{b \neq a} \sum_{s \in b}^{S_b} \left[\mathbf{F}^m(\mathbf{R}_s) + \mathbf{F}^n(\mathbf{R}_s) \right] \boldsymbol{\mu}_s^{\text{ind}}$$
$$- \sum_{b \neq a} \sum_{s \in b}^{S_b} \sum_{\alpha=x,y,z} \mu_{\alpha,s}^{\text{ind}}(\mathbf{F}) \sum_{pq \in a} t_{\alpha,pq}(\mathbf{R}_s) D_{pq}^a \tag{69}$$

and we obtain an effective Hamiltonian for the QM-region which is of the form

$$\hat{H}^{\text{eff}} = \hat{H}^a + \hat{v}^{\text{PE}}$$
$$= \hat{H}^a - \sum_{b \neq a} \sum_{s \in b}^{S_b} \sum_{|k|=0}^{K_s} \frac{(-1)^{|k|}}{k!} M_s^{(k)} \sum_{pq \in a} t_{pq}^{(k)}(\mathbf{R}_s) \hat{E}_{pq}^a$$
$$- \sum_{b \neq a} \sum_{s \in b}^{S_b} \sum_{\alpha=x,y,z} \mu_{\alpha,s}^{\text{ind}}(\mathbf{F}) \sum_{pq \in a} t_{\alpha,pq}(\mathbf{R}_s) \hat{E}_{pq}^a \tag{70}$$

Since the effective Hamiltonian depends on the induced dipole moments of the environment, which via Eq. 64, in turn, depends on the field of the QM-region, it is a nonlinear operator. Thus, the mutual polarization of the QM- and MM-regions leads to the Schrodinger equation requiring an iterative solution. Note that this arises only due to the induced dipoles, and that the electrostatic interactions do not lead to this complication.

At each step of the SCF procedure, the induced dipoles are updated according to Eq. 64, using the fields from the current guess to the wave function, in order to obtain the induction operator. Then, the induction operator is included in the next SCF step, and the procedure is repeated until convergence. Thus, both the induced dipoles and the QM wave function are fully variationally optimized.

3 PE Within Response Theory Framework

As outlined in the introductory part of this chapter, optical spectroscopy plays an important role in the investigation of molecular properties and the electronic structure of molecules [47]. The explicit calculation of excited states, however, is of considerable difficulty. DFT, for example, one of the most widely used tools currently at hand is unable to obtain excited states in a direct manner. A workaround is thus needed, which is conveniently found in response theory (which in the case of DFT led to the development of the widely used TD-DFT method). A distinctive feature of the PE model is that it has been designed for the calculation of molecular response properties, and therefore, has been formulated within a general response theory framework. Thus, it can be employed for the calculation of a wide range of spectroscopic properties, such as oscillator strengths, two-photon absorption cross-sections, as well as excitation energies and excited-state properties. By this method, also magnetic properties, such as shielding tensors can be obtained. More generally, any property that can be calculated based on the various levels of response can include environmental effects through the PE model. The PE model has also been formulated within a resonant-convergent response theory formalism, yielding equations that are valid over the whole frequency range, and in particular, where the conventional response theory approach outlined here shows divergences.

3.1 Introduction to Approximate State Response Theory

Response theory aims at the description of the change in a molecular property (the molecular "response") due to an external perturbation—it is thus perturbation theory by nature [47]. The perturbation can be an applied electric field (static or time-dependent) as in the case at hand, but the approach can be applied much more generally, enabling the calculation of molecular properties related to magnetic fields, changes in molecular geometry, and more.

We will outline the theory for single-determinant methods such as TD-HF [49] and TD-DFT [59]. For this, we consider the time-evolution of a molecular property A, which is obtained from the expectation value $\langle 0| \hat{A} |0\rangle$ of its time-independent corresponding operator $\hat{A}$. We then apply a periodic time-dependent field of the form

$$F_\alpha(t) = \sum_\omega F_\alpha^\omega e^{-i\omega t} e^{\epsilon t} \tag{71}$$

where the factor $e^{\epsilon t}$ ensures the slow switch-on of the field at $t \to -\infty$. The applied electric field couples to the QM-system via a coupling operator $\hat{V}_\alpha^\omega$, which can, e.g., be the dipole operator. The perturbation operator acting on the QM-system is thus

$$\hat{V}(t) = \sum_\omega \sum_\alpha \hat{V}_\alpha^\omega F_\alpha^\omega e^{-i\omega t} e^{\epsilon t} \tag{72}$$

and the requirement of a real-valued perturbation imposes the conditions that $F_\alpha^{-\omega} = [F_\alpha^\omega]^\dagger$ and $\hat{V}_\alpha^{-\omega} = [\hat{V}_\alpha^\omega]^\dagger = \hat{V}_\alpha^\omega$. Thus, the Hamiltonian of the QM-System becomes

$$\hat{H}(t) = \hat{H} + \hat{V}(t) = \hat{H} + \sum_\omega \sum_\alpha \hat{V}_\alpha^\omega F_\alpha^\omega e^{-i\omega t} e^{\epsilon t} \tag{73}$$

The wave function of the QM-system becomes time-dependent, and can be expanded in powers of the perturbation as

$$|\Psi(t)\rangle = |\Psi^{(0)}\rangle + |\Psi^{(1)}\rangle + |\Psi^{(2)}\rangle + |\Psi^{(3)}\rangle + \cdots \tag{74}$$

The expectation value of the molecular property A thus also becomes time-dependent and can be decomposed as

$$
\begin{aligned}
\langle\Psi(t)|\hat{A}|\Psi(t)\rangle = {} & \langle\Psi^{(0)}|\hat{A}|\Psi^{(0)}\rangle \\
& + \langle\Psi^{(1)}|\hat{A}|\Psi^{(0)}\rangle + \langle\Psi^{(0)}|\hat{A}|\Psi^{(1)}\rangle \\
& + \langle\Psi^{(2)}|\hat{A}|\Psi^{(0)}\rangle + \langle\Psi^{(1)}|\hat{A}|\Psi^{(1)}\rangle + \langle\Psi^{(0)}|\hat{A}|\Psi^{(2)}\rangle \\
& + \langle\Psi^{(3)}|\hat{A}|\Psi^{(0)}\rangle + \langle\Psi^{(2)}|\hat{A}|\Psi^{(1)}\rangle + \langle\Psi^{(1)}|\hat{A}|\Psi^{(2)}\rangle + \langle\Psi^{(0)}|\hat{A}|\Psi^{(3)}\rangle \\
& + \cdots
\end{aligned}
\tag{75}
$$

Collecting terms of the same power in the perturbation this becomes

$$\langle\Psi(t)|\hat{A}|\Psi(t)\rangle = \langle\Psi|\hat{A}|\Psi\rangle^{(0)} + \langle\Psi|\hat{A}|\Psi\rangle^{(1)} + \langle\Psi|\hat{A}|\Psi\rangle^{(2)} + \langle\Psi|\hat{A}|\Psi\rangle^{(3)} + \cdots \tag{76}$$

where the term $\langle\Psi(t)|\hat{A}|\Psi(t)\rangle^{(0)} = \langle\Psi^{(0)}|\hat{A}|\Psi^{(0)}\rangle$ is just the expectation value of the unperturbed system and the other terms constitute the corrections to this expectation value of first $(\langle\Psi(t)|\hat{A}|\Psi(t)\rangle^{(1)} = \langle\Psi^{(1)}|\hat{A}|\Psi^{(0)}\rangle + \langle\Psi^{(0)}|\hat{A}|\Psi^{(1)}\rangle)$, second $(\langle\Psi(t)|\hat{A}|\Psi(t)\rangle^{(0)} = \langle\Psi^{(2)}|\hat{A}|\Psi^{(0)}\rangle + \langle\Psi^{(1)}|\hat{A}|\Psi^{(1)}\rangle + \langle\Psi^{(0)}|\hat{A}|\Psi^{(2)}\rangle)$, and higher orders. We note the difference between order expansion in the wave function $|\Psi^{(n)}\rangle$

and in expectation values $\langle \Psi(t)| \hat{A} |\Psi(t)\rangle^{(n)}$. Having in mind this picture, we factor out the field in each of the terms of Eq. 76, and rewrite it as a power expansion in terms of the applied fields

$$
\begin{aligned}
\langle \Psi(t)| \hat{A} |\Psi(t)\rangle = \langle \hat{A}\rangle &+ \sum_{\omega_1} \sum_{\beta} \langle\langle \hat{A}; \hat{V}_\beta^{\omega_1}\rangle\rangle F_\beta^{\omega_1} e^{-i\omega_1 t} e^{\epsilon t} \\
&+ \frac{1}{2} \sum_{\omega_1 \omega_2} \sum_{\beta,\gamma} \langle\langle \hat{A}; \hat{V}_\beta^{\omega_1} \hat{V}_\gamma^{\omega_2}\rangle\rangle F_\beta^{\omega_1} F_\gamma^{\omega_2} e^{-i(\omega_1+\omega_2)t} e^{2\epsilon t} \\
&+ \frac{1}{6} \sum_{\omega_1 \omega_2 \omega_3} \sum_{\beta,\gamma,\delta} \langle\langle \hat{A}; \hat{V}_\beta^{\omega_1} \hat{V}_\gamma^{\omega_2} \hat{V}_\delta^{\omega_3}\rangle\rangle F_\beta^{\omega_1} F_\gamma^{\omega_2} F_\delta^{\omega_3} e^{-i(\omega_1+\omega_2+\omega_3)t} e^{3\epsilon t} \\
&+ \cdots
\end{aligned}
$$

$$(77)$$

The terms $\langle\langle \hat{A}; \hat{V}_\beta^{\omega_1}\rangle\rangle$, $\langle\langle \hat{A}; \hat{V}_\beta^{\omega_1} \hat{V}_\gamma^{\omega_2}\rangle\rangle$, $\langle\langle \hat{A}; \hat{V}_\beta^{\omega_1} \hat{V}_\gamma^{\omega_2} \hat{V}_\delta^{\omega_3}\rangle\rangle$, etc., that originate from this approach, are referred to as the linear, quadratic, cubic, etc. *response functions*, since they describe the response of the property of the QM-system to the external perturbation. We now turn to the question on how to obtain these response functions and shall explain this at the example of the linear response function. Generalization to higher-order response functions is essentially straightforward. In addressing this question, we shall use an exponential parametrization of the wave function of the form

$$
|\widetilde{\Psi}(\kappa)\rangle = e^{-i\hat{\kappa}} |\Psi\rangle \tag{78}
$$

which stems from the creation and annihilation operators obeying the relations

$$
\begin{aligned}
\tilde{a}_p^\dagger &= e^{-i\hat{\kappa}} \hat{a}_p^\dagger e^{i\hat{\kappa}} \\
\tilde{a}_q &= e^{-i\hat{\kappa}} \hat{a}_q e^{i\hat{\kappa}}
\end{aligned} \tag{79}
$$

The hermitian operator $\hat{\kappa}$ is defined as

$$
\hat{\kappa} = \sum_{pq} \kappa_{pq} \hat{a}_p^\dagger \hat{a}_q; \quad \hat{\kappa}^\dagger = \hat{\kappa} \tag{80}
$$

The parameters κ_{pq} are called the orbital rotation parameters. The advantage of this parametrization is that orthonormality is conserved, as the operator $e^{-i\hat{\kappa}}$ is unitary.

We need to introduce one additional aspect before we continue. It is convenient to extract out a phase factor $\phi(t)$ of the time-dependent wave function $\Psi(t)$ and rewrite it in the form

$$
|\Psi(t)\rangle = e^{-i\phi} |\overline{\Psi}(t)\rangle \tag{81}
$$

where $\overline{\Psi}(t)$ is called the phase-isolated wave function [9]. The division is made unique by the requirements that $\phi(t)$ be real and that the phase of the projection of

$\overline{\Psi}(t)$ onto $\Psi(t)$ be zero. In case of the absence of an external field, these requirements would lead back to the usual division

$$|\Psi(t)\rangle = e^{-i\phi} |\overline{\Psi}\rangle = e^{-iE_0t/\hbar} |0\rangle \tag{82}$$

which illustrates that the latter requirement also ensures that in perturbation theory, the zeroth-order contribution will be time-independent and equal to the eigenfunction of the time-independent, unperturbed, Hamiltonian. The division is useful since, when one is only interested in quantum mechanical averages, there is no need to determine the phase function, thereby simplifying the procedure.

In the following, we will show two different ways of deriving expressions for response functions within an explicit parametrization—the Ehrenfest and quasi-energy approaches. When discussing the extension of response functions to the polarizable embedding scheme, we will only consider the quasi-energy approach but include here for completeness also the Ehrenfest derivation.

3.2 Ehrenfest Approach

The Ehrenfest theorem provides an expression for the time derivative of the expectation value of an operator $\hat{A}$ [49]

$$\frac{\partial}{\partial t} \langle \Psi(t)|\hat{A}|\Psi(t)\rangle = \left\langle \frac{\partial \Psi(t)}{\partial t} \middle| \hat{A} \middle| \Psi(t) \right\rangle + \left\langle \Psi(t) \middle| \hat{A} \middle| \frac{\partial \Psi(t)}{\partial t} \right\rangle + \left\langle \Psi(t) \middle| \frac{\partial \hat{A}}{\partial t} \middle| \Psi(t) \right\rangle$$

$$= \frac{1}{i\hbar} \langle \Psi(t)|[\hat{A}, \hat{H}(t)]|\Psi(t)\rangle + \left\langle \Psi(t) \middle| \frac{\partial \hat{A}}{\partial t} \middle| \Psi(t) \right\rangle \tag{83}$$

where we substituted in the time-dependent Schrödinger equation for the time derivative of the wave function. If $\hat{A}$ does not include time differentiation, it is possible to substitute $\Psi(t)$ by its phase-isolated counterpart $\overline{\Psi}(t)$, which leads to

$$\frac{\partial}{\partial t} \langle \overline{\Psi}(t)|\hat{A}|\overline{\Psi}(t)\rangle - \left\langle \Psi(t) \middle| \frac{\partial \hat{A}}{\partial t} \middle| \Psi(t) \right\rangle = \frac{1}{i\hbar} \langle \overline{\Psi}(t)|[\hat{A}, \hat{H}_0 + \hat{V}(t)]|\overline{\Psi}(t)\rangle$$

$$\tag{84}$$

This equation can be solved by using time-dependent perturbation theory. For this, the wave function is expanded in terms of the perturbation by employing

$$\hat{\kappa}(t) = \hat{\kappa}^{(1)}(t) + \hat{\kappa}^{(2)}(t) + \hat{\kappa}^{(3)}(t) + \cdots \tag{85}$$

where the zeroth-order term drops out since the wave function is already variationally optimized. In exact state theory, exploiting Eqs. 84 and 85, by making the special choice for $\hat{A}$ to be the time-independent state-transition operator $|0\rangle \langle n|$, enables one to obtain a differential equation for the desired orbital rotation parameters. In approximate state theory, however, things are a bit more complicated.

A general one-electron operator in the basis of the time-independent molecular orbitals of the reference determinant can, in second quantization, be written as

$$\hat{A} = \sum_{pq} A_{pq} \hat{a}_p^\dagger \hat{a}_q = \sum_{pq} A_{pq} \hat{E}_{pq} \tag{86}$$

with $\hat{E}_{pq}$ being a (de)excitation operator, much in analogy to the state-transition operator $|0\rangle \langle n|$ in exact state theory. Proceeding analogously to exact state theory and making the special choice of the time-independent operator $\hat{E}_{pq}$ for $\hat{A}$ is not the best way to proceed, as the general operator $\hat{A}$ also contains (de)excitation operators $\hat{E}_{pq}$ for occupied-occupied and unoccupied-unoccupied transitions, which are redundant, and therefore, missing in the parametrization of single-determinant methods. Thus, if the orbital rotation parameters were determined in this incomplete set of excitation operators, this would lead to the problem that the Ehrenfest theorem would, in general, not be fulfilled for a general operator $\hat{A}$.

As can be shown [47], this problem can be circumvented by parametrizing the wave function in terms of the *time-dependent* molecular orbitals by using the time transformed creation and annihilation operators (see Eq. 79), which give rise to the time transformed (de)excitation operators

$$\tilde{\hat{E}}_{pq} = e^{-i\hat{\kappa}} \hat{E}_{pq} e^{i\hat{\kappa}} \tag{87}$$

A general operator $\hat{A}$ in this time-dependent basis takes the form

$$\hat{A} = \sum_{pq} A_{pq} \tilde{\hat{a}}_p^\dagger \tilde{\hat{a}}_q = \sum_{pq} A'_{pq} \tilde{\hat{E}}_{pq} \tag{88}$$

Having worked this out, we now make the special choice $\hat{A} = \tilde{\hat{E}}_{pq}$ in Eq. 84, yielding

$$\frac{1}{i\hbar} \left(\langle \overline{\Psi} | [\tilde{\hat{E}}_{pq}, \hat{H}_0] | \overline{\Psi} \rangle + \langle \overline{\Psi} | [\tilde{\hat{E}}_{pq}, \hat{V}(t)] | \overline{\Psi} \rangle \right) = \frac{\partial}{\partial t} \langle \overline{\Psi} | \tilde{\hat{E}}_{pq} | \overline{\Psi} \rangle - \langle \overline{\Psi} | \frac{\partial \tilde{\hat{E}}_{pq}}{\partial t} | \overline{\Psi} \rangle \tag{89}$$

We now evaluate each of the terms in Eq. 89, by substituting in Eqs. 78 and 87, and then expanding the exponential operators in a Baker-Campbell-Hausdorff (BCH) expansions which is of the form

$$e^{-i\hat{\kappa}} \hat{A} e^{i\hat{\kappa}} = \hat{A} + i[\hat{\kappa}, \hat{A}] - \frac{1}{2}[\hat{\kappa}[\hat{\kappa}, \hat{A}]] - \frac{i}{6}[\hat{\kappa}[\hat{\kappa}[\hat{\kappa}, \hat{A}]]] + \cdots \tag{90}$$

In doing so we shall collect all (de)excitation operators in a vector $\hat{\mathbf{q}}$ and all orbital rotation parameters in a vector κ. In this form, the definition of the operator $\hat{\kappa}$ reads

$$\hat{\kappa} = \hat{\mathbf{q}}^\dagger \kappa \tag{91}$$

3.3 Linear Response

For the determination of linear response properties we collect all terms of first order in the perturbation in Eq. 89. We thus arrive at

$$\frac{1}{i\hbar} \left(\langle 0|i[\hat{\mathbf{q}}, [\hat{\mathbf{q}}^\dagger \kappa_{\omega_1}^{(1)}, \hat{H}_0]]|0\rangle + \langle 0|[\hat{\mathbf{q}}, \hat{V}^\omega]|0\rangle \right) e^{-i\omega_1 t} = -\langle 0| \left[\hat{\mathbf{q}}, i\hat{\mathbf{q}}^\dagger \frac{\partial}{\partial t} \kappa_{\omega_1}^{(1)} e^{-i\omega_1 t} \right] |0\rangle$$

$$\frac{1}{\hbar} \left(\langle 0|[\hat{\mathbf{q}}, [\hat{\mathbf{q}}^\dagger \kappa_{\omega_1}^{(1)}, \hat{H}_0]]|0\rangle + i\langle 0|[\hat{\mathbf{q}}, \hat{V}^\omega]|0\rangle \right) = -i(-i\omega_1)\langle 0| \left[\hat{\mathbf{q}}, \hat{\mathbf{q}}^\dagger \kappa_{\omega_1}^{(1)} \right] |0\rangle$$

$$\langle 0|[\hat{\mathbf{q}}, [\hat{\mathbf{q}}^\dagger \kappa_{\omega_1}^{(1)}, \hat{H}_0]]|0\rangle + \hbar\omega_1 \langle 0| \left[\hat{\mathbf{q}}, \hat{\mathbf{q}}^\dagger \kappa_{\omega_1}^{(1)} \right] |0\rangle = -i\langle 0|[\hat{\mathbf{q}}, \hat{V}^\omega]|0\rangle$$

$$-\left(-\langle 0|[\hat{\mathbf{q}}, [\hat{\mathbf{q}}^\dagger, \hat{H}_0]]|0\rangle - \hbar\omega_1 \langle 0| \left[\hat{\mathbf{q}}, \hat{\mathbf{q}}^\dagger \right] |0\rangle \right) \kappa_{\omega_1}^{(1)} = -i\langle 0|[\hat{\mathbf{q}}, \hat{V}^\omega]|0\rangle \tag{92}$$

To bring out the matrix structure of Eq. 92, we define the matrices

$$\mathbf{E} = -\begin{pmatrix} \langle 0|[\hat{\mathbf{q}}, [\hat{\mathbf{q}}^\dagger, \hat{H}_0]]|0\rangle & \langle 0|[\hat{\mathbf{q}}, [\hat{\mathbf{q}}, \hat{H}_0]]|0\rangle \\ \langle 0|[\hat{\mathbf{q}}^\dagger, [\hat{\mathbf{q}}^\dagger, \hat{H}_0]]|0\rangle & \langle 0|[\hat{\mathbf{q}}^\dagger, [\hat{\mathbf{q}}, \hat{H}_0]]|0\rangle \end{pmatrix} \tag{93}$$

$$\mathbf{S} = \begin{pmatrix} \langle 0| \left[\hat{\mathbf{q}}, \hat{\mathbf{q}}^\dagger \right] |0\rangle & \langle 0| \left[\hat{\mathbf{q}}, \hat{\mathbf{q}} \right] |0\rangle \\ \langle 0| \left[\hat{\mathbf{q}}^\dagger, \hat{\mathbf{q}}^\dagger \right] |0\rangle & \langle 0| \left[\hat{\mathbf{q}}^\dagger, \hat{\mathbf{q}} \right] |0\rangle \end{pmatrix} \tag{94}$$

and the vector

$$\mathbf{V}^\omega = \begin{pmatrix} \langle 0|[\hat{\mathbf{q}}, \hat{V}^\omega]|0\rangle \\ \langle 0|[\hat{\mathbf{q}}^\dagger, \hat{V}^\omega]|0\rangle \end{pmatrix} \tag{95}$$

The general expectation value of an operator $\hat{A}$ can be expanded in orders of perturbation as follows:

$$\langle \overline{\Psi} | \hat{A} | \overline{\Psi} \rangle = \langle 0 | e^{-i\hat{\kappa}t} \hat{A} e^{i\hat{\kappa}t} | 0 \rangle$$

$$= \langle 0 | \hat{A} + i [\hat{\mathbf{q}}^\dagger \kappa(t), \hat{A}] - \frac{1}{2} [\hat{\mathbf{q}}^\dagger \kappa [\hat{\mathbf{q}}^\dagger \kappa, \hat{A}]] + \cdots | 0 \rangle$$

$$= \langle 0 | \hat{A} | 0 \rangle$$

$$+ i \sum_{\omega_1} \langle 0 | [\hat{\mathbf{q}}^\dagger, \hat{A}] | 0 \rangle \kappa^{(1)}_{\omega_1} e^{-i\omega_1 t} - \frac{1}{2} \sum_{\omega_1, \omega_2} \langle 0 | [\hat{\mathbf{q}}^\dagger \kappa^{(1)}_{\omega_1} [\hat{\mathbf{q}}^\dagger \kappa^{(1)}_{\omega_2}, \hat{A}]] | 0 \rangle \quad (96)$$

$$+ i \sum_{\omega_1, \omega_2} \langle 0 | [\hat{\mathbf{q}}^\dagger, \hat{A}] | 0 \rangle \kappa^{(2)}_{\omega_1, \omega_2} e^{-i(\omega_1 + \omega_2)t}$$

$$+ \cdots$$

Substituting $\hat{A} = \hat{H}_0$, we see that the matrix $\mathbf{E}$ has the form of an electronic Hessian. Using the definitions in Eqs. 93, 94 and 95, Eq. 92 becomes

$$(\mathbf{E} - \hbar\omega_1 \mathbf{S}) \, \kappa^{(1)}_{\omega_1} = i \mathbf{V}^\omega \quad (97)$$

An expression for a general linear response property can be obtained from Eq. 96. We have

$$\langle\langle \hat{A}; \hat{V}^\omega \rangle\rangle = i \langle 0 | [\hat{\mathbf{q}}^\dagger, \hat{A}] | 0 \rangle \kappa^{(1)}_{\omega_1} \quad (98)$$

and combining this with Eq. 97, we get

$$\langle\langle \hat{A}; \hat{V}^\omega \rangle\rangle = -\mathbf{A}^\dagger \, (\mathbf{E} - \hbar\omega_1 \mathbf{S})^{-1} \, \mathbf{V}^\omega \quad (99)$$

where we defined the vector $\mathbf{A}$ as

$$\mathbf{A} = \begin{pmatrix} \langle 0 | [\hat{\mathbf{q}}, \hat{A}] | 0 \rangle \\ \langle 0 | [\hat{\mathbf{q}}^\dagger, \hat{A}] | 0 \rangle \end{pmatrix}, \qquad \mathbf{A}^\dagger = - \left(\langle 0 | [\hat{\mathbf{q}}^\dagger, \hat{A}] | 0 \rangle \;\; \langle 0 | [\hat{\mathbf{q}}, \hat{A}] | 0 \rangle \right) \quad (100)$$

3.4 Quasi-energy Approach

In the previous section, we introduced the concept of the phase-isolated wave function (Eq. 81), and within the so-called quasi-energy formulation of response theory, this phase-isolated wave function plays a very central role. The derivations in the coming section follows Refs. [9, 47]. Based on the phase-isolated wave function we define the quasi-energy ($Q(t)$) as

$$Q(t) = \left\langle \overline{\Psi} \left| \left[\hat{H} - i\hbar \frac{\partial}{\partial t} \right] \right| \overline{\Psi} \right\rangle \quad (101)$$

The useful thing about this concept is that it can be used as the starting point for a time-dependent variational principle. A variation in the phase-isolated wave function must fulfill the normalization condition

$$\langle \overline{\Psi} | \overline{\Psi} \rangle = 1 \quad \Rightarrow \quad \delta \langle \overline{\Psi} | \overline{\Psi} \rangle = \langle \delta \overline{\Psi} | \overline{\Psi} \rangle + \langle \overline{\Psi} | \delta \overline{\Psi} \rangle = 0 \tag{102}$$

We also need the relation

$$\left\langle \overline{\Psi}_1 \left| \left[\hat{H} - i\hbar \frac{\partial}{\partial t} \right] \right| \overline{\Psi}_2 \right\rangle = \left\langle \overline{\Psi}_2 \left| \left[\hat{H} - i\hbar \frac{\partial}{\partial t} \right] \right| \overline{\Psi}_1 \right\rangle^* - i\hbar \frac{\partial}{\partial t} \langle \overline{\Psi}_1 | \overline{\Psi}_2 \rangle \tag{103}$$

We can now use these two equations to derive an expression for the variation in the quasi-energy δQ due to a variation in the wave function $\delta \overline{\Psi}$

$$
\begin{aligned}
\delta Q &= \left\langle \delta \overline{\Psi} \left| \left[\hat{H} - i\hbar \frac{\partial}{\partial t} \right] \right| \overline{\Psi} \right\rangle + \left\langle \overline{\Psi} \left| \left[\hat{H} - i\hbar \frac{\partial}{\partial t} \right] \right| \delta \overline{\Psi} \right\rangle \\
&= \left\langle \delta \overline{\Psi} \left| \left[\hat{H} - i\hbar \frac{\partial}{\partial t} \right] \right| \overline{\Psi} \right\rangle + \left\langle \delta \overline{\Psi} \left| \left[\hat{H} - i\hbar \frac{\partial}{\partial t} \right] \right| \overline{\Psi} \right\rangle^* - i\hbar \frac{\partial}{\partial t} \langle \overline{\Psi} | \delta \overline{\Psi} \rangle \\
&= Q(t) \left[\langle \delta \overline{\Psi} | \overline{\Psi} \rangle + \langle \delta \overline{\Psi} | \overline{\Psi} \rangle^* \right] - i\hbar \frac{\partial}{\partial t} \langle \overline{\Psi} | \delta \overline{\Psi} \rangle \\
&= -i\hbar \frac{\partial}{\partial t} \langle \overline{\Psi} | \delta \overline{\Psi} \rangle
\end{aligned}
\tag{104}
$$

and obtain

$$\delta Q + i\hbar \frac{\partial}{\partial t} \langle \overline{\Psi} | \delta \overline{\Psi} \rangle = 0 \tag{105}$$

It is possible to get rid of the second term by employing a mathematical trick: For any periodic function with period T

$$\frac{1}{T} \int_0^T \dot{f}(t) \mathrm{d}t = \frac{1}{T} [f(T) - f(0)] = \frac{1}{T} [f(0) - f(0)] = 0 \tag{106}$$

since $f(t + T) = f(t)$. Performing this time-averaging over Eq. 105, it becomes

$$\frac{1}{T} \int_t^{t+T} \delta Q \mathrm{d}t' + \frac{i\hbar}{T} \int_t^{t+T} \frac{\partial}{\partial t} \langle \overline{\Psi} | \delta \overline{\Psi} \rangle \mathrm{d}t' = \delta Q_T = 0 \tag{107}$$

where we defined the time-averaged quasi-energy

$$Q_T = \frac{1}{T} \int_t^{t+T} Q(t') \mathrm{d}t' = \frac{1}{T} \int_t^{t+T} \left\langle \overline{\Psi} \left| \left[\hat{H} - i\hbar \frac{\partial}{\partial t'} \right] \right| \overline{\Psi} \right\rangle \mathrm{d}t' \tag{108}$$

If we take the variation to be of the form

$$\delta\overline{\Psi} = \left|\frac{d\overline{\Psi}}{dF^\omega}\right\rangle \tag{109}$$

we can easily obtain an expression for the derivative of the quasi-energy with respect to the amplitude dF^ω of a component of the field in the time-dependent perturbation (compare Eq. 72)

$$
\begin{aligned}
\frac{\partial Q(t)}{\partial F^\omega} &= \frac{\partial}{\partial F^\omega}\left\langle\overline{\Psi}\left|\left[\hat{H} - i\hbar\frac{\partial}{\partial t}\right]\right|\Psi\right\rangle \\
&= \left\langle\frac{\partial\overline{\Psi}}{\partial F^\omega}\left|\left[\hat{H} - i\hbar\frac{\partial}{\partial t}\right]\right|\Psi\right\rangle + \left\langle\overline{\Psi}\left|\left[\hat{H} - i\hbar\frac{\partial}{\partial t}\right]\right|\frac{\partial\Psi}{\partial F^\omega}\right\rangle + \left\langle\overline{\Psi}\left|\frac{\partial\hat{H}}{\partial F^\omega}\right|\Psi\right\rangle \\
&= \delta Q + \left\langle\overline{\Psi}\left|\frac{\partial\hat{H}}{\partial F^\omega}\right|\Psi\right\rangle \\
&= -i\hbar\frac{\partial}{\partial t}\langle\overline{\Psi}|\frac{d\overline{\Psi}}{dF^\omega}\rangle + \left\langle\overline{\Psi}\left|\frac{\partial\hat{H}}{\partial F^\omega}\right|\Psi\right\rangle
\end{aligned}
\tag{110}
$$

This equation is known as the time-dependent Hellman-Feynman theorem. If we compare it to its time-independent counterpart

$$\frac{dE}{dF} = \left\langle\overline{\Psi}\left|\frac{\partial\hat{H}}{\partial F}\right|\Psi\right\rangle \tag{111}$$

it is immediately clear that there appears an additional term in the time-dependent version. Q_T is, in principle, a time-dependent function, but in the steady-state region, only the fundamental frequency of the field and its harmonics remain, essentially removing any time dependence. Therefore, Q_T can be seen as a function of F^ω and Q_T can be expended in terms of the field amplitudes F^ω of the perturbation around $F^\omega = 0$ as follows[9]

$$
\begin{aligned}
Q_T = {} & E_0 \\
& + \sum_{\omega_1}\sum_{\alpha}\left.\frac{dQ_T}{dF_\alpha^{\omega_1}}\right|_{F^\omega=0}F_\alpha^{\omega_1} \\
& + \frac{1}{2}\sum_{\omega_1,\omega_2}\sum_{\alpha,\beta}\left.\frac{d^2Q_T}{dF_\alpha^{\omega_1}dF_\beta^{\omega_2}}\right|_{F^\omega=0}F_\alpha^{\omega_1}F_\beta^{\omega_2} \\
& + \frac{1}{6}\sum_{\omega_1,\omega_2;\omega_3}\sum_{\alpha,\beta,\gamma}\left.\frac{d^3Q_T}{dF_\alpha^{\omega_1}dF_\beta^{\omega_2}dF_\gamma^{\omega_3}}\right|_{F^\omega=0}F_\alpha^{\omega_1}F_\beta^{\omega_2}F_\gamma^{\omega_3} \\
& + \cdots
\end{aligned}
\tag{112}
$$

To evaluate the first derivative, we perform the time-averaging over Eq. 110,

$$\frac{\mathrm{d}Q_T}{\mathrm{d}F_\alpha^\omega} = \frac{1}{T}\sum_\omega \int_t^{t+T}\left\langle \overline{\Psi}(t')\left|\frac{\partial \hat{H}}{\partial F_\alpha^\omega}\right|\overline{\Psi}(t')\right\rangle \mathrm{d}t'$$

$$= \frac{1}{T}\int_t^{t+T}\sum_\omega\left\langle \overline{\Psi}(t')\left|\hat{V}_\alpha^\omega\right|\overline{\Psi}(t')\right\rangle e^{-i\omega_1 t}e^{\epsilon t}\mathrm{d}t' \tag{113}$$

where we substituted in the expression for the Hamiltonian (Eq. 73). We now compare the expansion of the time-averaged quasi-energy to the expansion of an expectation value in time (Eq. 77), and choose the operator $\hat{A}$ as $\sum_{\omega_1}\hat{V}_\alpha^{\omega_1}e^{-i\omega_1 t}e^{\epsilon t}$ with α being the Cartesian component. The expansion will take the form

$$\sum_{\omega_1}\langle\Psi(t)|\,\hat{V}_\alpha^{\omega_1}\,|\Psi(t)\rangle\, e^{-i\omega_1 t}e^{\epsilon t}$$

$$= \sum_{\omega_1}\langle 0|\,\hat{V}_\alpha^{\omega_1}\,|0\rangle\, e^{-i\omega_1 t}e^{\epsilon t}$$

$$+ \sum_{\omega_1,\omega_2}\sum_\beta\langle\langle\hat{V}_\alpha^{\omega_1};\hat{V}_\beta^{\omega_2}\rangle\rangle F_\beta^{\omega_2}e^{-i(\omega_1+\omega_2)t}e^{2\epsilon t}$$

$$+ \frac{1}{2}\sum_{\omega_1,\omega_2\omega_3}\sum_{\beta,\gamma}\langle\langle\hat{V}_\alpha^{\omega_1};\hat{V}_\beta^{\omega_2}\hat{V}_\gamma^{\omega_3}\rangle\rangle F_\beta^{\omega_2}F_\gamma^{\omega_3}e^{-i(\omega_1+\omega_2+\omega_3)t}e^{3\epsilon t} \tag{114}$$

$$+ \frac{1}{6}\sum_{\omega_1,\omega_2\omega_3\omega_4}\sum_{\beta,\gamma,\delta}\langle\langle\hat{V}_\alpha^{\omega_1};\hat{V}_\beta^{\omega_2}\hat{V}_\gamma^{\omega_3}\hat{V}_\delta^{\omega_4}\rangle\rangle F_\beta^{\omega_2}F_\gamma^{\omega_3}F_\delta^{\omega_4}e^{-i(\omega_1+\omega_2+\omega_3+\omega_4)t}e^{4\epsilon t}$$

$$+ \cdots$$

Time-averaging Eq. 114 returns the equivalent to Eq. 110. In the limit of zero field strength all terms but the first vanish and thus, combining the Eqs. 110 and 114, we obtain

$$\left.\frac{\mathrm{d}Q_T}{\mathrm{d}F_\alpha^{\omega_1}}\right|_{F^\omega=0} = \frac{1}{T}\sum_{\omega_1}\int_t^{t+T}\langle 0|\,\hat{V}_\alpha^{\omega_1}\,|0\rangle\, e^{-i\omega_1 t'}e^{\epsilon t}\mathrm{d}t'$$

$$= \sum_{\omega_1}\langle 0|\,\hat{V}_\alpha^{\omega_1}\,|0\rangle\,\delta_{\omega_1,0} \tag{115}$$

Thus, only the static component $\omega_1 = 0$ remains. This is intuitive in that for a time-independent state $|0\rangle$, the absorption or emission of a photon of frequency $\omega_1 \neq 0$ is not allowed due to energy conservation. Taking the second derivative of Eq. 114, we obtain similarly for the second derivative of the time-averaged quasi-energy at zero field strength

$$\frac{d^2 Q_T}{dF_\alpha^{\omega_1} dF_\beta^{\omega_2}}\bigg|_{F^\omega=0} = \frac{1}{T}\int_t^{t+T} \frac{d}{dF_\beta^{\omega_2}}\bigg|_{F^\omega=0}\left[\sum_{\omega_1}\langle\Psi(t)|\hat{V}_\alpha^{\omega_1}|\Psi(t)\rangle e^{-i\omega_1 t}e^{\epsilon t}\right]$$

$$= \sum_{\omega_1,\omega_2}\langle\langle\hat{V}_\alpha^{\omega_1};\hat{V}_\beta^{\omega_2}\rangle\rangle\delta_{(\omega_1+\omega_2),0}$$

(116)

The higher derivatives are obtained in an analogous way, and hence, Eq. 112, becomes

$$Q_T = E_0$$
$$+ \sum_{\omega_1}\sum_\alpha \langle 0|\hat{V}_\alpha^{\omega_1}|0\rangle\delta_{\omega_1,0}F_\alpha^{\omega_1}$$
$$+ \frac{1}{2}\sum_{\omega_1,\omega_2}\sum_{\alpha,\beta}\langle\langle\hat{V}_\alpha^{\omega_1};\hat{V}_\beta^{\omega_2}\rangle\rangle\delta_{(\omega_1+\omega_2),0}F_\alpha^{\omega_1}F_\beta^{\omega_2}$$

(117)

$$+ \frac{1}{6}\sum_{\omega_1,\omega_2,\omega_3}\sum_{\alpha,\beta,\gamma}\langle\langle\hat{V}_\alpha^{\omega_1};\hat{V}_\beta^{\omega_2}\hat{V}_\gamma^{\omega_3}\rangle\rangle\delta_{(\omega_1+\omega_2+\omega_3),0}F_\alpha^{\omega_1}F_\beta^{\omega_2}F_\gamma^{\omega_3}$$

$$+ \cdots$$

We now address the question of how to obtain explicit expressions for the derivatives of the time-averaged quasi-energy. Due to the exponential parametrization of the wave function, the time-averaged quasi-energy (Eq. 108), is a function

$$Q_T = Q_T\left(\kappa^{(1)}(\omega_1), [\kappa^{(1)}(-\omega_1)]^*, \kappa^{(2)}(\omega_1,\omega_2), [\kappa^{(2)}(-\omega_1,-\omega_2)]^*, \ldots, F_\alpha^{\omega_1}, F_\beta^{\omega_2}, \ldots\right)$$

(118)

of the field strengths F^ω and the Fourier transforms of the orbital rotation coefficients $\kappa^{(k)}(\omega)$

$$\kappa_n(t) = \kappa_n^{(1)}(t) + \kappa_n^{(2)}(t) + \kappa_n^{(3)}(t) + \cdots$$
$$= \sum_{\omega_1}\kappa_n^{(1)}(\omega_1)e^{-i\omega_1 t} + \sum_{\omega_1,\omega_2}\kappa_n^{(2)}(\omega_1,\omega_2)e^{-i(\omega_1+\omega_2)t} + \cdots$$

(119)

with k the order of perturbation. Owing to the variational principle established earlier for Q_T (Eq. 107), the first derivatives with respect to the $\kappa^{(n)}(\omega)$ vanish

$$\frac{\partial Q_T}{\partial \kappa_n^{(1)}}\bigg|_{F^\omega} = 0; \quad \frac{\partial Q_T}{\partial [\kappa_n^{(1)}]^*}\bigg|_{F^\omega} = 0; \quad \frac{\partial Q_T}{\partial \kappa_n^{(2)}}\bigg|_{F^\omega} = 0; \quad \frac{\partial Q_T}{\partial [\kappa_n^{(2)}]^*}\bigg|_{F^\omega} = 0; \quad (120)$$

for all values of the field amplitudes F^ω at all frequencies ω. Thus, the wave function parameters are functions of the field strengths, and we need to determine the respective response of these parameters. This is done by differentiating Eq. 120, with respect to the field strengths, where the differential is evaluated at $F^\omega = 0$ for consistency with Eq. 112. From this results the set of response equations [47]

$$
\frac{d}{dF_\alpha^\omega} \frac{\partial Q_T}{\partial \kappa_n^{(k)}} \bigg|_{F^\omega=0} = \frac{\partial^2 Q_T}{\partial F_\alpha^\omega \partial \kappa_n^{(k)}} \bigg|_{F^\omega=0}
$$

$$
+ \sum_m \left[\frac{\partial^2 Q_T}{\partial \kappa_n^{(k)} \partial \kappa_m^{(k)}} \frac{\partial \kappa_m^{(k)}}{\partial F_\alpha^\omega} + \frac{\partial^2 Q_T}{\partial \kappa_n^{(k)} \partial [\kappa_m^{(k)}]^*} \frac{\partial [\kappa_m^{(k)}]^*}{\partial F_\alpha^\omega} \right] \bigg|_{F^\omega=0} = 0
$$

$$
\frac{d}{dF_\alpha^\omega} \frac{\partial Q_T}{\partial [\kappa_n^{(k)}]^*} \bigg|_{F^\omega=0} = \frac{\partial^2 Q_T}{\partial F_\alpha^\omega \partial [\kappa_n^{(k)}]^*} \bigg|_{F^\omega=0}
$$

$$
+ \sum_m \left[\frac{\partial^2 Q_T}{\partial [\kappa_n^{(k)}]^* \partial \kappa_m^{(k)}} \frac{\partial \kappa_m^{(k)}}{\partial F_\alpha^\omega} + \frac{\partial^2 Q_T}{\partial [\kappa_n^{(k)}]^* \partial [\kappa_m^{(k)}]^*} \frac{\partial [\kappa_m^{(k)}]^*}{\partial F_\alpha^\omega} \right] \bigg|_{F^\omega=0} = 0
$$

$$(121)$$

from which the responses of the orbital rotation parameters to the field strengths

$$
\frac{\partial \kappa_m^{(k)}}{\partial F_\alpha^\omega}, \quad \frac{\partial [\kappa_m^{(k)}]^*}{\partial F_\alpha^\omega}
\tag{122}
$$

can be calculated. Higher-order responses of the orbital rotation parameters can be obtained by performing additional differentiations in the same way as shown. The response equations in Eq. 121, can be cast into matrix form [47]

$$
\boldsymbol{Q}_T^{2;0} \boldsymbol{P}^\alpha + \boldsymbol{Q}_T^{1;\alpha} = 0
$$

$$
\boldsymbol{P}^\alpha = - \left[\boldsymbol{Q}_T^{2;0} \right]^{-1} \boldsymbol{Q}_T^{1;\alpha}
\tag{123}
$$

where

$$
\boldsymbol{Q}_T^{2;0} = \begin{pmatrix} \frac{\partial^2 Q_T}{\partial [\kappa_n^{(k)}]^* \partial \kappa_m^{(k)}} \big|_{F^\omega=0} & \frac{\partial^2 Q_T}{\partial [\kappa_n^{(k)}]^* \partial [\kappa_m^{(k)}]^*} \big|_{F^\omega=0} \\ \frac{\partial^2 Q_T}{\partial \kappa_n^{(k)} \partial \kappa_m^{(k)}} \big|_{F^\omega=0} & \frac{\partial^2 Q_T}{\partial \kappa_n^{(k)} \partial [\kappa_m^{(k)}]^*} \big|_{F^\omega=0} \end{pmatrix}
\tag{124}
$$

$$
\boldsymbol{Q}_T^{1;\alpha} = \begin{pmatrix} \frac{\partial^2 Q_T}{\partial F_\alpha^\omega \partial [\kappa_n^{(k)}]^*} \big|_{F^\omega=0} \\ \frac{\partial^2 Q_T}{\partial F_\alpha^\omega \partial \kappa_n^{(k)}} \big|_{F^\omega=0} \end{pmatrix} ; \quad \boldsymbol{P}^\alpha = \begin{pmatrix} \frac{\partial \kappa_m^{(k)}}{\partial F_\alpha^\omega} \big|_{F^\omega=0} \\ \frac{\partial [\kappa_m^{(k)}]^*}{\partial F_\alpha^\omega} \big|_{F^\omega=0} \end{pmatrix}
\tag{125}
$$

We now need to calculate the derivatives of Q_T in Eqs. 124 and 125, in order to be able to evaluate Eq. 123, for the responses $\boldsymbol{P}^\alpha$ of the orbital rotation parameters to the field strengths. This can be done by making use of a BCH-expansion of the time-averaged quasi-energy

$$Q_T = \frac{1}{T} \int_t^{t+T} \langle \overline{\Psi} | \hat{H}_0 + \hat{V}(t) - i\hbar \frac{\partial}{\partial t} | \overline{\Psi} \rangle dt'$$

$$= \frac{1}{T} \int_t^{t+T} \langle 0 | e^{i\hat{\kappa}} \hat{Q} e^{-i\hat{\kappa}} | 0 \rangle dt' \qquad (126)$$

$$= \frac{1}{T} \int_t^{t+T} \left(\langle 0 | \hat{Q} | 0 \rangle + i \langle 0 | [\hat{\kappa}, \hat{Q}] | 0 \rangle - \frac{1}{2} \langle 0 | [\hat{\kappa} [\hat{\kappa}, \hat{Q}]] | 0 \rangle + \cdots \right) dt'$$

where we introduced $\hat{Q}$ as the quasi-energy operator.

3.5 Linear Response

From Eq. 116, the linear response function is given by the second-order derivative of Q_T with respect to the field strength.

$$\frac{d^2 Q_T}{dF_\alpha^{\omega_1} dF_\beta^{\omega_2}} \bigg|_{F^\omega = 0} = \sum_{\omega_1, \omega_2} \langle\langle \hat{V}_\alpha^{\omega_1}; \hat{V}_\beta^{\omega_2} \rangle\rangle \delta_{(\omega_1 + \omega_2), 0} \qquad (127)$$

Writing out the first-order derivative of Q_T with respect to $F_\alpha^{\omega_1}$ by using chain-rule differentiation we obtain

$$\frac{d^2 Q_T}{dF_\alpha^{\omega_1} dF_\beta^{\omega_2}} = \left[\frac{\partial^2 Q_T}{\partial F_\alpha^{\omega_1} \partial \kappa_n} \frac{\partial \kappa_n}{\partial F_\beta^{\omega_2}} + \frac{\partial^2 Q_T}{\partial F_\alpha^{\omega_1} \partial \kappa_n^*} \frac{\partial \kappa_n^*}{\partial F_\beta^{\omega_2}} \right] \qquad (128)$$

which can conveniently be rewritten in matrix notation [47]

$$\frac{d^2 Q_T}{dF_\alpha^{\omega_1} dF_\beta^{\omega_2}} \bigg|_{F_\omega = 0} = [Q_T^{1;\alpha}]_-^\dagger P^\beta$$

$$= -[Q_T^{1;\alpha}]_-^\dagger \left[Q_T^{2;0} \right]^{-1} Q_T^{1;\beta} \qquad (129)$$

where P^β was obtained from the linear response equations. Working out finally the explicit expressions for the vectors and matrix in the above equation, and defining the Hessian and overlap matrix as in the case of the Ehrenfest derivation, allows the linear response function to be written in the same way as obtained in the Ehrenfest approach, i.e., through the use of the electronic Hessian and overlap matrices

$$\frac{d^2 Q_T}{dF_\alpha^{\omega_1} dF_\beta^{\omega_2}} = -[Q_T^{1;\alpha}]_-^\dagger \left[Q_T^{2;0} \right]^{-1} Q_T^{1;\beta}$$

$$= -[V^{\omega_1}]_-^\dagger (E - \hbar\omega_2 S)^{-1} V^{\omega_2} \delta_{(\omega_1 + \omega_2), 0} \qquad (130)$$

3.6   Effect of the PE Model

Having established the basics of the concept of response theory, we now turn to the description of the PE model in the response theory framework [38]. In the case of an externally applied time-dependent electric field, the electron density will respond accordingly, while the response of the nuclear configuration will, in general, be too slow to follow the perturbation (for an isolated molecule, this is the well-known Franck-Condon-principle). Thus, we naturally arrive at the description of vertical (de-)excitations, leaving the molecular geometry intact, which, in the terminology of embedding models is referred to as the nonequilibrium regime. In the PE model, the fixed orientational polarization of the environment from the unperturbed ground-state calculation is described by the respective permanent multipole moments and induced dipole moments, while the electronic polarization of the environment arising from the applied electric field is included through additional induced dipole moments.

An important aspect of the PE model in regard to response theory is that we need to account for the fact that the field that is actually acting on the embedded quantum region—the local field—is not equal to the externally applied electric field (which would be experienced by the QM-region in isolation [37]). It is the superposition of the externally applied field and the fields generated by the permanent and induced charge distributions of the environment. Within a distributed multipole expansion, the expression for this, evaluated at point $\mathbf{R}_s$ within the QM-region, becomes

$$\mathbf{F}^{\mathrm{LF}}(\mathbf{R}_s) = \mathbf{F}^{\mathrm{ext}}(\mathbf{R}_s) + \mathbf{F}^m(\mathbf{R}_s) + \sum_{s'=1}^{S} \mathbf{T}_{ss'}^{(2)} \boldsymbol{\mu}_{s'}^{\mathrm{ind}}(\tilde{\mathbf{F}}) \tag{131}$$

where $\mathbf{F}^{\mathrm{ext}}$ denotes the externally applied electric field, $\mathbf{F}^m$ the field originating by the permanent multipoles of the environment and $\boldsymbol{\mu}_s^{\mathrm{ind}}(\tilde{\mathbf{F}})$ the induced dipoles of the environment. The induced dipoles are created by the superposition of the fields from the quantum region, the permanent multipoles and the externally applied electric field (compare Eq. 63)

$$\tilde{\mathbf{F}}(\mathbf{R}_s) = \mathbf{F}^n(\mathbf{R}_s) + \langle \tilde{a}| \hat{\mathbf{F}}^e(\mathbf{R}_s) |\tilde{a}\rangle + \mathbf{F}^m(\mathbf{R}_s) + \mathbf{F}^{\mathrm{ext}}(\mathbf{R}_s) \tag{132}$$

We shall proceed by incorporating the local field effects in the quasi-energy formulation of response theory. As previously discussed, the quasi-energy originates from applying the time-dependent Schrödinger equation to the wave function partitioned in the phase-isolated wave function and the phase function. For the composite system, this reads

$$Q(t) = \hbar \dot{\phi}(t); \quad Q(t) = \left\langle \overline{\Psi}_{\mathrm{sys}}(t) \left| \left[\hat{H}_{\mathrm{sys}} + \hat{V}(t) - i\hbar \frac{\partial}{\partial t} \right] \right| \overline{\Psi}_{\mathrm{sys}}(t) \right\rangle \tag{133}$$

Since we only determine the wave function for the QM-region of the system, we repartition the quasi-energy according to

$$Q(t) = \left\langle \bar{\tilde{a}}(t) \left| \left[\hat{H}_0 + \hat{v}^{es} + \hat{V}(t) - i\hbar\frac{\partial}{\partial t} \right] \right| \tilde{a}(t) \right\rangle - \frac{1}{2}\sum_{s,s'}^{S} \tilde{\mathbf{F}}(\mathbf{R}_s)^T \mathbf{B}_{ss'}\tilde{\mathbf{F}}(\mathbf{R}_s') + \tilde{\epsilon}_{env}$$

(134)

The response functions are obtained from differentiating the quasi-energy (Eq. 101), after performing the time-averaging. Thus, the linear response function for the composite system reads [37]

$$\frac{d^2 Q_T}{dF_\alpha^{-\omega}dF_\beta^\omega}\bigg|_{F^\omega=0} = \alpha_{\alpha,\beta}^{env}(-\omega,\omega) + \sum_{\omega_1,\omega_2} \langle\langle \hat{V}_\alpha^{-\omega}; \hat{V}_\beta^\omega \rangle\rangle$$

(135)

where we used that the Kronecker delta-functional (compare Eq. 116), is only nonzero for $\omega_1 = -\omega_2$ in the case of linear response. $\alpha_{\alpha,\beta}^{env}(-\omega,\omega)$ is an explicit contribution to the linear polarizability from the isolated environment (the $\tilde{\epsilon}^{env}$ term in Eq. 134). The second term of Eq. 135, comprises the interactions between the QM-region and the environment and can thus be seen as the linear response function of the embedded system. It is thus suited for determining transition properties of the latter.

We now turn to the evaluation of the linear response function of the embedded system. Due to the introduction of $\alpha_{\alpha,\beta}^{env}(-\omega,\omega)$, we only need to consider the first two terms of Eq. 134. The evaluation of the second derivative of the field follows the process for an isolated system closely. Differentiating the first term two times with respect to the externally applied field leaves only the second derivatives in κ for the parts of the term in $\hat{H}_0 + \hat{v}^{es}$ and in the time differentiation operator $\frac{\partial}{\partial t}$. The overlap matrix $\mathbf{S}$ arises from the latter and remains unchanged

$$\mathbf{S} = \begin{pmatrix} \langle a|\left[\hat{q}, \hat{q}^\dagger\right]|a\rangle & \langle a|\left[\hat{q}, \hat{q}\right]|a\rangle \\ \langle a|\left[\hat{q}^\dagger, \hat{q}^\dagger\right]|a\rangle & \langle a|\left[\hat{q}^\dagger, \hat{q}\right]|a\rangle \end{pmatrix}$$

(136)

Following through all the steps for the Hamiltonian already displayed, but with $\hat{H}_0$ substituted by $\hat{H}_0 + \hat{v}^{es}$ leads to the matrix

$$\mathbf{E}^{(es)} = -\begin{pmatrix} \langle a|[\hat{q}, [\hat{q}^\dagger, \hat{H}_0 + \hat{v}^{es}]]|a\rangle & \langle a|[\hat{q}, [\hat{q}, \hat{H}_0 + \hat{v}^{es}]]|a\rangle \\ \langle a|[\hat{q}^\dagger, [\hat{q}^\dagger, \hat{H}_0 + \hat{v}^{es}]]|a\rangle & \langle a|[\hat{q}^\dagger, [\hat{q}, \hat{H}_0 + \hat{v}^{es}]]|a\rangle \end{pmatrix}$$

(137)

which, as we shall show shortly, is here only a part of the total electronic Hessian.

Forming the second derivative with respect to the external field strength of the perturbation part of the first term returns the unchanged property gradient $\mathbf{V}^\omega$, which is one part to the total property gradient.

$$\mathbf{V}^\omega = \begin{pmatrix} \langle a|[\hat{q}, \hat{V}^\omega]|a\rangle \\ \langle a|[\hat{q}^\dagger, \hat{V}^\omega]|a\rangle \end{pmatrix}$$

(138)

We now turn to the evaluation of the second term in Eq. 134, which can be simplified by remembering, that we have to take the second derivative of the time-averaged quasi-energy with respect to the externally applied field F^ω. If we write it out in total (using Eq. 132), we get

$$-\frac{1}{2}\sum_{s,s'}^{S}\tilde{\mathbf{F}}(\mathbf{R}_s)^T\mathbf{B}_{ss'}\tilde{\mathbf{F}}(\mathbf{R}'_s)$$

$$=-\frac{1}{2}\sum_{s,s'}^{S}\left[\mathbf{F}^n+\left\langle\tilde{a}\right|\hat{\mathbf{F}}^e\left|\tilde{a}\right\rangle+\mathbf{F}^m+\mathbf{F}^\omega\right]\mathbf{B}_{ss'}\left[\mathbf{F}^n+\left\langle\tilde{a}\right|\hat{\mathbf{F}}^e\left|\tilde{a}\right\rangle+\mathbf{F}^m+\mathbf{F}^\omega\right]$$

(139)

where it is easy to see that after two-fold differentiation with respect to F^ω only two terms remain,

$$-\frac{\partial^2}{\partial\kappa_n^{(1)}\partial\kappa_m^{(1)}}\left\langle\tilde{a}\right|\hat{\mathbf{F}}^e\left|\tilde{a}\right\rangle\mathbf{B}_{ss'}\left\langle\tilde{a}\right|\hat{\mathbf{F}}^e\left|\tilde{a}\right\rangle$$

(140)

and

$$-\frac{\partial^2}{\partial\kappa_n^{(1)}\partial\mathbf{F}^\omega}\mathbf{F}^\omega\mathbf{B}_{ss'}\left\langle\tilde{a}\right|\hat{\mathbf{F}}^e\left|\tilde{a}\right\rangle$$

(141)

Evaluating the first term by applying the BCH-expansion leads to a two-fold differentiation with respect to the orbital rotation parameters κ

$$\frac{\partial^2}{\partial\kappa_n^{(1)}\partial\kappa_m^{(1)}}\left\langle\tilde{a}\right|\hat{\mathbf{F}}^e\left|\tilde{a}\right\rangle\mathbf{B}_{ss'}\left\langle\tilde{a}\right|\hat{\mathbf{F}}^e\left|\tilde{a}\right\rangle=\mathbf{E}^{(\mathrm{dyn})}+\mathbf{E}^{(\mathrm{ind})}$$

(142)

where

$$\mathbf{E}^{(\mathrm{dyn})}=-\begin{pmatrix}\langle a|\left[\hat{q},\hat{\mathbf{F}}^{e\dagger}\right]|a\rangle^T\mathbf{B}_{ss'}\langle a|\left[\hat{q},\hat{\mathbf{F}}^{e\dagger}\right]|a\rangle & \langle a|\left[\hat{q},\hat{\mathbf{F}}^e\right]|a\rangle^T\mathbf{B}_{ss'}\langle a|\left[\hat{q},\hat{\mathbf{F}}^e\right]|a\rangle\\ \langle a|\left[\hat{q}^\dagger,\hat{\mathbf{F}}^{e\dagger}\right]|a\rangle^T\mathbf{B}_{ss'}\langle a|\left[\hat{q}^\dagger,\hat{\mathbf{F}}^{e\dagger}\right]|a\rangle & \langle a|\left[\hat{q}^\dagger,\hat{\mathbf{F}}^e\right]|a\rangle^T\mathbf{B}_{ss'}\langle a|\left[\hat{q}^\dagger,\hat{\mathbf{F}}^e\right]|a\rangle\end{pmatrix}$$

(143)

and

$$\mathbf{E}^{(\mathrm{ind})}=-\begin{pmatrix}\langle a|[\hat{q}[\hat{q}^\dagger,\hat{\mathbf{F}}^e]]|a\rangle^T\mathbf{B}_{ss'}\langle a|\hat{\mathbf{F}}^e|a\rangle & \langle a|[\hat{q}[\hat{q},\hat{\mathbf{F}}^e]]|a\rangle^T\mathbf{B}_{ss'}\langle a|\hat{\mathbf{F}}^e|a\rangle\\ \langle a|[\hat{q}^\dagger[\hat{q}^\dagger,\hat{\mathbf{F}}^e]]|a\rangle^T\mathbf{B}_{ss'}\langle a|\hat{\mathbf{F}}^e|a\rangle & \langle a|[\hat{q}^\dagger[\hat{q},\hat{\mathbf{F}}^e]]|a\rangle^T\mathbf{B}_{ss'}\langle a|\hat{\mathbf{F}}^e|a\rangle\end{pmatrix}$$

$$=-\begin{pmatrix}\langle a|[\hat{q}[\hat{q}^\dagger,\mu^{\mathrm{ind}}\hat{\mathbf{F}}^e]]|a\rangle^T & \langle a|[\hat{q}[\hat{q},\mu^{\mathrm{ind}}\hat{\mathbf{F}}^e]]|a\rangle^T\\ \langle a|[\hat{q}^\dagger[\hat{q}^\dagger,\mu^{\mathrm{ind}}\hat{\mathbf{F}}^e]]|a\rangle^T & \langle a|[\hat{q}^\dagger[\hat{q},\mu^{\mathrm{ind}}\hat{\mathbf{F}}^e]]|a\rangle^T\end{pmatrix}$$

$$=-\begin{pmatrix}\langle a|[\hat{q}[\hat{q}^\dagger,\hat{v}^{\mathrm{ind}}]]|a\rangle^T & \langle a|[\hat{q}[\hat{q},\hat{v}^{\mathrm{ind}}]]|a\rangle^T\\ \langle a|[\hat{q}^\dagger[\hat{q}^\dagger,\hat{v}^{\mathrm{ind}}]]|a\rangle^T & \langle a|[\hat{q}^\dagger[\hat{q},\hat{v}^{\mathrm{ind}}]]|a\rangle^T\end{pmatrix}$$

(144)

It can be seen that adding together matrices $\mathbf{E}^{(\mathrm{es})}$ and $\mathbf{E}^{(\mathrm{ind})}$ yields the matrix

$$\mathbf{E}^{(\text{stat})} = -\begin{pmatrix} \langle \overline{a}|[\hat{\mathbf{q}}, [\hat{\mathbf{q}}^{\dagger}, \hat{H}_0 + \hat{v}^{\text{PE}}]]|\overline{a}\rangle & \langle \overline{a}|[\hat{\mathbf{q}}, [\hat{\mathbf{q}}, \hat{H}_0 + \hat{v}^{\text{PE}}]]|\overline{a}\rangle \\ \langle \overline{a}|[\hat{\mathbf{q}}^{\dagger}, [\hat{\mathbf{q}}^{\dagger}, \hat{H}_0 + \hat{v}^{\text{PE}}]]|\overline{a}\rangle & \langle \overline{a}|[\hat{\mathbf{q}}^{\dagger}, [\hat{\mathbf{q}}, \hat{H}_0 + \hat{v}^{\text{PE}}]]|\overline{a}\rangle \end{pmatrix} \tag{145}$$

The last part of the second derivatives of Eq. 134, with respect to the external electric field we need to evaluate is Eq. 141. Inserting the BCH-expansion this returns

$$\mathbf{V}^{\omega}_{\text{corr}} = -\begin{pmatrix} \mathbf{B}_{ss'}\langle \overline{a}|\left[\hat{\mathbf{q}}, \hat{\mathbf{F}}^{e}(\mathbf{R}_{s'})\right]|\overline{a}\rangle \\ \mathbf{B}_{ss'}\langle \overline{a}|\left[\hat{\mathbf{q}}^{\dagger}, \hat{\mathbf{F}}^{e}(\mathbf{R}_{s'})\right]|\overline{a}\rangle \end{pmatrix} \tag{146}$$

which returns an additional contribution to the property gradient. We can now add the two contributions to the electronic Hessian and the two terms obtained for the property gradient to obtain

$$\mathbf{E} = \mathbf{E}^{(\text{stat})} + \mathbf{E}^{(\text{dyn})} \tag{147}$$

and

$$\overline{\mathbf{V}}^{\omega} = \mathbf{V}^{\omega} + \mathbf{V}^{\omega}_{\text{corr}} \tag{148}$$

Looking back at Eqs. 131 and 132, it can be seen that the local field can be divided into two contributions: The so-called *reaction field* originating from the dipole polarization in Eq. 131, induced by the contribution of the QM-region to $\tilde{\mathbf{F}}$ (the first two terms in Eq. 132) and the *effective external field*, EEF [37]. The latter arises since an externally applied electric field with frequency ω will induce an additional electric field of the same frequency in the environment. Thus, it consists of the applied external field and the field from the polarization of the environment caused by the external field itself (the contribution to the polarization in Eq. 131, from the last term in Eq. 132), and takes the form

$$\mathbf{F}^{\text{EEF},\omega}(\mathbf{R}_s) = \mathbf{F}^{\omega}(\mathbf{R}_s) + \sum_{s'=1}^{S} \mathbf{T}^{(2)}_{ss'} \boldsymbol{\mu}^{\text{ind}}_{s'}(\mathbf{F}^{\omega}) \tag{149}$$

In this picture, the two contributions $\mathbf{E}^{(\text{stat})}$ and $\mathbf{E}^{(\text{dyn})}$ to the electronic Hessian model the effects of the reaction-field contribution to the local field, i.e., the impact of the direct coupling between the quantum region and the environment. $\mathbf{E}^{(\text{stat})}$ and $\mathbf{E}^{(\text{dyn})}$ can be seen as the static and dynamic contributions to the electronic Hessian, respectively. The static term $\mathbf{E}^{(\text{stat})}$ corresponds to the situation in which the environment polarization is kept frozen in the polarization induced by the ground state of the QM-region and is not allowed to respond to the time-dependent changes in the electronic density of the QM-region induced by the applied external field. This (linear) environmental response is taken into account by the dynamic term $\mathbf{E}^{(\text{dyn})}$ and describes the polarization induced in the environment due to the periodic changes in the electronic density of the QM-region caused by the external field, yielding a dynamic contribution to the reaction field.

The remaining part of the local field (the effective external field) describes the implications of the direct environmental response to the applied external field. Thus, the property gradient $\overline{\mathbf{V}}^{\omega}$ (for simplicity only one element of the matrix is shown here)

$$\overline{\mathbf{V}}^{\omega} = \langle \overline{a} | [\hat{\mathbf{q}}, \hat{V}^{\omega}] | \overline{a} \rangle - \mathbf{B}_{ss'} \langle \overline{a} | \left[\hat{\mathbf{q}}, \hat{\mathbf{F}}^{e}(\mathbf{R}_{s'}) \right] | \overline{a} \rangle \tag{150}$$

can be seen as caused by the effective external field, where the first term in Eq. 150, from the applied external field, and the second term is caused by the environmental response. The connection to the physical picture of the induced dipoles caused by the applied time-dependent electric field interacting with the electric field of the QM-region is not directly obvious in Eq. 150, but is easily seen in Eq. 141.

Making use of the above derivations, we recover the linear response equations

$$\langle \langle \hat{V}_{\alpha}^{-\omega}; \hat{V}_{\beta}^{\omega} \rangle \rangle = -[\overline{\mathbf{V}}^{-\omega}]_{-}^{\dagger}(\mathbf{E} - \hbar\omega\mathbf{S})^{-1}\overline{\mathbf{V}}^{\omega} \tag{151}$$

but with altered property gradient $\overline{\mathbf{V}}^{\omega}$ and electronic Hessian $\mathbf{E}$.

3.7   Connection to Experimental Properties

Having derived the general expressions for the linear response functions for single-determinant wave functions, we shall now show how the response theory framework can be put use to obtain molecular properties of interest [47]. The properties obtained straightforwardly way are the response functions themselves. The respective property obtained depends on the form of the perturbation that is used. When coupling electric fields to the QM-system via the dipole operator, one obtains as the linear response function the dipole-dipole polarizability α, and the hyperpolarizabilities $\beta, \gamma, \ldots$ for the higher-order response functions. Moreover, there is no need to restrict ourselves to electric fields and the corresponding derivatives as perturbations. Applying, e.g., a magnetic field as the perturbation, provides the dipole magnetizability ξ from the linear response function and hypermagnetizabilities from higher-order response functions. By this approach, shielding tensors and coupling constants relevant for the prediction of NMR-spectra can be obtained.

In addition to these molecular properties, there is a lot more to be gained from the approach. Additional properties are "hidden" in the response functions. Hence, we now turn to a particularly useful set of molecular properties to be gained from response theory, namely molecular excitation energies. Deriving the general expression for the linear response function in an exact state framework, one arrives at

$$\langle\langle \hat{A}; \hat{V}_{\beta}^{\omega} \rangle\rangle = -\frac{1}{\hbar} \sum_{n}' \left[\frac{\langle 0|\hat{A}|n\rangle\langle n|\hat{V}_{\beta}^{\omega}|0\rangle}{\omega_{n0} - \omega} + \frac{\langle 0|\hat{V}_{\beta}^{\omega}|n\rangle\langle n|\hat{A}|0\rangle}{\omega_{n0} + \omega} \right] \tag{152}$$

where ω_{n0} are the excitation energies and ω the frequency of the applied field. It is immediately clear that the linear response function will have poles if $\omega_{n0} = \omega$, and this provides access to excitation energies.

We now proceed to look for a similar expression in the approximate state framework. Comparing Eq. 152 to Eq. 99, we see that the linear response function takes a similar form in both the exact and approximate state formulations. With that in mind, it is interesting to look at the solutions of the generalized eigenvalue equation

$$\mathbf{E}\mathbf{X}_e = \lambda_e \mathbf{S}\mathbf{X}_e \tag{153}$$

To proceed, we evaluate the matrix elements of $\mathbf{E}$ and $\mathbf{S}$ for the linear response equation obtained for a single-determinant description of a molecule in gas-phase. By doing so, one arrives at

$$\mathbf{E} = \begin{pmatrix} \mathbf{A} & \mathbf{B} \\ \mathbf{B}^* & \mathbf{A}^* \end{pmatrix}; \quad \mathbf{S} = \begin{pmatrix} 1 & 0 \\ 0 & -1 \end{pmatrix} \tag{154}$$

where, with ab denoting occupied and rs denoting unoccupied spinorbitals, for Hartree-Fock

$$A_{ar,bs} = (\epsilon_r - \epsilon_a)\delta_{ab}\delta_{rs} + \langle rb||as\rangle; \quad B_{ar,bs} = \langle rs||ab\rangle \tag{155}$$

and for (global hybrid) DFT functionals

$$A_{ar,bs} = (\epsilon_r - \epsilon_a)\delta_{ab}\delta_{rs} + \langle rb|as\rangle - c_{\text{HF}}\langle rb|sa\rangle + (1 - c_{\text{HF}})\langle rb|f_x|as\rangle + \langle rb|f_c|as\rangle$$
$$B_{ar,bs} = \langle rs|ab\rangle - c_{\text{HF}} + \langle rs|ba\rangle + (1 - c_{\text{HF}})\langle rs|f_x|ab\rangle + \langle rs|f_c|ab\rangle \tag{156}$$

with

$$\langle rb|f_x|as\rangle = \int \phi_r^*(r)\phi_b^*(r') \frac{\partial^2 E_x}{\partial\rho(r)\partial\rho(r')}\phi_a(r)\phi_s(r')\mathrm{d}^3r\,\mathrm{d}^3r'$$
$$\langle rb|f_c|as\rangle = \int \phi_r^*(r)\phi_b^*(r') \frac{\partial^2 E_c}{\partial\rho(r)\partial\rho(r')}\phi_a(r)\phi_s(r')\mathrm{d}^3r\,\mathrm{d}^3r' \tag{157}$$

where E_x and E_c are the exchange and correlation energies, respectively. Substituting in these expressions, Eq. 153, becomes the well-known TD-HF and TD-DFT equations

$$\begin{pmatrix} \mathbf{A} & \mathbf{B} \\ \mathbf{B}^* & \mathbf{A}^* \end{pmatrix} \begin{pmatrix} \mathbf{Y}_e \\ \mathbf{Z}_e \end{pmatrix} = \lambda_e \begin{pmatrix} 1 & 0 \\ 0 & -1 \end{pmatrix} \begin{pmatrix} \mathbf{Y}_e \\ \mathbf{Z}_e \end{pmatrix}, \quad e = -n, \dots, -1, 1, \dots, n \tag{158}$$

where the λ_e are the excitation energies, as we shall now show.

The eigenvalue equation can be solved by diagonalizing the matrix $\mathbf{S}^{-1}\mathbf{E}$. The solutions to Eq. 158, come in pairs, $\mathbf{X}_e = (\mathbf{Y}_e, \mathbf{Z}_e)$ corresponding to eigenvalue λ_e

and $\mathbf{X}_{-e} = (\mathbf{Z}_e^*, \mathbf{Y}_e^*)$ corresponding to eigenvalue $-\lambda_e$. Collecting these eigenvalues in a matrix $\mathbf{X}$

$$\mathbf{X} = \begin{pmatrix} \mathbf{Y}_e & \mathbf{Z}_e^* \\ \mathbf{Z}_e & \mathbf{Y}_e^* \end{pmatrix} \tag{159}$$

the matrix $\mathbf{S}^{-1}\mathbf{E}$ can be diagonalized according to

$$\mathbf{X}_e^{-1}\mathbf{S}^{-1}\mathbf{E}\mathbf{X}_e = \begin{pmatrix} \lambda & 0 \\ 0 & -\lambda \end{pmatrix} \tag{160}$$

It can be shown that matrix $\mathbf{X}$ also diagonalizes matrices $\mathbf{E}$ and $\mathbf{S}$ individually according to

$$\mathbf{X}_e^{\dagger}\mathbf{E}\mathbf{X}_e = \begin{pmatrix} \lambda & 0 \\ 0 & \lambda \end{pmatrix}, \quad \mathbf{X}_e^{\dagger}\mathbf{S}\mathbf{X}_e = \begin{pmatrix} 1 & 0 \\ 0 & -1 \end{pmatrix} \tag{161}$$

Thus, the inverse $(\mathbf{E} - \hbar\omega\mathbf{S})^{-1}$ in Eq. 99, is easily formed according to

$$(\mathbf{E} - \hbar\omega\mathbf{S})^{-1} = \mathbf{X}[\mathbf{X}^{\dagger}(\mathbf{E} - \hbar\omega\mathbf{S})\mathbf{X}]^{-1}\mathbf{X}^{\dagger} \tag{162}$$

and Eq. 99, takes the form

$$\langle\langle \hat{A}; \hat{V}_{\beta}^{\omega} \rangle\rangle = -[\mathbf{A}]_{-}^{\dagger} \sum_{e>0} \left(\frac{\mathbf{X}_e\mathbf{X}_e^{\dagger}}{\lambda_e - \hbar\omega} + \frac{\mathbf{X}_{-e}\mathbf{X}_{-e}^{\dagger}}{\lambda_e + \hbar\omega} \right) \mathbf{V}^{\omega} \tag{163}$$

which resembles the exact state linear response function (Eq. 152), and thereby leads to the interpretation of the eigenvalues in Eq. 153, as the excitation energies. Regarding the interpretation of the eigenvectors $\mathbf{X}_e$, care needs to be taken as it can, in general, not be ascribed to the excited state wave function.

The derived picture for TD-HF and TD-DFT for a molecule in gas-phase is readily extended to the PE model. Eigenvalue equation 160, is then solved for the modified electronic Hessian and property gradient, yielding the excitation energies and corresponding eigenvectors for the embedded quantum region.

Another useful technique to obtain molecular properties from response functions is residue analysis. For any function $f(z)$ of the form

$$f(z) = \frac{g(z)}{(z - a)^n} \tag{164}$$

the residue of pole a of order n is defined as

$$\text{Res}(f, a) = \frac{1}{(n - 1)!} \lim_{z \to a} \frac{d^{n-1}}{dz^{n-1}}[(z - a)^n f(z)] \tag{165}$$

which means that for poles a of order one the residue becomes equal to $g(a)$. Thus, calculating the residue of Eq. 163, as

$$\lim_{\hbar\omega\to\lambda_e}(\lambda_e - \hbar\omega)\langle\langle\hat{A}; \hat{V}_\beta^\omega\rangle\rangle = -[\mathbf{A}]_-^\dagger\mathbf{X}_e\mathbf{X}_e^\dagger\mathbf{V}^\omega \tag{166}$$

and comparing it to the residue of the exact state picture (Eq. 152),

$$\lim_{\hbar\omega\to\hbar\omega_{n0}}(\hbar\omega_{n0} - \hbar\omega)\langle\langle\hat{A}; \hat{V}_\beta^\omega\rangle\rangle = \langle 0|\hat{A}|n\rangle\langle n|\hat{V}_\beta^\omega|0\rangle \tag{167}$$

we obtain the association of $[\mathbf{A}]_-^\dagger\mathbf{X}_e$ and $\mathbf{X}_e^\dagger\mathbf{V}^\omega$ with $\langle 0|\hat{A}|n\rangle$ and $\langle n|\hat{V}_\beta^\omega|0\rangle$, respectively. In the case of an applied electric field, these expectation values become the dipole transition moments, which are related to the one-photon absorption cross-sections by

$$\begin{aligned}\sigma^{\text{OPA}} &= \frac{4\pi^2 e^2}{\hbar c}\omega_e g(\omega; \omega_e, \gamma_e)\left(\frac{1}{3}\sum_\alpha[\mathbf{V}^{-\omega_e}]_-^\dagger\mathbf{X}_{e,\alpha}\mathbf{X}_{e,\alpha}^\dagger\mathbf{V}^{\omega_e}\right)\\ &= \frac{4\pi^2 e^2}{\hbar c}\omega g(\omega; \omega_e, \gamma_e)\langle\delta^{\text{OPA}}\rangle\end{aligned} \tag{168}$$

where $g(\omega; \omega_e, \gamma_e)$ is a line-shape function parameterized by a broadening factor γ_e, and where the summation over α indicates a summation over Cartesian components.

From Eq. 160, it is also clear that neglecting $\mathbf{E}^{\text{dyn}}$ in the calculation of excitation energies and transition moments in the PE model will affect both. On the other hand, neglecting the correction to the property gradient introduced by the PE model (i.e., the effective external field) will not affect the obtained excitation energies, but will still affect the obtained oscillator strengths and multi-photon cross-sections, since the property gradient does not appear in the equations needed for the determination of the former.

Employing the concept of residue analysis on the higher-order response functions calculated within the linear approximation for a perturbing electric field enables one to obtain various properties associated either with light absorption or with the excited states themselves. The concept can also be generalized to perturbing field gradients or magnetic fields, etc. Concerning properties associated with light absorption, transition matrix elements for two and multiple photon absorption can be obtained, which are related to two and multiple photon absorption cross-sections. Two-photon absorption matrix elements are calculated from residues of the first-order hyperpolarizability (first-order nonlinear response), and assuming linearly polarized light with parallel polarization, the resulting relation to the two-photon absorption cross-section, $\langle\delta^{\text{TPA}}\rangle$, is

$$\sigma^{\text{TPA}} = \frac{8\pi^3 e^4}{\hbar^2 c^2} \omega_e^2 2\, g(\omega; \omega_e, \gamma_e) \langle \delta^{\text{TPA}} \rangle; \quad \langle \delta^{\text{TPA}} \rangle = \frac{1}{15} \sum_{\alpha\beta} (2M_{\alpha\beta}M_{\alpha\beta}^* + M_{\alpha\alpha}M_{\beta\beta}^*)$$

$$(169)$$

where $M_{\alpha,\beta}$ are the two-photon transition matrix elements. Three-photon absorption cross-sections are obtained from residues of the second-order hyperpolarizability and so on.

4   Environmental Effects on Nile Red—A Case Study Using Polarizable Embedding

In this section, we aim to discuss by an example some of the practical considerations when performing computational investigations using the PE model. However, not all important aspects can be discussed here, especially not the precise syntax of the setup of the calculations, and we, therefore, refer the reader to Ref. [68], a rather comprising tutorial review, for additional information.

The example we look at in this section is the particular case of the calculation of excitation energies and two-photon absorption cross-sections of Nile red in different environments, namely in gas-phase, water and β-lactoglobulin (BLG), a protein that constitutes a major component of milk. Nile red (see Fig. 1), is an uncharged hydrophobic dye that strongly fluoresces in the red spectral region in hydrophobic environments, and is, therefore, used in microbiology as an optical probe for the detection of lipids. It also exhibits significant solvatochromic behavior. Two-photon absorption spectroscopy shows great potential for biological applications, amongst

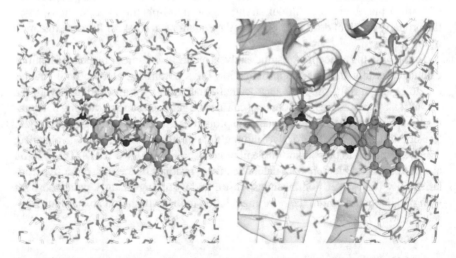

Fig. 1 The Nile red molecule in two environments, water (left), and the BLG protein (right)

other things due to the low energy of the photons used. Due to the useful practical applications of Nile red the investigation of its two-photon absorption properties are of high interest.

4.1 Computational Details

4.1.1 Molecular Dynamics

Snapshots of Nile red in water and in the BLG protein were generated using molecular dynamics. A two-step procedure was used for this purpose: First a classical molecular dynamics simulation was run in order to address configurational sampling, followed by several short QM/MM MD trajectories to obtain high-quality structures of the Nile red molecule within the specific environment. The procedure used in generating the snapshots is identical to the one described in Ref. [61], except that we have simulated several independent QM/MM MD trajectories and extracted a much larger number of snapshots for this study.

The crystal structure of BLG was obtained from the Protein Data Bank (PDB ID: 1BOO [82]) and prepared using the Protein Preparation Wizard from the Schrödinger Suite [60] in order to add missing atoms. The existing palmitate ligand was removed, and N- and C- terminals were capped. Disulfide bonds were created between cysteines 106 and 118, and cysteines 66 and 160. Protonation states were assigned using PROPKA [54, 66] according to a pH of 7.0, and the protein was finally minimized in order to remove any bad contacts. The Nile red molecule was then docked into the binding pocket of the BLG protein with the Glide program [14, 15, 21], using the XP protocol. The initial ligand–receptor complex was solvated in a rectangular box of 19684 water molecules, along with 8 Na^+ ions in order to neutralize the system. For Nile red in water, a box containing 5675 water molecules and a single Nile red molecule was created with the leap utility from the Amber program suite [8, 80]. Water molecules were modeled with the TIP3P force field [30], while the Nile red molecule was modeled by the GAFF force field [76]. The protein was described by the ff14SB force field [40]. Charges of the Nile red molecule were assigned using the RESP procedure [5], in the Antechamber program [75], based on an electrostatic potential evaluated at a HF/6-31G* level of theory, which was computed with the Gaussian program [16].

The assembled systems were then equilibrated in a multi-stage procedure with the Gromacs program [1], version 2016.4. Following 2000 steps of the steepest descent minimizer, the systems were heated to a temperature of 303 K under constant volume for 50 ps using a time-step of 2 fs. Next, NPT production runs were carried out for each of the systems. Long-range electrostatics were treated using the Particle Mesh Ewald method [11], (with a short-range cutoff of 12 Å). Bonds involving hydrogens were restrained using the LINCS procedure [25]. The temperature was controlled toward 303 K using the Nose-Hoover thermostat [26, 48], while the pressure was controlled toward 1.0 atm using the Parrinello-Rahman barostat [55]. For Nile red in water, a 300 ns production run was used, whereas a longer 800 ns production was used for Nile red in BLG. In both cases, the first 50 ns were discarded as equilibration.

Starting from equally spaced points (sampled every 10 ns) of the classical MD trajectory of Nile red in water, 25 independent QM/MM MD runs were carried out using the NAMD [56]/Orca [45] QM/MM interface [42]. The QM-region consisted of the Nile red molecule, which was treated at a M06L [85]/def2-SVP [79] level of theory. The calculations were accelerated using the RI approximation with the def2/J auxillary fitting set [78]. The QM/MM interaction was calculated according to CHELPG charges [7], which were updated at every time-step. For each trajectory, the system was first minimized for 100 steps, followed by 3000 steps of dynamics (total of 1.5 ps). Ten snapshots were extracted from each trajectory, for a total of 250 snapshots. Likewise, 75 independent QM/MM MD trajectories of Nile red in BLG were started from equally spaced points of the classical MD. For each trajectory, the system was first minimized for 100 steps, followed by 3000 steps of dynamics. Two snapshots were extracted from each trajectory, for a total of 150 snapshots.

4.1.2 Embedding Parameters

For each of the extracted snapshots, embedding parameters were calculated using the PyFraME program [51]. For Nile red in water, a shell of the nearest 3000 water molecules was included in the embedding potential. Distributed multipole moments (up to and including quadrupoles) and anisotropic dipole–dipole polarizabilities were derived using the LoProp approach [17, 74], based on CAM-B3LYP [83]/Loprop-6-31+G* calculations. For Nile red in BLG, the protein, ions and 10,000 water molecules were included in the embedding potential. Protein residues and water within 7 Å of the Nile red molecule were described with LoProp-derived parameters. The remaining water molecules and ions were described by the isotropic SEP potential [6], while the remaining protein residues were described using the CP^3 potential [58]. The parameters in both of these potentials consist of charges and isotropic polarizabilities, and the entire environment is thus polarizable.

4.1.3 Property Calculations

Excitation energies and associated one- and two-photon strengths of Nile red were computed using the Dalton program [3], at the CAM-B3LYP [83]/6-31+G* level of theory. For the excitation energies and oscillator strengths, the four lowest excited states were included. For the two-photon cross-section, only the lowest excited state was considered. Interactions between induced dipoles were damped in order to avoid issues with over-polarization, and the EEF model [37], was employed in order to treat local field effects. All spectra were generated assuming Lorentzian broadening function with a FWHM of 0.1 eV. In addition to the embedding calculations, the properties were also calculated in the gas-phase. For consistency, a structure of Nile red optimized at a M06L [85]/def2-SVP [79], level of theory was used for these calculations, while the actual property calculations were conducted using CAM-B3LYP [83]/6-31+G*.

4.2 General Remarks

Before turning to the discussion of the example calculations, some general aspects regarding calculations employing the PE model should be mentioned [68]. The PE model is not restricted to any specific kind of environment but may be used on an equal footing for environments of rather different types, such as solvents, proteins, DNA, and lipids. Of course, the practical setup varies depending on the environment composition, in particular with regards to fragmentation. The fragmentation of a solvent environment is straightforward, while the fragmentation of a protein is more involved as it is necessary to cut covalent bonds in the definition of the fragments. Fortunately, the required procedures have been automated [51]. A similar issue arises from the partitioning of the supersystem into the QM-region and the environment. This division is again straightforward if the QM-region is not covalently linked to the environment, which is the case for solute–solvent systems. However, for covalently bonded systems, the quantum-classical interface becomes a delicate matter and needs careful consideration due to the risk of, e.g., over-polarization or electron spill-out [57, 63]. For these reasons, several interfacing schemes have been developed, usually relying on a simple hydrogen link-atom approach. We will not discuss this issue further here, but instead, refer the reader to the discussion in the recent tutorial review in Ref. [68].

Concerning available programs, the PE model is implemented in the PE library, which has been interfaced with the Dalton and Dirac programs. The implementation in the PE library currently allows up to fifth-order multipoles, which allows for highly accurate reproduction of the electrostatic potential of the environmental fragments. Independently on the PE library, the PE model has also been implemented in other electronic-structure programs, such as PE-RI-CC2 in TURBOMOLE [27, 62], and in Psi4 [73], PySCF [70], and QChem [34] via the CPPE library [61]. Finally, the possibility to combine the PE model with continuum solvation models is notable [46]. In this approach, the environment is split into inner (treated with the PE) and outer shells (treated with a continuum model). This can be useful, e.g., when modeling an optical probe embedded in a solvated DNA environment, where the DNA molecule would be reasonably treated with the PE model and the solvation effects can be included with the continuum model. The big advantage of this approach is that the need for the sampling of the configurations in the outer shell (e.g., the solvent) is removed, leading to significant reductions in the computational cost.

For a fully polarizable environment, directly solving the induction problem in Eq. 34, (i.e., solving a system of linear equations) formally scales cubically with the number of polarizable sites. By using iterative solution methods instead, the computational cost in standard PE implementations is quadratic with respect to the number of environment sites, since each of the environment sites accumulate fields from all remaining environment sites (i.e., each of the N environment sites require computation of the fields from the remaining $N - 1$ sites). Practically, this limits the sizes of systems which can reasonably be studied at around 50,000 environment sites. For nonpolarizable embedding calculations of *response* properties, linear scaling is, in principle, easily achieved, since interaction energy between permanent multipoles (the evaluation of which is quadratically scaling) can be ignored, which leaves only a linear (and non-iterative) amount of operator contributions. To treat larger systems in a fully polarizable, reduced-scaling computational schemes are required, such as the Fast Multipole Method implementation reported in Ref. [35]. One of the gains in using a polarizable formulation of the environment is the possibility of employing smaller quantum regions than in standard QM/MM, while still getting accurate results [44].

We finally remark on the procedure used to generate the structures used in the PE calculations, i.e., the sampling protocol. As shown in Ref. [31], even for small and rather simple molecules, using standard force fields to generate structures to be used in subsequent embedding calculations may lead to inaccurate, or in some cases, even qualitative wrong, results. This issue can be solved by using specially made force fields for the molecules in question, or by using classical molecular dynamics followed by a number of short QM/MM based molecular simulations, which is the strategy we follow in this presentation.

4.3 One-Photon Spectra

The computed UV/vis spectra of Nile red in gas-phase, water, or protein environments is shown in Fig. 2. The gas-phase spectrum shows a single bright peak at 2.95 eV, with a small secondary peak at 3.8 eV. When placed in an environment, the bright peak remains, but with a diminished maximal intensity, and the higher-energy peak turns into a broad shoulder toward higher energies.

When placed in an environment (water or the BLG protein), the peak position is red-shifted by about 0.21 eV compared to the gas-phase. The peak position is almost identical between the water and protein environments. Likewise, when external field effects are not taken into account, the intensities are also almost identical. Only when including EEF effects, differences between the two environments becomes appreciable. In both cases, the inclusion leads to reductions in the intensity compared to using a plain PE model, but with the enhanced screening occurring in the protein environment. Thus, this example calculation underlines the importance of including local field effects (EEF) when modeling absorption spectra in heterogeneous environments.

Fig. 2 One-photon spectrum of Nile red in gas-phase (black line), water (blue lines) or BLG (red lines) environments. Solid lines indicate that the EEF model was included. The spectra were normalized relative to the maximum intensity of the gas-phase spectrum

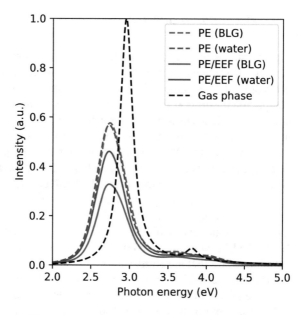

Fig. 3 Two-photon spectrum of Nile red in gas-phase (black line), water (blue lines) or BLG (red lines) environments. Solid lines indicate that the EEF model was included

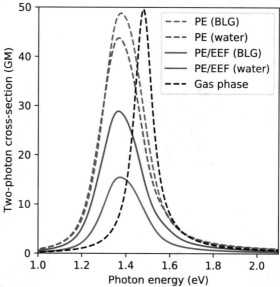

4.4 Two-Photon Spectra

Figure 3 shows computed two-photon absorption spectra of the Nile red molecule in various environments. For both the water and BLG environments, the peak position is red-shifted (by about 0.1 eV) compared to the vacuum result in agreement with the one-photon absorption spectra. The inclusion of sampling in the water and protein environments leads to noticeably broader peaks in the condensed phase, even though the same broadening function (and line-width) was employed for the individual transitions as in vacuum. When in an environment, the peak positions are almost identical when comparing water (1.365 eV) to BLG (1.37 eV). The peak position does not change depending on whether or not EEF effects are included since the excitation energies are unchanged by the inclusion thereof. Thus, the largest differences are found in the shape and intensity of the spectra. When including EEF effects, the intensity is significantly higher when in water (28.9 GM) than in the protein (15.7 GM). The EEF intensities are both lowered compared to the plain PE results, which have intensities that are similar to the vacuum result (43.8 and 48.8 GM for water and BLG, respectively). Importantly, the difference between the intensities in the two environments is much smaller, and further, the intensities are even in the opposite order. Thus, as for the case of one-photon absorption, including local field effects

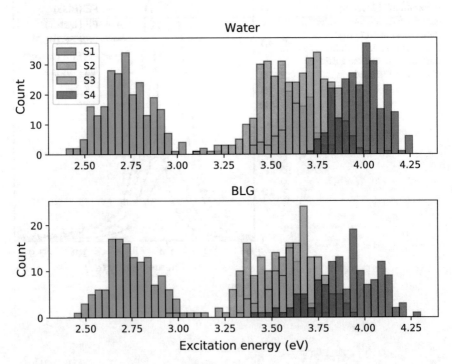

Fig. 4 Distribution of excitation energies for the four lowest excited states of Nile red in water (top) and the BLG protein (bottom)

is of significant importance when modeling two-photon absorption spectra in heterogeneous environments, and neglect hereof may even lead to qualitatively wrong predictions.

4.5 Importance of Sampling

When using discrete approaches to represent the environment, sampling of environmental configurations has to be included explicitly. Figure 4 shows distributions of the four lowest excitation energies of Nile red in water or protein environments. Clearly, including proper statistical sampling is important. For example, the lowest excitation energy ranges, depending on the snapshot, between 2.4 and 3.2 eV. Selecting any single snapshot can, therefore, lead to wrong conclusions. From Fig. 4, it is seen that the distribution of the excitation energies span the same energy range independent of the environment, i.e., rather similar distributions are obtained in water or in the protein. This picture changes when considering similar distributions for the two-photon absorption, i.e., Fig. 5. Here, a much broader distribution is obtained for the case of the water environment indicating a significant influence of the environment on this

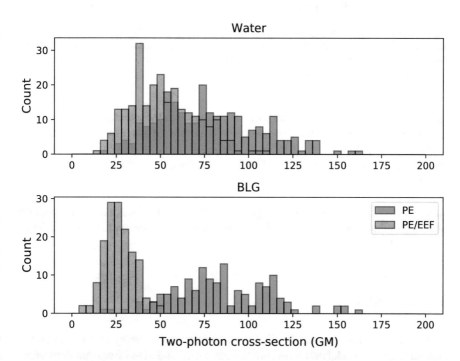

Fig. 5 Distribution of the two-photon cross-section for the lowest excited state of Nile red in water (top) and the BLG protein (bottom)

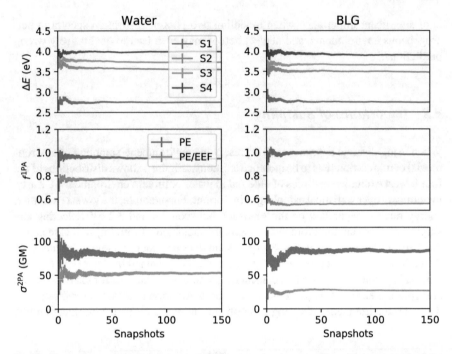

Fig. 6 Convergence of the four lowest excitation energies (top panel), oscillator strength of the lowest excited state (middle panel), and two-photon cross-section of the lowest excited state (bottom) panel of Nile red in water (left panels) or the BLG protein (right panels). Error bars were estimated from the standard error of the mean

specific nonlinear property. Finally, the convergence in the average excitation energies, oscillator strengths, and two-photon cross-sections with respect to the number of included snapshots is shown in Fig. 6. Among the three properties, the two-photon cross-section is clearly the most difficult to converge, since this property is highly sensitive to local structural changes in the environment. In conclusion, we find that especially nonlinear properties—like the two-photon absorption cross-section—are very sensitive to both structural sampling and local field effects, and that modeling of such nonlinear properties, therefore, puts special demands on the computational procedure.

5   Conclusion

In this contribution, we have detailed the background and theory of the polarizable embedding (PE) model and have shown how it can be derived from standard intermolecular perturbation theory. We have reviewed the basic principles of approximate state linear response theory and have shown its connection to experimental properties and time-dependent Hartree-Fock and density functional theory. We have, furthermore, demonstrated how the PE model can be formulated within a response theory framework and which contributions to the response functions arise due to the inclusion of environmental effects. Finally, we have showcased a few applications of the PE model to realistic chemical systems and underlined the need for a proper account of both structural dynamics, as well as consideration of local field effects.

Acknowledgements Computations/simulations for the work described herein were supported by the DeIC National HPC Centre, SDU.

References

1. Abraham MJ, Murtola T, Schulz R, Páll S, Smith JC, Hess B, Lindahl E (2015) Gromacs: high performance molecular simulations through multi-level parallelism from laptops to supercomputers. SoftwareX 1:19–25
2. Ahlström P, Wallqvist A, Engström S, Jönsson B (1989) A molecular dynamics study of polarizable water. Mol Phys 68(3):563–581
3. Aidas K, Angeli C, Bak KL, Bakken V, Bast R, Boman L, Christiansen O, Cimiraglia R, Coriani S, Dahle P, Dalskov EK, Ekström U, Enevoldsen T, Eriksen JJ, Ettenhuber P, Fernández B, Ferrighi L, Fliegl H, Frediani L, Hald K, Halkier A, Hättig C, Heiberg H, Helgaker T, Hennum AC, Hettema H, Hjertenæs E, Høst S, Høyvik IM, Iozzi MF, Jansík B, Jensen HJAa, Jonsson D, Jørgensen P, Kauczor J, Kirpekar S, Kjærgaard T, Klopper W, Knecht S, Kobayashi R, Koch H, Kongsted J, Krapp A, Kristensen K, Ligabue A, Lutnæs OB, Melo JI, Mikkelsen KV, Myhre RH, Neiss C, Nielsen CB, Norman P, Olsen J, Olsen JMH, Osted A, Packer MJ, Pawlowski F, Pedersen TB, Provasi PF, Reine S, Rinkevicius Z, Ruden TA, Ruud K, Rybkin VV, Sałek P, Samson CCM, de Merás AS, Saue T, Sauer SPA, Schimmelpfennig B, Sneskov K, Steindal AH, Sylvester-Hvid KO, Taylor PR, Teale AM, Tellgren EI, Tew DP, Thorvaldsen AJ, Thøgersen L, Vahtras O, Watson MA, Wilson DJD, Ziolkowski M, Ågren H (2014) The Dalton quantum chemistry program system. WIREs Comput Mol Sci 4(3):269–284
4. Barone V, Cossi M (1998) Quantum calculation of molecular energies and energy gradients in solution by a conductor solvent model. J Phys Chem A 102(11):1995–2001
5. Bayly CI, Cieplak P, Cornell W, Kollman PA (1993) A well-behaved electrostatic potential based method using charge restraints for deriving atomic charges: the RESP model. J Phys Chem 97(40):10269–10280
6. Beerepoot MTP, Steindal AH, List NH, Kongsted J, Olsen JMH (2016) Averaged solvent embedding potential parameters for multiscale modeling of molecular properties. J Chem Theory Comput 12(4):1684–1695
7. Breneman CM, Wiberg KB (1990) Determining atom-centered monopoles from molecular electrostatic potentials. The need for high sampling density in formamide conformational analysis. J Comput Chem 11(3):361–373
8. Case DA, Ben-Shalom IY, Brozell SR, Cerutti DS, Cheatham TE III, Cruzeiro VWD, Darden TA, Duke RE, Ghoreishi D, Gilson MK, Gohlke H, Goetz AW, Greene D, Harris R, Homeyer N, Izadi S, Kovalenko A, Kurtzman T, Lee TS, LeGrand S, Li P, Lin C, Liu J, Luchko T, Luo R, Mermelstein DJ, Merz KM, Miao Y, Monard G, Nguyen C, Nguyen H, Omelyan I, Onufriev A, Pan F, Qi R, Roe DR, Roitberg A, Sagui C, Schott-Verdugo S, Shen J, Simmerling CL, Smith J, Salomon-Ferrer R, Swails J, Walker RC, Wang J, Wei H, Wolf RM, Wu X, Xiao L, York DM, Kollman PA (2018) Amber 2018. University of California, San Francisco
9. Christiansen O, Poul J, Hättig C (1998) Response functions from Fourier component variational perturbation theory applied to a time-averaged quasienergy. Int J Quantum Chem 68(1):1–52

10. Curutchet C, Muñoz-Losa A, Monti S, Kongsted J, Scholes GD, Mennucci B (2009) Electronic energy transfer in condensed phase studied by a polarizable QM/MM model. J Chem Theory Comput 5(7):1838–1848
11. Darden T, York D, Pedersen L (1993) Particle mesh Ewald: an $N \log(N)$ method for Ewald sums in large systems. J Chem Phys 98(12):10089–10092
12. Eriksen JJ, Sauer S, Mikkelsen KV, Jensen HJA, Kongsted J (2012) On the importance of excited state dynamic response electron correlation in polarizable embedding methods. J Comput Chem 33(25):2012–2022
13. Field MJ, Bash PA, Karplus M (1990) A combined quantum mechanical and molecular mechanical potential for molecular dynamics simulations. J Comput Chem 11(6):700–733
14. Friesner RA, Banks JL, Murphy RB, Halgren TA, Klicic JJ, Mainz DT, Repasky MP, Knoll EH, Shelley M, Perry JK, et al (2004) Glide: a new approach for rapid, accurate docking and scoring. 1. Method and assessment of docking accuracy. J Med Chem 47(7):1739–1749
15. Friesner RA, Murphy RB, Repasky MP, Frye LL, Greenwood JR, Halgren TA, Sanschagrin PC, Mainz DT (2006) Extra precision glide: docking and scoring incorporating a model of hydrophobic enclosure for protein-ligand complexes. J Med Chem 49(21):6177–6196
16. Frisch MJ, Trucks GW, Schlegel HB, Scuseria GE, Robb MA, Cheeseman JR, Scalmani G, Barone V, Mennucci B, Petersson GA, Nakatsuji H, Caricato M, Li X, Hratchian HP, Izmaylov AF, Bloino J, Zheng G, Sonnenberg JL, Hada M, Ehara M, Toyota K, Fukuda R, Hasegawa J, Ishida M, Nakajima T, Honda Y, Kitao O, Nakai H, Vreven T, Montgomery JA Jr, Peralta JE, Ogliaro F, Bearpark M, Heyd JJ, Brothers E, Kudin KN, Staroverov VN, Kobayashi R, Normand J, Raghavachari K, Rendell A, Burant JC, Iyengar SS, Tomasi J, Cossi M, Rega N, Millam JM, Klene M, Knox JE, Cross JB, Bakken V, Adamo C, Jaramillo J, Gomperts R, Stratmann RE, Yazyev O, Austin AJ, Cammi R, Pomelli C, Ochterski JW, Martin RL, Morokuma K, Zakrzewski VG, Voth GA, Salvador P, Dannenberg JJ, Dapprich S, Daniels AD, Farkas Ö, Foresman JB, Ortiz JV, Cioslowski J, Fox DJ (2009) Gaussian 09 revision d.01. Gaussian Inc. Wallingford CT 2009
17. Gagliardi L, Lindh R, Karlström G (2004) Local properties of quantum chemical systems: the LoProp approach. J Chem Phys 121(10):4494–4500
18. Goez A, Neugebauer J (2017) Embedding methods in quantum chemistry. In: Frontiers of quantum chemistry. Springer, Singapore, pp 139–179
19. Gomes ASP, Jacob CR (2012) Quantum-chemical embedding methods for treating local electronic excitations in complex chemical systems. Annu Rep Prog Chem Sect C: Phys Chem 108:222
20. Gordon MS, Fedorov DG, Pruitt SR, Slipchenko LV (2011) Fragmentation methods: a route to accurate calculations on large systems. Chem Rev 112(1):632–672
21. Halgren TA, Murphy RB, Friesner RA, Beard HS, Frye LL, Pollard WT, Banks JL (2004) Glide: a new approach for rapid, accurate docking and scoring. 2. Enrichment factors in database screening. J Med Chem 47(7):1750–1759
22. Hedegård ED, List NH, Jensen HJA, Kongsted J (2013) The multi-configuration self-consistent field method within a polarizable embedded framework. J Chem Phys 139(4):044101
23. Hedegård ED, Olsen JMH, Knecht S, Kongsted J, Jensen HJA (2015) Polarizable embedding with a multiconfiguration short-range density functional theory linear response method. J Chem Phys 142(11):114113
24. Hedegård ED, Bast R, Kongsted J, Olsen JMH, Jensen HJA (2017) Relativistic polarizable embedding. J Chem Theory Comput 13(6):2870–2880
25. Hess B, Bekker H, Berendsen HJC, Fraaije JGEM (1997) LINCS: a linear constraint solver for molecular simulations. J Comput Chem 18(12):1463–1472
26. Hoover WG (1985) Canonical dynamics: equilibrium phase-space distributions. Phys Rev A 31(3):1695
27. Hršak D, Marefat Khah A, Christiansen O, Hättig C (2015) Polarizable embedded RI-CC2 method for two-photon absorption calculations. J Chem Theory Comput 11(8):3669–3678
28. Jacob CR, Neugebauer J (2014) Subsystem density-functional theory. WIREs Comput Mol Sci 4(4):325–362

29. Jensen L, Van Duijnen PT, Snijders JG (2003) A discrete solvent reaction field model within density functional theory. J Chem Phys 118(2):514–521
30. Jorgensen WL, Chandrasekhar J, Madura JD, Impey RW, Klein ML (1983) Comparison of simple potential functions for simulating liquid water. J Chem Phys 79(2):926–935
31. Kjellgren ER, Olsen JMH, Kongsted J (2018) Importance of accurate structures for quantum chemistry embedding methods: which strategy is better? J Chem Theory Comput 14(8):4309–4319
32. Klamt A (1995) Conductor-like screening model for real solvents: a new approach to the quantitative calculation of solvation phenomena. J Phys Chem 99(7):2224–2235
33. Klamt A, Schüürmann G (1993) COSMO: a new approach to dielectric screening in solvents with explicit expressions for the screening energy and its gradient. J Chem Soc Perkin Trans 2(5):799–805
34. Krylov AI, Gill PMW (2013) Q-chem: an engine for innovation. WIREs Comput Mol Sci 3(3):317–326
35. Lipparini F (2019) General linear scaling implementation of polarizable embedding schemes. J Chem Theory Comput 15(8):4312–4317
36. Lipparini F, Cappelli C, Barone V (2012) Linear response theory and electronic transition energies for a fully polarizable QM/classical Hamiltonian. J Chem Theory Comput 8(11):4153–4165
37. List NH, Jensen HJA, Kongsted J (2016a) Local electric fields and molecular properties in heterogeneous environments through polarizable embedding. Phys Chem Chem Phys 18(15):10070–10080
38. List NH, Olsen JMH, Kongsted J (2016b) Excited states in large molecular systems through polarizable embedding. Phys Chem Chem Phys 18(30):20234–20250
39. List NH, Norman P, Kongsted J, Jensen HJA (2017) A quantum-mechanical perspective on linear response theory within polarizable embedding. J Chem Phys 146(23):234101
40. Maier JA, Martinez C, Kasavajhala K, Wickstrom L, Hauser KE, Simmerling C (2015) ff14SB: improving the accuracy of protein side chain and backbone parameters from ff99SB. J Chem Theory Comput 11(8):3696–3713
41. Marini A, Munõz-Losa A, Biancardi A, Mennucci B (2010) What is solvatochromism? J Phys Chem B 114(51):17128–17135
42. Melo MCR, Bernardi RC, Rudack T, Scheurer M, Riplinger C, Phillips JC, Maia JDC, Rocha GB, Ribeiro JV, Stone JE, Neese F, Schulten K, Luthey-Schulten Z (2018) Namd goes quantum: an integrative suite for hybrid simulations. Nat Methods 15(5):351
43. Murugan NA, Kongsted J, Rinkevicius Z, Ågren H (2012) Color modeling of protein optical probes. Phys Chem Chem Phys 14(3):1107–1112
44. Nåbo LJ, Olsen JMH, Martínez TJ, Kongsted J (2017) The quality of the embedding potential is decisive for minimal quantum region size in embedding calculations: the case of the green fluorescent protein. J Chem Theory Comput 13(12):6230–6236
45. Neese F (2018) Software update: the ORCA program system, version 4.0. WIREs Comput Mol Sci 8(1):e1327
46. Nørby MS, Steinmann C, Olsen JMH, Li H, Kongsted J (2016) Computational approach for studying optical properties of DNA systems in solution. J Chem Theory Comput 12(10):5050–5057
47. Norman P, Ruud K, Saue T (2018) Principles and practices of molecular properties: theory, modeling, and simulations. Wiley
48. Nosé S (1984) A unified formulation of the constant temperature molecular dynamics methods. J Chem Phys 81(1):511–519
49. Olsen J, Jørgensen P (1985) Linear and nonlinear response functions for an exact state and for an MCSCF state. J Chem Phys 82(7):3235–3264
50. Olsen JMH (2013) Development of quantum chemical methods towards rationalization and optimal design of photoactive proteins. PhD thesis, University of Southern Denmark. https://doi.org/10.6084/m9.figshare.156852

51. Olsen JMH (2018) PyFraME: python tools for fragment-based multiscale embedding. https://doi.org/10.5281/zenodo.1443314
52. Olsen JMH, Kongsted J (2011) Molecular properties through polarizable embedding. In: Advances in quantum chemistry, vol 61. Elsevier, pp 107–143
53. Olsen JMH, Aidas K, Kongsted J (2010) Excited states in solution through polarizable embedding. J Chem Theory Comput 6(12):3721–3734
54. Olsson MHM, Søndergaard CR, Rostkowski M, Jensen JH (2011) PROPKA3: consistent treatment of internal and surface residues in empirical pKa predictions. J Chem Theory Comput 7(2):525–537
55. Parrinello M, Rahman A (1981) Polymorphic transitions in single crystals: a new molecular dynamics method. J Appl Phys 52(12):7182–7190
56. Phillips JC, Braun R, Wang W, Gumbart J, Tajkhorshid E, Villa E, Chipot C, Skeel RD, Kale L, Schulten K (2005) Scalable molecular dynamics with NAMD. J Comput Chem 26(16):1781–1802
57. Reinholdt P, Kongsted J, Olsen JMH (2017) Polarizable density embedding: a solution to the electron spill-out problem in multiscale modeling. J Phys Chem Lett 8(23):5949–5958
58. Reinholdt P, Kjellgren ER, Steinmann C, Olsen JMH (2019) Cost-effective potential for accurate polarizable embedding calculations in protein environments. J Chem Theory Comput. https://doi.org/10.1021/acs.jctc.9b00616
59. Sałek P, Vahtras O, Helgaker T, Ågren H (2002) Density-functional theory of linear and nonlinear time-dependent molecular properties. J Chem Phys 117(21):9630–9645
60. Sastry GM, Adzhigirey M, Day T, Annabhimoju R, Sherman W (2013) Protein and ligand preparation: parameters, protocols, and influence on virtual screening enrichments. J Comput-Aided Mol Des 27(3):221–234
61. Scheurer M, Reinholdt P, Kjellgren ER, Olsen JMH, Dreuw A, Kongsted J (2019) CPPE: an open-source C++ and python library for polarizable embedding. J Chem Theory Comput 15(11):6154–6163
62. Schwabe T, Sneskov K, Olsen JMH, Kongsted J, Christiansen O, Hättig C (2012) PERI-CC2: a polarizable embedded RI-CC2 method. J Chem Theory Comput 8(9):3274–3283
63. Senn HM, Thiel W (2009) QM/MM methods for biomolecular systems. Angew Chem Int Ed 48(7):1198–1229
64. Singh UC, Kollman PA (1986) A combined ab initio quantum mechanical and molecular mechanical method for carrying out simulations on complex molecular systems: applications to the CH_3Cl^+ Cl^- exchange reaction and gas phase protonation of polyethers. J Comput Chem 7(6):718–730
65. Sneskov K, Schwabe T, Kongsted J, Christiansen O (2011) The polarizable embedding coupled cluster method. J Chem Phys 134(10):104108
66. Søndergaard CR, Olsson MHM, Rostkowski M, Jensen JH (2011) Improved treatment of ligands and coupling effects in empirical calculation and rationalization of pKa values. J Chem Theory Comput 7(7):2284–2295
67. Stefanovich EV, Truong TN (1995) Optimized atomic radii for quantum dielectric continuum solvation models. Chem Phys Lett 244(1):65–74
68. Steinmann C, Reinholdt P, Nørby MS, Kongsted J, Olsen JMH (2018) Response properties of embedded molecules through the polarizable embedding model. Int J Quantum Chem 119(1):e25717
69. Stone A (2013) The theory of intermolecular forces. OUP Oxford
70. Sun Q, Berkelbach TC, Blunt NS, Booth GH, Guo S, Li Z, Liu J, McClain JD, Sayfutyarova ER, Sharma S et al (2018) PySCF: the python-based simulations of chemistry framework. WIREs Comput Mol Sci 8(1):e1340
71. Tomasi J, Mennucci B, Cammi R (2005) Quantum mechanical continuum solvation models. Chem Rev 105(8):2999–3094
72. Truong TN, Stefanovich EV (1995) A new method for incorporating solvent effect into the classical, ab initio molecular orbital and density functional theory frameworks for arbitrary shape cavity. Chem Phys Lett 240(4):253–260

73. Turney JM, Simmonett AC, Parrish RM, Hohenstein EG, Evangelista FA, Fermann JT, Mintz BJ, Burns LA, Wilke JJ, Abrams ML et al (2012) Psi4: an open-source ab initio electronic structure program. WIREs Comput Mol Sci 2(4):556–565
74. Vahtras O (2014) LoProp for Dalton. https://doi.org/10.5281/zenodo.13276
75. Wang J, Wang W, Kollman PA, Case DA (2001) Antechamber: an accessory software package for molecular mechanical calculations. J Am Chem Soc 222:U403
76. Wang J, Wolf RM, Caldwell JW, Kollman PA, Case DA (2004) Development and testing of a general amber force field. J Comput Chem 25(9):1157–1174
77. Warshel A, Levitt M (1976) Theoretical studies of enzymatic reactions: dielectric, electrostatic and steric stabilization of the carbonium ion in the reaction of lysozyme. J Mol Biol 103:227–249
78. Weigend F (2006) Accurate coulomb-fitting basis sets for H to Rn. Phys Chem Chem Phys 8(9):1057–1065
79. Weigend F, Ahlrichs R (2005) Balanced basis sets of split valence, triple zeta valence and quadruple zeta valence quality for H to Rn: design and assessment of accuracy. Phys Chem Chem Phys 7(18):3297–3305
80. Weiner PK, Kollman PA (1981) Amber: assisted model building with energy refinement. A general program for modeling molecules and their interactions. J Comput Chem 2(3):287–303
81. Wesolowski TA, Shedge S, Zhou X (2015) Frozen-density embedding strategy for multilevel simulations of electronic structure. Chem Rev 115(12):5891–5928
82. Wu SY, Pérez MD, Puyol P, Sawyer L (1999) β-lactoglobulin binds palmitate within its central cavity. J Biol Chem 274(1):170–174
83. Yanai T, Tew DP, Handy NC (2004) A new hybrid exchange-correlation functional using the coulomb-attenuating method (CAM-B3LYP). Chem Phys Lett 393(1–3):51–57
84. Yoo S, Zahariev F, Sok S, Gordon MS (2008) Solvent effects on optical properties of molecules: a combined time-dependent density functional theory/effective fragment potential approach. J Chem Phys 129(14):144112
85. Zhao Y, Truhlar DG (2006) A new local density functional for main-group thermochemistry, transition metal bonding, thermochemical kinetics, and noncovalent interactions. J Chem Phys 125(19):194101

Computational Studies of Photochemistry in Phytochrome Proteins

Jonathan R. Church, Aditya G. Rao, Avishai Barnoy, Christian Wiebeler, and Igor Schapiro

Abstract In this book chapter, the computational research which has been performed when studying the photochemistry of phytochrome proteins is reviewed. Phytochromes represent a large family of ubiquitous photoreceptor proteins that are considered for various biotechnological applications. A comprehensive understanding of their photochemical properties is essential for their utilization. A brief introduction to phytochromes is given followed by an array of computational studies that cover different aspects of their photochemical properties. This includes the choice of different quantum chemical methods which have been used to calculate the excitation energies, as well as how to accurately model the protein environmental itself and its interaction with the chromophore. Several benchmarking studies with various quantum chemical methods used in excitation energy calculations and their comparisons are summarized. Further benchmarks include the determination of the correct structure of the chromophores found in the protein. Another issue which has been encountered in these studies is accounting for the high dimensionality of the protein environment in calculation of excitation energies. Several procedures used to alleviate this problem, including conformational sampling and the optimization procedures, are discussed. Eventually, new insights from computer simulations such as the resolution of the chromophore conformation or its protonation state are highlighted, augmented by the molecular mechanism for spectral tuning.

Keywords Phytochromes · CBCRs · QM/MM · Conformational sampling · Excitation energies

1 Introduction

Phytochromes (Phy) constitute a superfamily of photoreceptor proteins found across a wide range of life domains such as plants, algae, and cyanobacteria. These proteins

J. R. Church · A. G. Rao · A. Barnoy · C. Wiebeler · I. Schapiro (✉)
Fritz Haber Center for Molecular Dynamics Research Institute of Chemistry, The Hebrew University of Jerusalem, 9190401 Jerusalem, Israel
e-mail: Igor.Schapiro@mail.huji.ac.il

© Springer Nature Switzerland AG 2021
T. Andruniów and M. Olivucci (eds.), *QM/MM Studies of Light-responsive Biological Systems*, Challenges and Advances in Computational Chemistry and Physics 31, https://doi.org/10.1007/978-3-030-57721-6_4

play an important role in biological systems such as circadian rhythm regulation and chromatic acclimation in cyanobacteria [1–5]. They are regarded as promising candidates for in vivo imaging because of their ability to absorb and emit light that falls in the near-infrared region while also being non-toxic to the biological environment [6–9]. Although the exact protein architecture can vary among phytochromes, a key characteristic is a conserved GAF domain which contains a covalently bound linear bilin chromophore. The bilin chromophore inside the GAF domain is linked to the protein environment through a conserved cysteine residue. Bilin chromophores, which are derivatives of heme, are linear tetrapyrroles with two propionate functional groups attached to the B and C pyrrole rings. The chromophore molecule bound in the GAF domain can range from either biliverdin (BV), phytochromobilin (PΦB), phycocyanobilin (PCB) to phycoviolobilin (PVB) (Fig. 1) [6].

Each of these chromophores is highly conjugated making them adapt at light absorption in the visible range. Under biological conditions BV can be produced from heme by heme oxygenase (HO) [10]. In bacterial systems, including cyanobacteria, both PΦB and PCB can be produced from BV through reduction by ferredoxin-dependent bilin reductase (FDBR) [6, 10–13].

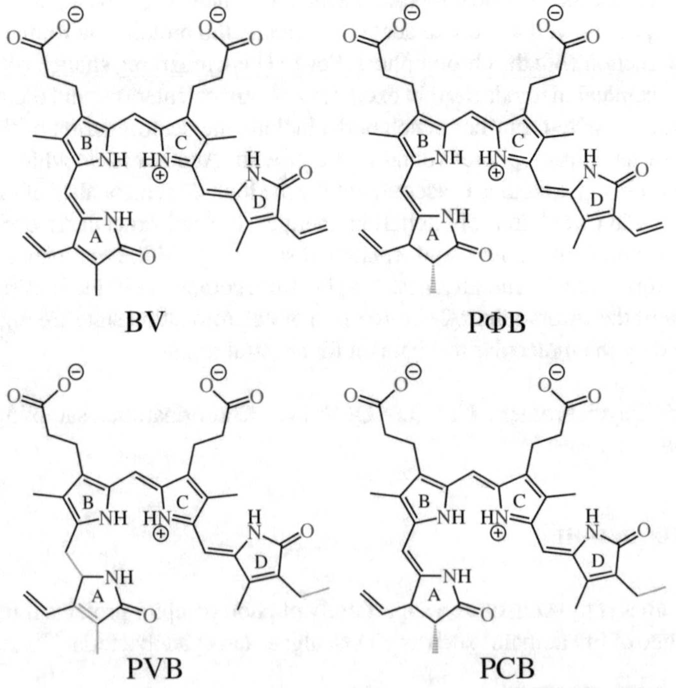

Fig. 1 Bilin chromophores found in biological systems: biliverdin (BV), phytochromobilin (PΦB), phycocyanobilin (PCB), and phycoviolobilin (PVB). The major differences between the chromophores are highlighted in orange

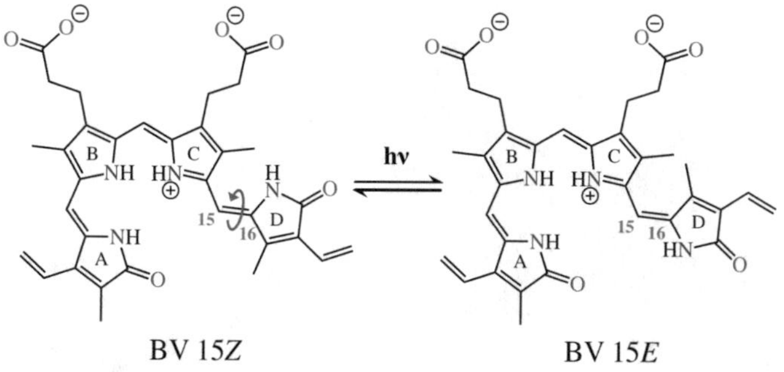

Fig. 2 Photoisomerization between the 15Z to 15E conformers of biliverdin. The isomerization of the $C_{15} = C_{16}$ bond leads to a flip of the D-ring

Upon light absorption the bilin molecules will undergo a double bond isomerization [14–19]. The isomerizing double bond $C_{15} = C_{16}$ is located in the methine bridge between the rings C and D. Therefore, the bilins can be switched between the 15Z and 15E configurations. These two configurations are associated with a change in color which in the most common case is an interconversion between a red absorbing state (P_r) and a far-red absorbing state (P_{fr}) (Fig. 2).

The photosensory region in phytochromes may contain PAS and PHY domains in addition to the GAF domain. For example, the photosensory region of canonical phytochromes is comprised of a three-domain architecture, which includes PAS-GAF-PHY domains. PAS domains are sensor modules thought to play a role in sensing light, small molecules, and oxygen [20]. The PAS domain is also the smallest domain of the sensory portion of phytochromes. GAF domains are somewhat structurally similar to PAS domains, but the GAF domain contains additional helices which can play a role in the dimer formation for some phytochromes [11, 21–23]. The GAF and PAS domain of the sensory region interacts through a knotting mechanism, in which a loop or "lasso" structure produced by the GAF domain has several N-terminal residues of the PAS domain passed through it. The PHY domain contains an α-helix which is connected to the GAF domain through an additional α-helix. This yields a long continuous helix which spans the length of the sensory region of the phytochrome [20]. Additionally, the PHY domain is thought to play a role in shielding the chromophore in the GAF domain. A conserved loop or "tongue" from the PHY domain has been found to block the binding pocket of the GAF domain which is thought to protect the bound chromophore from unwanted solvent interactions [11, 21, 22, 24, 25]. The loop from the PHY domain which approaches the binding pocket often contains several conserved aromatic residues which are thought to play a role in stabilizing the P_r and P_{fr} states of the chromophore [24]. For example, in the *Deinococcus radiodurans* phytochrome protein, *Dr*BphP, there are two tyrosine residues within the PHY loop which are thought to help stabilize the bound bilin molecule in the binding pocket (Fig. 3) [12, 26].

Similarities between phytochrome containing systems have led to the classification of three primary groups of phytochromes. Essen and coworkers define them as

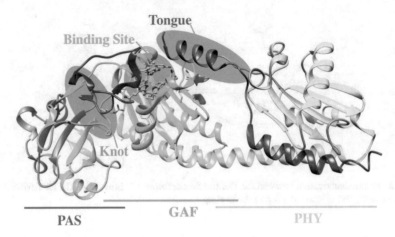

Fig. 3 Three domain architecture (PAS-GAF-PHY) of the photosensory region found inside the bacteriophytochrome *Deinococcus radiodurans*. This structure was experimentally determined by Takala et al. using X-ray crystallography (PDB code 4o01) [27]. Dashed lines near the tongue region indicate residues which were not resolved in the experimentally determined structure. The bilin chromophore is shown in a ball and stick representation, bound in the binding pocket of the GAF domain

Group I through III, while Lagarias and coworkers define it as knotted phytochromes, knotless phytochromes and cyanobacteriochromes (CBCRs) [10]. The photosensory portion of canonical phytochromes, Group I, is characterized as having a three-domain architecture comprised of PAS-GAF-PHY domains. This subset of phytochromes are typically found in bacteria, algae, and fungal species [6]. The phytochromes found within cyanobacteria are typically classified as either Group II or III. The phytochromes of Group II are structurally similar to those found within Group I with the exception of a missing PAS domain, shortening their architecture to GAF-PHY. Group III differs from the previous groups in that they only contain a GAF domain that can occur standalone or in tandem. CBCRs are also currently known to only exist within cyanobacteria [5, 28–31]. Unlike in plants which typically have a single GAF domain, there are cases where cyanobacteria can have multiple photosensory GAF domains in tandem, each with different spectral properties [10, 32, 33].

In Groups I and II, the 15Z conformation typically corresponds to the red absorbing state (P$_r$) and 15E corresponds to the red-shifted metastable photoproduct (P$_{fr}$) [1, 10]. In bathyphytochromes the order is reversed, because their 15E conformation is dark stable and the 15Z conformer corresponds to the metastable photoproduct [10]. The simplified architecture of Group III phytochromes allows for more unique interactions between the chromophore and the surrounding environment than those of Groups II and I. These additional interactions are thought to play a role in enabling the chromophore to absorb light at a wider range of wavelengths ranging from ultraviolet to infrared [29, 34–39]. CBCRs are further divided into four subtypes

Fig. 4 GAF domain of the cyanobacteriochrome (CBCR) AnPixJ. This protein structure was determined using X-ray crystallography by Narikawa et al. (PDB code 3W2Z) [39]. A bilin chromophore is shown as a ball and stick model inside the binding pocket

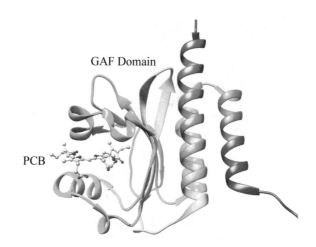

according to their spectral properties, photochemical reactivity, and conserved structural elements. Two of these subtypes display opposite photochromicity, i.e., the 15Z and 15E isomers have a red-to-green photocycle for one subtype and green-to-red for the other subtypes [34, 39–41]. The red-to-green conversion shows that for CBCRs, the 15E isomer does not always correspond to the state which absorbs lower energy light [6]. Narikawa et al. have used X-ray crystallography to resolve the first structure of a CBCR which exhibits this red/green photocycle, namely the P_r form of AnPixJ (Fig. 4) from *Anabaena* sp. PCC 7120 [42]. More recently Xu et al. [33] have solved the crystal structures of both the red-absorbing P_r and the green-absorbing P_g forms of Slr1393g3 from *Synechocystis* sp. PCC6803. This progress made it possible to reveal structural elements regulating the photochromicity and pointed out that both peripheral rings A and D are the main cause of the shift to the green.

Additional subtypes of CBCRs originate from systems which have an additional cysteine linking residue that can form an adduct with the chromophore and are known as insert-CYS and DXCF. This linkage separates the conjugated π-system of the chromophore in two separated, short π-systems [29, 43, 44]. Therefore, the shortened conjugation leads to absorption in the near-UV to blue regions of the spectrum [29, 43]. TePixJ, a DXCF-type CBCR, is an example of a CBCR which displays this additional thiol linkage [37, 42, 45]. TePixJ has been found to cycle between a blue absorbing resting state to a photoproduct which absorbs in the green region [42]. Structural studies to understand the photoconversion in TePixJ are currently in progress [46].

Despite the biological relevance of the phytochromes and the large body of experimental work, the number of computational studies on phytochromes is still relatively sparse within the literature compared to rhodopsins, photoactive yellow proteins and green fluorescent proteins [47–49]. This is surprising because computational methods have played a key role in the understanding of other photoreceptor families, e.g., in rhodopsin [50]. There are two main reasons for the small number of computation studies on phytochromes: (i) being that the bilin chromophores themselves are quite

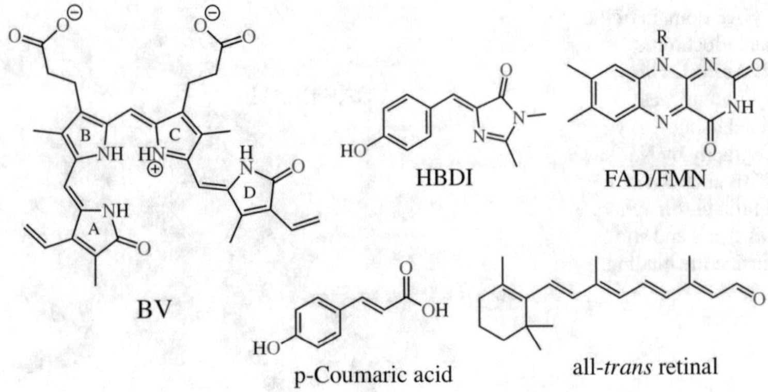

Fig. 5 Comparison of chromophores from the main photoreceptor protein families. HBDI is the chromophore in Green Fluorescent Protein. FAD/FMN are the chromophores in flavin proteins. Retinal is the chromophore in rhodopsins. P-coumaric acid is the chromophore found within Photoactive Yellow Protein

large in size making excited-state calculations using high levels of theory computationally very demanding (Fig. 5) and (ii) the availability of the crystal structures that serve as a starting point for simulations [1, 51]. The protein environment plays a crucial role in the photochemistry of photoreceptor proteins in general and specifically in phytochromes. The protein can contribute to the spectral tuning mechanism by stabilizing the chromophore in a constrained conformation which would affect its absorption, e.g., in the red/green CBCRs [37, 52–54]. In the photoisomerization the protein selects the double bond of the methine bridge between C and D rings, which is in contrast to the double bond between the B and C rings in the gas-phase.

The absorption and CD spectra of phytochromes are characterized by two intense spectral bands, namely Soret (or B) and Q-bands (Fig. 6). The name of each of these absorption bands stems from the nomenclature used for porphyrins. The B-band originates from the electronically excited state with B symmetry that is an allowed transition [55]. It is more common to refer to it as the Soret band in honor of Jacques-Louis Soret who characterized this band [56]. The Q-band term was coined by Platt [57–59] because it is usually a quasi-allowed electronic transition in porphyrins. The Q-band absorbs in the visible region, whereas the Soret band absorbs in the near-UV region of the electromagnetic spectrum.

The Q-band can often exhibit a shoulder at shorter wavelengths. The origin of this shoulder can be attributed to either structural or conformational ground-state heterogeneity or to vibronic progressions in a homogeneous ground-state structure. The cyanobacterial phytochrome Cph1 is one of the best studied phytochromes to address this question. Experimental evidence in support of the vibronic progression model has been provided on the basis of RR spectroscopy by the Mathies group [61–63] and more recently by Bizimana et al. [64] using transient absorption experiments. Spillane et al. were able to identify the shoulder as a vibronic band on the basis of Raman intensity analysis [61]. Other experiments support the heterogeneous model,

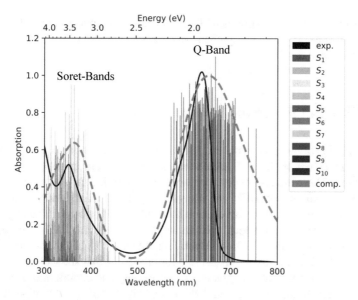

Fig. 6 The experimental UV/Vis absorption spectrum of the truncated Group II phytochrome all2699g1 in the P_r form (solid black) superimposed with a theoretical spectrum computed from 100 structures (dash gray). The lines indicate excited energies from individual snapshots. For each snapshot 10 excited states were calculated [60]

e.g., from NMR studies by Song et al. [18, 65–67]. More recently, RR studies by Hildebrandt and coworkers have revealed a pH dependence of the heterogeneity which could reconcile contradicting experiments [68]. Computational studies could help to explain the origin of the shoulder, however, it would require to account for vibronic coupling which is challenging for an entire protein. Recently, Macaluso et al. have done such simulations for the bacterial phytochrome *Dr*BphP and confirmed the vibronic nature of the Q-band shoulder [69].

Both Q and Soret bands are dominated by excitations of π type character. Generally, the Q-band is characterized by an electronic transition from the HOMO to the LUMO orbitals and the Soret band is composed of several electronic states that are characterized by transitions involving the HOMO-1, HOMO, LUMO and LUMO + 1 orbitals. In porphyrins, the orbitals involved in the major spectral bands are described using the four orbital model proposed by Martin Gouterman [70].

2 Methodology

In the following section, we describe the methodology that has been used to calculate the excitation energies and spectra of phytochromes from studies found in the literature. This section comprises different electronic structure methods as well as methods to account for environmental effects and sampling techniques. The goal of this section is to provide a comprehensive overview without being overly technical in the procedural details.

2.1 Choice of Electronic Structure Method

The calculation of the ground and excited state energies is the main focus of this book chapter. In this section, we will discuss the choice of the appropriate quantum chemical method. There are two major approaches to determine the electronic structure: density functional theory (DFT) and wave function theory. The former approach is a popular cost-effective choice for organic molecules like the bilins. The major families of DFT functionals are local spin-density approximation (LSDA), generalized gradient approximation (GGA), and Hybrid functionals [71]. LSDAs are exact for the homogeneous electron gas and inaccurate for properties that show density inhomogeneity. GGAs incorporate the gradient of the density and exhibit improved performance over LSDAs. However, both LSDAs and GGAs suffer from self-interaction error. Global hybrid (GH) functionals reduce this error by incorporating a constant amount of Hartree–Fock exchange to the DFT functional.

In order to use DFT to obtain the excitation energies of molecules, a time-dependent (TD) extension has been developed (TD-DFT) [72]. Within TD-DFT, there are different functionals from which one can choose to compute spectroscopic properties. The spectra of many light-absorbing proteins have been generated using TD-DFT with a variety of functionals including MPW1PW91 [73], B3LYP [74, 75], and BLYP [76]. The above-mentioned functionals were developed for the ground state. When these functionals were employed in the framework of TD-DFT, they were found to give a poor description of charge transfer states and double excitations [77–80]. While the latter is not of particular importance for the low-lying excited state, the charge transfer is a common feature in biological chromophores [80]. To mitigate this shortcoming, range-separated hybrid functionals (RSH) were developed [81]. RSHs partition the interelectronic interaction into short- and long-range components and employ the HF exchange for the long-range component and local DFT exchange for the short-range component. RSHs such as CAM-B3LYP and ωB97 have been employed for calculation of excitation energies in phytochromes [76]. Those functionals have range separation parameters that are fixed. However, it is also possible to tune these parameters for a specific molecule [81].

Wave function methods present an alternative approach to DFT. They can be further subdivided into single reference and multireference wave function methods. Accurate excitation energies have been obtained for organic molecules using ab initio single reference methods such as second-order approximate coupled-cluster singles and doubles model CC2 or the second-order algebraic-diagrammatic construction scheme ADC(2) [82, 83]. The ADC(2) method has favorable computational cost and good quality description of excited states. The ADC(2) method is also advantageous because it is both size-consistent and formulates the eigenvalue equation as a Hermitian matrix which is suitable for derivation of properties [83, 84]. The CC2 method has also been shown to be efficient for obtaining an accurate description of excited states. The accuracy of CC2 was assessed previously in the benchmark studies by Thiel and coworkers using 28 organic molecules [85]. Recently, an assessment of the ADC(2) method was made using the same set of molecules [86]. The singlet excitation energies of CC2 method were found to have a 0.29 ± 0.28 eV (mean error $\pm$ standard deviation) for 103 vertical excitation energies compared to the theoretical benchmarks, while ADC(2) produced 0.22 ± 0.38 eV for 104 vertical excitation energies. Similarly, for triplets a comparison to 63 vertical excitation energies showed a deviation of 0.17 ± 0.13 eV and 0.12 ± 0.16 eV for CC2 and ADC(2), respectively. The two wave function-based methods were used in the "resolution of the identity" (RI) formulation, which is an order of magnitude more efficient in comparison to the conventional algorithms [87].

Multireference methods have the potential to obtain accurate spectra by accounting for the static and dynamic electron correlation [88]. They are in particular suitable for excited-state reaction pathways that might involve the characterization of crossing such as conical intersections (CIXs) [89]. Conical intersections are particularly important in order to accurately model non-radiative processes in light-absorbing molecules. These intersections are topological features that serve as reaction funnels between electronic states with the same spin multiplicity. They are associated with ultrafast events and are ubiquitous in photochemical studies because the topology controls the outcome of the reaction. Since at least two states are degenerate in such a CIX, one needs at least two configurations to obtain a flexible wave function that can describe these states. The lowest approach in the hierarchy of these multireference methods is the complete-active space self-consistent-field (CASSCF) method which can be found in most computational packages. A prerequisite is to select an active space which is composed of electrons and orbitals that are relevant for the description of the excited states. Within the active space, all possible electronic configurations are generated, also known as full configuration interaction (FCI). The number of configurations increases exponentially with the size of the active space and, therefore, using this method presents a challenge for phytochromes and CBCRs. The number of all π-like electrons and orbitals that are required to describe the low-lying excited state is 32 electrons in 29 orbitals for BV and 28 electrons in 25 orbitals for PCB (Fig. 7).

The CASSCF method requires a correction to yield quantitative results because this method neglects dynamic electron correlation and only accounts for the static electron correlation. This correction can be accounted for by variational or perturbative methods. The former can be achieved by multireference CI (MR-CI) which, however, is very computationally demanding. While the spectroscopically oriented MR-CI (SORCI) methods are significantly more efficient for excitation energies [76], the more feasible method is the multireference perturbation theory (MR-PT). Among the methods are CASPT2, XMCQDPT2, and NEVPT2. The first two have been already applied to phytochromes [51, 54, 90].

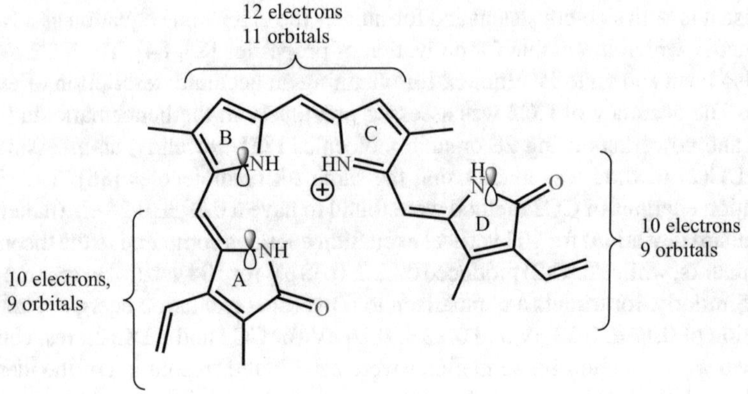

Fig. 7 Molecular orbitals of the conjugated π-system that constitute the active space for the BV chromophore. The displayed model is truncated and doesn't include the propionate side chains

2.2 Description of the Protein Environment

Despite the ongoing development of efficient electronic structure methodology and the increase of the computational power of modern computer hardware, it is not possible to calculate the full phytochrome using quantum chemical methods. Hence, some studies have relied on the calculation of the isolated chromophore [28, 91]. While it provided qualitative results and helped to understand circular dichroism spectra, the effect of the protein remained unknown.

A rudimentary way to model the effects of the protein environment is by using implicit solvation that models the solvent as a continuum. They have been successfully applied to describe chromophores in solution. The most commonly used implicit solvent model is the polarizable continuum model (PCM) [92]. In this model, the solute is constructed as series of interlocking spheres placed into a cavity. The cavity is surrounded by a polarizable dielectric medium replicating the solvent and is characterized by a dielectric constant. For proteins, the dielectric constant of the continuum is generally taken to be 4.0. Gonzalez and coworkers have used PCM models to calculate excitation energies in phytochromes [93]. However, such approaches are unable to describe specific solute–solvent interactions such as salt-bridges and H-bonds.

The protein environment consists of amino acids that have neutral, polar, or charged sidechains which require a more accurate description. A cost-effective method of describing such a heterogeneous environment is to use multiscale modeling such as the hybrid quantum mechanics/molecular mechanics (QM/MM) approach. This approach is in particular suitable for photoreceptor proteins such as phytochromes because it exploits the local nature of the electronic excitation. Typically, the excitation is limited to the chromophore and its immediate environment. In this approach, the protein can be partitioned into a photoactive region described using QM and an inactive region treated at a lower level of resolution using the efficient force-field-based molecular mechanics (MM). Further, the energy of the full system is evaluated using an additive scheme, as proposed originally by Warshel and Karplus. In this scheme, the total energy is evaluated by summing up the individual energy contributions from the QM and MM regions and adding the interaction between the two regions. Alternatively, a "subtractive scheme" has been developed by Morokuma and coworkers known as "Our own N-layered Integrated molecular Orbital and molecular Mechanics" (ONIOM) [94]. Hence, the critical step is to choose the boundary when partitioning the system into QM and MM regions. This selection is initially based on chemical intuition and then expanded stepwise, starting from the minimal model which corresponds to the chromophore with a QM/MM boundary [52, 76]. The choice is then validated by calculations of excitation energies until the results are converged. It is worth noting that the QM/MM method can also be used for the determination of other properties, e.g., the calculation of Raman spectra of phytochromes [74, 95].

The interactions between the QM and MM regions can be accounted for in different ways. The simplest way is mechanical embedding which only considers van der Waals interaction between the QM and MM regions. However, a more accurate embedding procedure allows for electrostatic or polarizable interactions [96–100]. This QM/MM methodology has enabled researchers to accurately model the absorption and emission spectra of several types of proteins containing bound chromophores [47, 48, 101–105].

2.3 Conformational Sampling

The description of the entire phytochrome using the QM/MM scheme allows the construction of a realistic model in terms of chromophore–protein interactions. However, the calculation of an absorption spectrum for such a high dimensional system becomes more complicated compared to an isolated chromophore in the gas-phase. For an isolated chromophore a spectrum is determined from a single conformation of the chromophore corresponding to the ground-state equilibrium geometry which can be obtained from a geometry optimization (Fig. 8A) [106–108]. Thus, such a single structure provides a limited set of excitation energies which result in a so-called stick spectrum. The height of the stick is determined from the oscillator strength that is related to the experimental extinction coefficient. Often, the resulting stick spectrum, as shown in Fig. 8B, can then be further modified by Gaussian broadening each stick to resemble an experimental spectrum [109, 110].

This approach is appealing because it requires only one geometry. Hence, it has been successfully applied for many decades [111, 112]. However, this approximation will only yield reliable results for systems with low-dimensional potential energy surfaces (PES). For small systems, the global minimum of the surface can be obtained during structural relaxation giving rise to one dominant conformation of the chromophore [113, 114]. However, for systems with high-dimensionality ensuring the localization of a minimum is impossible given large number of degrees of freedom in a protein. In fact, multiple minima can exist and contribute to the spectra [115,

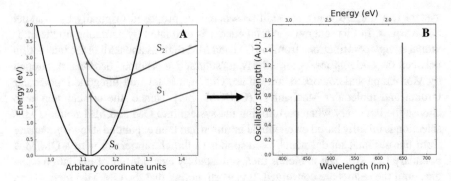

Fig. 8 **A**) Vertical excitation from a ground state minimum geometry to calculate the two lowest excited states, **B**) A stick spectrum for the two excited states shown in **A**

116]. This fact arises from the higher dimensionality of the potential energy surface where multiple conformations are easily accessible at a given temperature and may contribute to the absorption spectrum.

To address this issue, conformational sampling [52, 60, 117, 118] can be used. In this approach, many structures of the system are generated in order to probe different conformations of the chromophore and the protein environment [52, 119]. This procedure is in line with the Ergodic hypothesis, [120, 121] which states that for a statistical ensemble the major contributing structures of the chromophore will be sampled more than those that are less important, resulting in a better description of the absorption spectrum.

There are many realizations of the conformational sampling. Typically, the first step after preparing the crystal structure for simulation is to run a molecular dynamics simulation to equilibrate the protein and to explore the potential energy surface. This can be achieved by using a classical force field or QM/MM [100, 122] for dynamics. The former allows longer timescales and, therefore, more sampling to be covered [119, 123, 124, 125]. However, classical force fields require parameters for the chromophore which are not always readily available. Examining the long classical trajectories can also reveal important structural changes and different subpopulations, unraveling portions of the ground-state PES that may result in different excitation energies. In the second step, the resulting trajectory can then be sampled in fixed time intervals and for each sampled structure a vertical excitation is calculated. The generated spectra are then averaged to generate the stationary spectrum of the protein. There are two procedures that can be applied to the ensemble of snapshots in order to obtain an accurate absorption spectrum: 1) the snapshots can be directly taken from the resulting trajectory for excitation energy calculation or 2) they can be further optimized by using the QM/MM method. The excitation energies obtained from the ensemble of conformers can then be convoluted in order to generate a theoretical estimation of the spectrum [126, 127]. Although similar to the Gaussian broadening of a stick spectrum discussed earlier, this method differs from the prior due to the sampling of multiple conformers per excited state transition.

3   Results

3.1   Benchmark Studies

In 2004, Durbeej et al. [128] tested the effect of the methyl and propionate groups on the excitation energy in phytochromobilin (PΦB), the chromophore in plant phytochromes. Using TD-DFT with B3LYP/6-31G*, they found that the truncation of both groups had only a small effect (<0.10 eV) on the excitation energy and the spectral shift caused by the Z-to-E isomerization. In the follow-up study, the effect of enlarging the basis was benchmarked and resulted in changes below 0.05 eV [129]. Further tests were done using the PBE0 functional and the wave function methods CIS, CIS(D), and symmetry-adapted-cluster CI (SAC-CI) [130]. In 2011, Strambi and Durbeej have extended the benchmark to modern functionals including range-separated hybrids (RSHs) [131].

In order to accurately reproduce the absorption and emission spectrum of bilins in solution, Falklöf and Durbeej [132] benchmarked a series of quantum chemical methods on sterically locked linear tetrapyrroles. Ten different DFT functionals from different families were assessed by calculating the excitation energies and then comparing them to experimental absorption maxima. Solvent effects were incorporated using the polarizable continuum (PCM) approach during geometry optimization and in the calculation of excitation energies at the TD-DFT level in state-specific regime. The energies obtained from the molecular structures optimized in an implicit PCM solvent model were found to consistently shift the excitation energies by 0.02 eV relative to those optimized with explicit water molecules. Within the state-specific regime, the predicted excitation energies were found to overestimate the experimental values regardless of the amount of HF exchange included. Generalized gradient approximation (GGA)-based functionals were found to perform the best from the functional types tested, having a deviation of 0.15 eV. Whereas the RSH functionals performed the worst with a deviation of 0.4–0.5 eV in absorption and 0.7 eV in emission. Global hybrids were found to produce errors that fell between RSH and GGA functionals. Therefore, the authors have argued that including HF exchange results in larger deviation from experimental values because the charge distribution of S_0 and S_1 states is similar with no apparent charge transfer character. It was also found that TD-DFT performs better than the tested wave function methods CIS, CIS(D), CC2, and CASPT2 based on comparison with experiment.

In an extension of their previous work, Falklöf and Durbeej [105] examined the effect of interactions between the protein and the chromophore on the absorption spectrum using the hybrid QM/MM scheme. The crystal structure of DrBphP bacteriophytochrome [105] was used for this purpose, which contains the BV chromophore in a ZZZ-ssa conformation. A two-layer ONIOM (QM/MM) approach was employed to accurately predict the Soret and Q-band absorption maxima using TD-DFT. Also here different density functionals were assessed as part of the QM region. The effect of the functional choice on the structural relaxation as well as the effect on the excitation energies was tested. To this end, the authors determined the Q-band maximum using several functionals and then determined the error produced by each functional with respect to the experimental value at 700 nm. The GGA functionals were found to have the best agreement with experiment followed by global hybrids. These functional types produced deviations from experiment of 0.15 eV and 0.3 eV, respectively. The highest deviations were produced by RSH functionals which exhibited a deviation of 0.5 eV. A similar behavior was observed for the Soret band whose experimental peak is at 380 nm, in line with their previous work [105]. The RSH functionals performed the worst with deviations ≥ 0.3 eV while the rest of the functionals exhibited deviations less than 0.13 eV. The authors have attributed the cause of these large deviations in the predicted Q-band to little or no contribution arising from charge-transfer transitions. They also compared the predicted TD-DFT spectrum of the isolated chromophore with those computed using the SAC-CI method and found that TD-DFT yielded similar results. Expansion of the QM region by inclusion of Asp207 and His260 residues resulted in minor deviations of 0.02–0.08 eV and 0.05 eV of the predicted Q and Soret bands, respectively.

The influence of structural relaxation on the absorption maxima was probed by relaxing the chromophore in a stepwise manner starting from the geometrical parameters of the chromophore (bonds, angles, and torsions) to the full binding site within the protein upon which a convergence of the predicted absorption was observed. The unrelaxed crystal structure reproduced the experimental values the best and the authors explained it by a cancelation of errors. Compared to predicted values from the crystal structure, the bond and torsional relaxations blue-shift the absorption by 0.1 eV each. The angle relaxation and inclusion of residues within the QM region do not shift the absorption maxima.

When comparing the Q-band absorption maxima obtained from QM/MM and solvent model based on density (SMD) descriptions of the protein environment, the specific short-range interactions not considered in the SMD model produce a blue-shifted Q-band, whereas the bulk long-range electrostatic interactions red-shift the Q-band. By computing the protein-induced absorption shift, the authors have concluded that the protein environment hypsochromically shifts the Q-band absorption by 0.2 eV and this effect is more accurately reproduced by the hybrid functionals than the GGA functionals.

Recently Wiebeler and Schapiro [76] have benchmarked quantum chemical methods for PCB. They assessed the performance of ADC(2) and other wave function methods in addition to the approaches discussed above. Also, semi-empirical methods were tested for geometry relaxation in order to find a suitable method for long QM/MM sampling (discussed in section *Spectrum calculation by QM/MM Sampling*). The semi-empirical DFTB2+D method [133, 134, 135] was found to yield similar PCB structures to computationally more demanding ab initio methods [76].

3.2 Structural Elucidation Based on Quantum Chemical Calculations

Quantum chemical calculations give an accurate estimate of the Q and Soret absorption band positions and their intensities. Since these parameters inherently depend on the structure of the chromophore, the calculations become a tool to derive the configuration of the chromophore. This has been demonstrated in the studies by Matute et al. [75, 93] where the electronic absorption spectra of bilin chromophores were computed in different orientations, protonation states, and/or solvation in order to predict its exact configuration in cyanobacterial [75] and plant phytochromes [93]. However, recent spectroscopic studies have revealed that the chromophore can adopt more than one conformation when bound to the protein, therefore, impeding the structural characterization on the basis of UV/Vis absorption spectra [68]. The presence of multiple conformations, also known as heterogeneity, can significantly influence the quantum yield of isomerization as demonstrated in the study by Heyne et al. [136]. Additional insights into the conformation can be obtained from circular dichroism (CD) spectroscopy. This method can be used to distinguish chiral elements in molecular systems.

3.2.1 Resolving the Conformation Using Circular Dichroism Spectra

The CD spectrum of the Cph1 and Tlr0924 protein structures were studied by Rockwell et al. [91]. The spectra were first obtained using experimental methods and then calculated using theoretical methods for comparison. To this end, gas-phase calculations on reduced models of the chromophore adducts were used. The reduced models were created by replacing the propionate side chains with methyl groups and the thioether linkage was replaced with a hydrogen cap. The resulting models were then optimized using B3LYP(VWN5)/6–31+G* level of theory. The vertical excitation energies were calculated for the 15 lowest states using TD-DFT with the BLYP functional and 6–311+G* basis set. While examining the calculated spectra, these authors found that the facial orientation of the D-ring of the chromophore can change the sign of the Q-band in the resulting CD spectrum. In one conformation, the D-ring is above the central B-C ring plane and while in the other it is below (Fig. 9). Rockwell et al. [137] named these as α-facial (D-α_f) and β-facial (D-β_f) orientations, respectively.

They found that when the D-ring was D-α_f relative to the B and C pyrrole rings, the Q-band exhibited negative rotatory strength and the Soret band was positive. They also found that varying the facial disposition of the A-ring had no effect on the CD spectrum. For each facial disposition, good agreement was also found between the calculated excitation energies and the observed maximum wavelengths obtained experimentally.

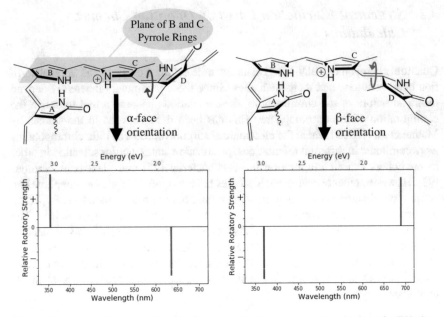

Fig. 9 The α-facial (D-α$_f$) and β-facial (D-β$_f$) orientations of the D-pyrrole ring of a BV chromophore. Here the A-ring is shown in an α-facial orientation for each structure, however, its disposition has no effect on the sign of the rotatory strength. Each facial orientation also has an effect on the sign of the rotatory strength on the Q and Soret bands as shown [137]

   In 2014, Heyne and coworkers [136] reported similar results in a joint spectroscopic-computational study. They used time-resolved vibrational spectroscopy to measure the dynamics. Heyne and coworkers found that there were two distinct D-ring carbonyl stretching bands resolved at 1701 cm^{-1} and 1708 cm^{-1}. These vibrational frequencies correspond to the D-α$_f$ and D-β$_f$ conformations of the chromophore respectively. Since the crystal structure of Cph1 consists of only the H-bonded D-α$_f$ geometry, a non-H-bonded D-β$_f$ conformation of the chromophore was prepared in the computational part of the study. Further, based on polarization resolved decay experiments, the authors have proposed that only a sub-population of D-α$_f$ chromophores isomerizes to form the first intermediate of the reaction pathway (Lumi-R). The D-β$_f$ chromophores are unsuccessful and return to the ground state without isomerization. The chromophore configuration in the intermediate was validated by the relative angle between vibrational transition dipole moment and electronically excited transition dipole moments were computed to be 61°, which is in close agreement to the experimentally measured value of 54°, confirming the configuration of Lumi-R to be that of ZZE-ssa. After determining the correct conformation of the intermediate structure, the authors attempted to find the correct facial orientation of the D-ring. In order to find the correct facial orientation, the authors computed the CD spectra for both D-α$_f$ and D-β$_f$ geometries using TD-DFT. The CD displayed a sign change in the rotatory strengths of both the Soret and Q-bands for both geometries and a two-fold intensity for the D-α_f geometry. Noting that a similar reversal is observed upon isomerization from P_r to P_{fr}, the authors have concluded that the isomerizing D-α_f population rotates α-facially passing through a barrier on the excited state and forms the Lumi-R photoproduct.

3.2.2 Probing Chromophore Conformation by Excitation Energies

To discern the correct configuration of PCB in cyanobacterial phytochrome Cph1, three different structures were investigated: semi-cyclic (*ZZZ-ssa*) conformations of BV and PCB and a stretched (*ZZZ-asa*) conformation of PCB (Fig. 10). For each structure, four models were constructed and used for spectrum calculation. These models served for a systematic approach to evaluate the effect of protonation and environment (gas-phase, water, and protein). Electronic spectra were computed on both unrelaxed and relaxed geometries by treating the chromophore using TD-DFT with a hybrid functional B3LYP and accounting for the surroundings by a polariz-able continuum. The authors have noted that the computed Q-band from the DFT optimized structures was systematically blue-shifted relative to the experimentally resolved structures obtained using X-ray diffraction. The authors have attributed the shift to steric effects of the protein, non-planarity of rings and higher helicity of the optimized structures compared to X-ray structures. The results from their findings are given below in Table 1.

The semi-cyclic (BV and PCB) and stretched (PCB) conformations were found to exhibit a bathochromic shift upon protonation. The shift relative to the experi-mental absorption was lowest for the semi-cyclic (BV) and largest for the stretched conformation (PCB). Thus, Matute et al. concluded that the chromophore must be

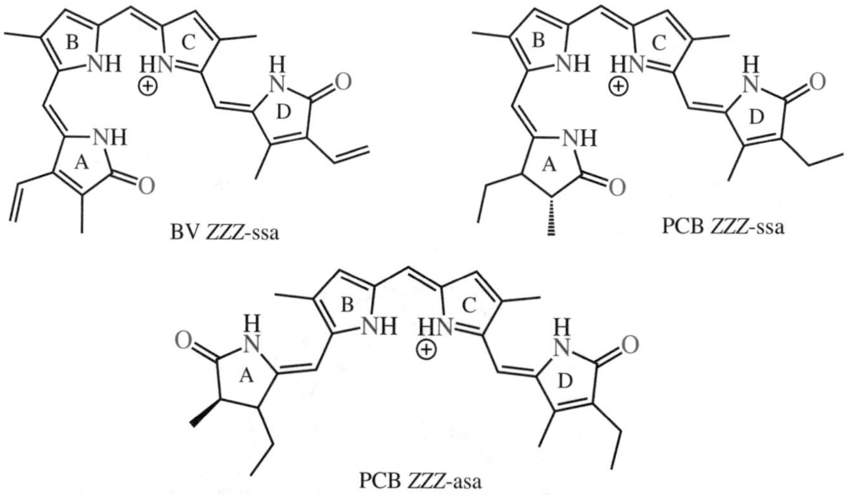

Fig. 10 Protonated BV and PCB chromophores used in the work by Matute and Gonzalez [93]

Table 1 Q-band excitation energies (in nm) of the BV and PCB chromophores calculated using TD-B3LYP/6-31G* in different environments

		Model I	Model II	Model III	Model IV	Expt
Protonated		No	Yes	Yes	Yes	–
Environment		Gas-phase	Gas-phase	Water	Protein	–
BV-ssa	X-ray	648	662	708	712	702 (BV in DrBphP)
	DFT	588	643	659	665	
PCB-ssa	X-ray	603	620	659	661	659 (PCB in Cph1)
	DFT	526	590	609	613	
	DFT		574			
PCB-asa	X-ray	528	590	606	614	618 (PCB in C-PC)
	DFT	508	559	584	588	
	DFT	539	541, 582			

protonated within the protein environment. The computed Q-band for the semi-cyclic (PCB) in both water and protein environment also closely matched the experimental Q-band at 659 nm. The sum of these effects allowed the authors to determine that PCB in Cph1 was most likely in the *ZZZ-ssa* conformation.

Follow up work by Matute and Gonzalez [75] examined conformations of the phytochromobilin (PΦB) chromophore found in plant phytochromes. Here three conformations of PΦB were chosen to compute the spectra, namely semi-cyclic (*ZZZ-ssa*), helical (*ZZZ-sss*), and stretched (*ZZZ-asa*) (Fig. 11). The semi-cyclic

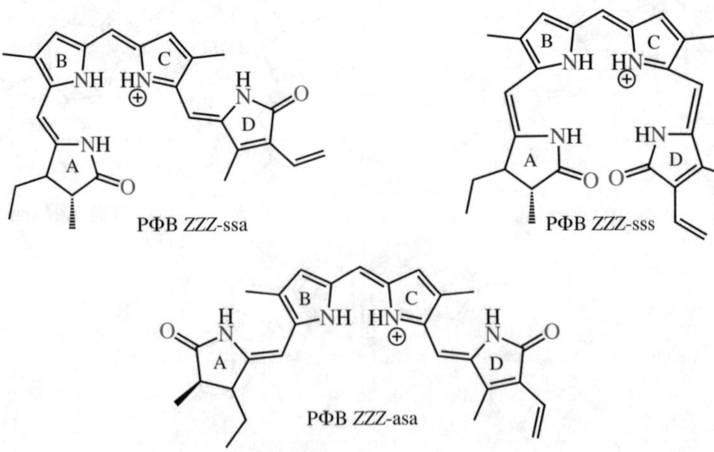

Fig. 11 Three conformations of PΦB used for quantum chemical calculations by Matute and Gonzalez [75]

Table 2 Calculated TD-DFT excitation energies (in nm) and Q/S ratios. Values in parentheses correspond to structures refined with a constraint optimization (frozen dihedrals). Values with h superscripts have been obtained using the BV template and values with i superscripts have been obtained using the PCB template

Model	Absorption max		Q/S ratio	
	Expt	TD-DFT	Expt	TD-DFT
BV-sss	377, 696, 710	385,712	0.29	0.26
PCB-sss	375, 692	368,675	0.33, 0.43	0.37
PΦB-sss	386, 700	384,701		0.34
BV-ssa	380, 702	417,712	2.69	1.99
PCB-ssa	380, 659	394,661	1.17	2.28
PΦB-ssa	380, 666	395 ^{h} (382)^h, 664 ^{h}(633)^h, 410^i(382)^i, 688^i(611)^i	1.36, 1.45	2.12 ^{h}(2.02)^h, 2.36^i(1.90)^i
PCB-asa	380, 618	385,614	4.1	5.26
PΦB-asa		398(385),640(619)		5.27 (4.26)

conformations were prepared by removing the propionates and partially relaxing each molecule *in vacuo* while also placing restraints on the dihedral angles. The approach resembled the one used for PCB in cyanobacterial phytochrome [93]. However, the authors have acknowledged that interactions between residues are not explicitly treated in PCM. The results are summarized in Table 2. Comparing the absorption maxima among the DFT optimized chemical structures, the theoretical prediction of the semi-cyclic-BV conformation was closest to the experimental value. However, the wavelength shift relative to the semi-cyclic PCB and the stretched conformation was only 0.07 eV. The authors note that the comparison of absorption maxima between relaxed geometries and the non-optimized crystal structures is difficult since the former show larger deviations relative to experiment values. Thus, to unequivocally assign the spectra to a specific configuration, the ratio of the oscillator strengths of the Q and Soret bands, known as the Q/S index, were determined for each conformer. A higher Q/S index indicates higher linearity of the tetrapyrrole, whereas a lower Q/S index indicates higher helicity. The semi-cyclic conformation was found to exhibit a value closest to the experiment when using this index. The authors, therefore, concluded that the semi-cyclic conformation of the chromophore is the most probable conformer present in the plant phytochrome.

3.3 Spectrum Calculation by QM/MM Sampling

3.3.1 Phytochromes

In an attempt to accurately compute the absorption spectrum of a phytochrome Polyakov et al. examined the spectrum produced by the protein-bound BV chromophores using TD-DFT as well as the XMCQDPT2 approach [51]. In this work, the authors used the crystal structure of a protein variant termed IFP4.1 (PDB code 5AJG), a derivative of DrBphP as a starting point for their simulations [90]. The authors also included protein effects in the absorption spectrum by calculating the excitation energies for two configurations of the chromophore within the protein environment using a QM/MM scheme. In each of their calculations, the bound BV chromophore was treated as being fully protonated, a decision based on previous studies [51]. The conformers used for the spectrum generation were obtained by first performing a classical MD trajectory for 10 ns and then optimizing the final snapshots. In the optimization, the QM region comprised of PBE0/6-31G* and the MM partition was treated using a classical force field. Vertical excitation energy calculations were then performed on the resulting structures using both ab initio multireference and DFT methods. This procedure was performed for both the red and far-red absorbing forms. The TD-DFT results were found to overestimate the energy gap of the Q-band of the P_r state by 0.2–0.3 eV [90, 138]. The XMCQDPT2/SA(10)-CASSCF(16,12)/cc-pVDZ approach produced more accurate predictions for both the Q and Soret bands when compared to experiment with differences of only 0.05 eV and 0.12 eV, respectively. Prior to this study, the work by Matute and Gonzalez [75, 93] as well as Falklöf and Durbeej [93] suggested that more accurate energies are obtained using non-optimized geometries from the crystal structure. This work showed that by using higher level methods the accuracy of calculated excitation energies can be improved for optimized structures [90].

The work by Modi et al. examined the red and far-red state of the photosensory region found within DrBphP [118]. Here the authors used the experimentally determined protein structures reported by Takala et al. [27] to model the BV chromophore in the bacteriophytochrome DrBphP (PDB codes 4O0P and 4O0I). In order to determine the most likely protonation state of the BV chromophore in the binding pocket, the authors calculated the absorption spectrum for various protonation states of the pyrrole-nitrogen atoms. Similar to the work by Polyakov et al. these authors used a QM/MM scheme while treating the QM region with XMCQDPT2/SA(5)-CASSCF(12,12)/cc-pVDZ. It should be mentioned that the size of the active space as well as the number of averaged states in the CASSCF wave function is smaller than the work performed by Polyakov et al. [90]. However, the spectra were computed for multiple conformations that were obtained by taking snapshots every nanosecond over a 100 ns classical trajectory. This sampling procedure differs from the one used by Polyakov et al. in that these authors took snapshots at room temperature with the protein in solution and did not perform additional minimizations. By comparing

Table 3 Computational details of the multireference methods used to compute properties in phytochromes taken from literature

	Chromophore	Method	Active space (elec,orb)	Number of roots	Type of calculation	Ref.
1	PΦB (gas phase)	CASSCF	(4,4)	-	Optimization	Borg et al. [130]
2	B and C rings of PΦB (gas phase)	CASSCF	(12,11)	2	Excitation energies	Durbeej et al. [140]
3	PΦB (gas phase)	CASSCF and RASSCF	(8,8) and (12,11)	–	Optimization	Altoè et al. [54]
		CASPT2	(10,11)	3	Excitation energies	
4	Locked BV (gas phase)	CASPT2	(14,14)	2	Excitation energies	Falklöf et al. [132]
5	BV (QM/MM)	XMCQDPT2	(16,12)	10	Excitation energies	Polyakov et al. [51]
6	BV (QM/MM)	XMCQDPT2	(12,12)	5	Excitation energies	Modi et al. [118]

to the experimental absorption spectra, these authors determined that the BV chromophore in the binding pocket is most likely fully protonated inside the protein. Subsequent work by Polyakov et al. also examined the various protonation states of BV and the resulting absorption spectrum in the IFP1.4 variant [90]. These authors found that the lowest energy structure of the BV chromophore was found to be the all-protonated version, and the relative energetics between the various conformers only ranged by 5–6 kcal/mol (Table 3). Recently, Maximowitsch and Domratcheva studied the origin of the spectral tuning in phytochromes [139]. They found an asparatate residue in the vicinity of the chromophore's D-ring to play a key role in red-shifting the photoproduct of various phytochromes that have this specific amino acid conserved.

3.3.2   Cyanobacteriochromes

Canonical phytochromes have red to far-red spectral tuning and, therefore, a somewhat limited spectral shift. Cyanobacteriochromes have a more diverse tuning mechanism. One representative of the red/green subfamily is Slr1393g3 from *Syneccocystis* sp. PCC6803. Recently, Xu et al. [33] have solved the crystal structure for both, the red and the green-absorbing forms of this protein. In conjunction with QM/MM simulations, it was shown that the large spectral shift from the red to the green is not directly induced by the protein or other environmental effects. Instead a distortion of the A- and D-rings out of the chromophore plane are mainly responsible for the large change. This protein was further studied by Wiebeler et al. [52] by including

QM/MM conformational sampling. To this end DFTB2+D/AMBER MD simulations were employed. The structure was prepared for the simulation in several steps that included restraints of the chromophore to its crystallographic position since reliable force field parameters were missing. Eventually, a DFTB2+ D/AMBER optimization without any restraints was performed as this method [133, 134, 135] had been found to yield similar PCB structures to computationally more demanding ab initio methods [76]. The spectra were generated from 100 equidistant snapshots from a 1 ns QM/MM MD trajectory. This choice of the method was motivated by the lack of the force field parameters which could alter the geometry, in line with similar findings for different chromophores in complex environments [141, 142, 143]. The resulting spectra allowed to deduce the origin of the spectral shift from a red- to a green-absorbing form in a cyanobacteriochrome. The calculations revealed that the effective conjugation length in the chromophore becomes shorter upon conversion from the red to the green form. This is related to the planarity of the entire chromophore. A large distortion is found for the terminal pyrrole rings A and D, however, the D-ring contributes stronger to the photoproduct tuning, despite a larger change in the twist of the A-ring. These findings implicate that the D-ring twist can be exploited to regulate absorption of the photoproduct. Hence, mutations that affect the D-ring twist can lead to rational tuning of the photoproduct absorption that allows tailoring of cyanobacteriochromes for biotechnological applications such as optogenetics and bioimaging.

4 Conclusion

Computational studies of photochemical properties of phytochromes have only recently emerged due to limitations. One is the lack of experimentally resolved crystal structures and the other is the large computational cost associated with the large size of phytochrome chromophores. Therefore, initial studies were limited to truncated chromophores and implicit solvent models. Nevertheless, it allowed to draw conclusions on important molecular properties. Several computational studies have aided the determination of the most likely protonation states of the each nitrogen of the pyrrole rings as well as the conformation by comparing to experimental values.

With the advance in crystallography protein structures of phytochromes became available. This allowed more accurate calculations. More recent studies employed hybrid QM/MM schemes, extended chromophore models in tandem with conformational sampling to accurately predict the absorption spectra of the stable and metastable states of Phytochromes. For the calculation of the properties, the choice of the quantum chemical method and the extent of its region is essential. On the methodology side, TD-DFT was the preferred choice for spectrum calculation due to its favorable efficiency and higher accuracy than wave function methods. For this purpose various functionals and basis sets have been benchmarked. With the advancement of computer hardware and algorithms more expensive ab initio methodology were applied and assessed. In particular the single reference method ADC(2) and the multireference method XMCQDPT2 were found to be in close agreement with experimental absorption maxima. This progress is paving the way to go beyond the calculation of spectra toward the simulation of nonadiabatic dynamics in order to study the photoisomerization process.

Acknowledgements This project is funded from European Research Council (ERC) under European Union's Horizon 2020 research and innovation program (grant agreement 678169 "PhotoMutant"). JC would like to acknowledge the Zuckerman Foundation for their support. CW acknowledges funding by the Deutsche Forschungsgemeinschaft (WI 4853/2-1 and 4853/1-1).

Bibliography

1. Nemukhin AV, Grigorenko BL, Khrenova MG, Krylov AI (2019) Computational challenges in modeling of representative bioimaging proteins: GFP-like proteins. Flav Phytochrom 123:6133–6149. https://doi.org/10.1021/acs.jpcb.9b00591
2. Wiltbank LB, Kehoe DM (2016) Two cyanobacterial photoreceptors regulate photosynthetic light harvesting by sensing teal, green, yellow, and red light. MBio 7. https://doi.org/10.1128/mBio.02130-15
3. Terauchi K, Montgomery BL, Grossman AR et al (2004) RcaE is a complementary chromatic adaptation photoreceptor required for green and red light responsiveness. Mol Microbiol 51:567–577. https://doi.org/10.1046/j.1365-2958.2003.03853.x
4. Wiltbank LB, Kehoe DM (2019) Diverse light responses of cyanobacteria mediated by phytochrome superfamily photoreceptors. Nat Rev Microbiol 17:37–50
5. Ikeuchi M, Ishizuka T (2008) Cyanobacteriochromes: a new superfamily of tetrapyrrole-binding photoreceptors in cyanobacteria. In: Photochemical and Photobiological Sciences. pp 1159–1167
6. Anders K, Essen LO (2015) The family of phytochrome-like photoreceptors: diverse, complex and multi-colored, but very useful. Curr Opin Struct Biol 35:7–16
7. Chernov KG, Redchuk TA, Omelina ES, Verkhusha VV (2017) Near-infrared fluorescent proteins, biosensors, and optogenetic tools engineered from phytochromes. Chem Rev 117:6423–6446. https://doi.org/10.1021/acs.chemrev.6b00700
8. Baloban M, Shcherbakova DM, Pletnev S et al (2017) Designing brighter near-infrared fluorescent proteins: insights from structural and biochemical studies. Chem Sci 8:4546–4557. https://doi.org/10.1039/c7sc00855d
9. Lehtivuori H, Bhattacharya S, Angenent-Mari NM et al (2015) Removal of chromophore-proximal polar atoms decreases water content and increases fluorescence in a near infrared phytofluor. Front Mol Biosci 2. https://doi.org/10.3389/fmolb.2015.00065
10. Rockwell NC, Lagarias JC (2019) Phytochrome evolution in 3D: deletion, duplication, and diversification. New Phytol nph.16240. https://doi.org/10.1111/nph.16240
11. Essen LO, Mailliet J, Hughes J (2008) The structure of a complete phytochrome sensory module in the Pr ground state. Proc Natl Acad Sci USA 105:14709–14714. https://doi.org/10.1073/pnas.0806477105
12. Yang X, Kuk J, Moffat K (2008) Crystal structure of Pseudomonas aeruginosa bacteriophytochrome: photoconversion and signal transduction. Proc Natl Acad Sci USA 105:14715–14720. https://doi.org/10.1073/pnas.0806718105
13. Burgie ES, Bussell AN, Walker JM et al (2014) Crystal structure of the photosensing module from a red/far-red light-absorbing plant phytochrome. Proc Natl Acad Sci USA 111:10179–10184. https://doi.org/10.1073/pnas.1403096111

14. Rohmer T, Lang C, Hughes J et al (2008) Light-induced chromophore activity and signal transduction in phytochromes observed by13C and15N magic-angle spinning NMR. Proc Natl Acad Sci USA 105:15229–15234. https://doi.org/10.1073/pnas.0805696105

15. Andel F, Lagarias JC, Mathies RA (1996) Resonance Raman analysis of chromophore structure in the Lumi-R photoproduct of phytochrome [†]. Biochemistry 35:15997–16008. https://doi.org/10.1021/bi962175k

16. Andel F, Murphy JT, Haas JA et al (2000) Probing the photoreaction mechanism of phytochrome through analysis of resonance Raman vibrational spectra of recombinant analogues [†]. Biochemistry 39:2667–2676. https://doi.org/10.1021/bi991688z

17. Foerstendorf H, Lamparter T, Hughes J et al (2007) The photoreactions of recombinant phytochrome from the cyanobacterium synechocystis: a low-temperature UV-Vis and FT-IR spectroscopic study. Photochem Photobiol 71:655–661. https://doi.org/10.1562/0031-8655(2000)0710655TPORPF2.0.CO2

18. Song C, Psakis G, Lang C et al (2011) Two ground state isoforms and a chromophore D-ring photoflip triggering extensive intramolecular changes in a canonical phytochrome. Proc Natl Acad Sci USA 108:3842–3847. https://doi.org/10.1073/pnas.1013377108

19. Yang X, Ren Z, Kuk J, Moffat K (2011) Temperature-scan cryocrystallography reveals reaction intermediates in bacteriophytochrome. Nature 479:428–431. https://doi.org/10.1038/nature10506

20. Auldridge ME, Forest KT (2011) Bacterial phytochromes: more than meets the light. Crit Rev Biochem Mol Biol 46:67–88. https://doi.org/10.3109/10409238.2010.546389

21. Wagner JR, Brunzelle JS, Forest KT, Vierstra RD (2005) A light-sensing knot revealed by the structure of the chromophore-binding domain of phytochrome. Nature 438:325–331. https://doi.org/10.1038/nature04118

22. Anders K, Daminelli-Widany G, Mroginski MA et al (2013) Structure of the cyanobacterial phytochrome 2 photosensor implies a tryptophan switch for phytochrome signaling. J Biol Chem 288:35714–35725. https://doi.org/10.1074/jbc.M113.510461

23. Bellini D, Papiz MZ (2012) Structure of a bacteriophytochrome and light-stimulated protomer swapping with a gene repressor. Structure 20:1436–1446. https://doi.org/10.1016/j.str.2012.06.002

24. Karniol B, Wagner JR, Walker JM, Vierstra RD (2005) Phylogenetic analysis of the phytochrome superfamily reveals distinct microbial subfamilies of photoreceptors. Biochem J 392:103–116. https://doi.org/10.1042/BJ20050826

25. Hughes J (2010) Phytochrome three-dimensional structures and functions. Biochem Soc Trans 38:710–716

26. Yang X, Kuk J, Moffat K (2009) Conformational differences between the Pfr and Pr states in Pseudomonas aeruginosa bacteriophytochrome. Proc Natl Acad Sci USA 106:15639–15644. https://doi.org/10.1073/pnas.0902178106

27. Takala H, Björling A, Berntsson O et al (2014) Signal amplification and transduction in phytochrome photosensors. Nature 509:245–248. https://doi.org/10.1038/nature13310

28. Rockwell NC, Lagarias JC (2010) A brief history of phytochromes. ChemPhysChem 11:1172–1180. https://doi.org/10.1002/cphc.200900894

29. Rockwell NC, Martin SS, Feoktistova K, Lagarias JC (2011) Diverse two-cysteine photocycles in phytochromes and cyanobacteriochromes. Proc Natl Acad Sci USA 108:11854–11859. https://doi.org/10.1073/pnas.1107844108

30. Song JY, Cho HS, Il CJ et al (2011) Near-UV cyanobacteriochrome signaling system elicits negative phototaxis in the cyanobacterium Synechocystis sp. PCC 6803. Proc Natl Acad Sci USA 108:10780–10785. https://doi.org/10.1073/pnas.1104242108

31. Hirose Y, Rockwell NC, Nishiyama K et al (2013) Green/red cyanobacteriochromes regulate complementary chromatic acclimation via a protochromic photocycle. Proc Natl Acad Sci USA 110:4974–4979. https://doi.org/10.1073/pnas.1302909110

32. Rockwell NC, Martin SS, Gan F et al (2015) NpR3784 is the prototype for a distinctive group of red/green cyanobacteriochromes using alternative Phe residues for photoproduct tuning. Photochem Photobiol Sci 14:258–269. https://doi.org/10.1039/c4pp00336e

33. Xu X, Port A, Wiebeler C et al (2020) Structural elements regulating the photochromicity in a cyanobacteriochrome. Proc Natl Acad Sci USA 117:2432–2440. https://doi.org/10.1073/pnas.1910208117

34. Narikawa R, Enomoto G, Ni-Ni-Win, et al (2014) A new type of dual-cys cyanobacteriochrome GAF domain found in cyanobacterium acaryochloris marina, which has an unusual red/blue reversible photoconversion cycle. Biochemistry 53:5051–5059. https://doi.org/10.1021/bi500376b

35. Hauck AFE, Hardman SJO, Kutta RJ et al (2014) The photoinitiated reaction pathway of full-length cyanobacteriochrome Tlr0924 monitored over 12 orders of magnitude. J Biol Chem 289:17747–17757. https://doi.org/10.1074/jbc.M114.566133

36. Velazquez Escobar F, Utesch T, Narikawa R et al (2013) Photoconversion mechanism of the second GAF domain of cyanobacteriochrome AnPixJ and the cofactor structure of its green-absorbing state. Biochemistry 52:4871–4880. https://doi.org/10.1021/bi400506a

37. Rockwell NC, Martin SS, Gulevich AG, Lagarias JC (2012) Phycoviolobilin formation and spectral tuning in the DXCF cyanobacteriochrome subfamily. Biochemistry 51:1449–1463. https://doi.org/10.1021/bi201783j

38. Enomoto G, Hirose Y, Narikawa R, Ikeuchi M (2012) Thiol-based photocycle of the blue and teal light-sensing cyanobacteriochrome Tlr 1999. Biochemistry 51:3050–3058. https://doi.org/10.1021/bi300020u

39. Narikawa R, Fukushima Y, Ishizuka T et al (2008) A novel photoactive GAF domain of cyanobacteriochrome AnPixJ that shows reversible green/red photoconversion. J Mol Biol 380:844–855. https://doi.org/10.1016/j.jmb.2008.05.035

40. Narikawa R, Nakajima T, Aono Y et al (2015) A biliverdin-binding cyanobacteriochrome from the chlorophyll d-bearing cyanobacterium Acaryochloris marina. Sci Rep 5. https://doi.org/10.1038/srep07950

41. Gottlieb SM, Kim PW, Chang C-W et al (2015) Conservation and diversity in the primary forward photodynamics of red/green cyanobacteriochromes. Biochemistry 54:1028–1042. https://doi.org/10.1021/bi5012755

42. Narikawa R, Ishizuka T, Muraki N et al (2013) Structures of cyanobacteriochromes from phototaxis regulators AnPixJ and TePixJ reveal general and specific photoconversion mechanism. Proc Natl Acad Sci U S A 110:918–923. https://doi.org/10.1073/pnas.1212098110

43. Rockwell NC, Martin SS, Lagarias JC (2015) Identification of DXCF cyanobacteriochrome lineages with predictable photocycles. Photochem Photobiol Sci 14:929–941. https://doi.org/10.1039/c4pp00486h

44. Burgie ES, Walker JM, Phillips GN, Vierstra RD (2013) A photo-labile thioether linkage to phycoviolobilin provides the foundation for the blue/green photocycles in DXCF-cyanobacteriochromes. Structure 21:88–97. https://doi.org/10.1016/j.str.2012.11.001

45. Ishizuka T, Kamiya A, Suzuki H et al (2011) The cyanobacteriochrome, TePixJ, isomerizes its own chromophore by converting phycocyanobilin to phycoviolobilin. Biochemistry 50:953–961. https://doi.org/10.1021/bi101626t

46. Sethe Burgie E, Clinger JA, Miller MD et al (2020) Photoreversible interconversion of a phytochrome photosensory module in the crystalline state. Proc Natl Acad Sci USA 117:300–307. https://doi.org/10.1073/pnas.1912041116

47. Boggio-Pasqua M, Burmeister CF, Robb MA, Groenhof G (2012) Photochemical reactions in biological systems: Probing the effect of the environment by means of hybrid quantum chemistry/molecular mechanics simulations. The Royal Society of Chemistry

48. Schäfer LV, Groenhof G, Klingen AR et al (2007) Photoswitching of the fluorescent protein asFP595: mechanism, proton pathways, and absorption spectra. Angew Chemie Int Ed 46:530–536. https://doi.org/10.1002/anie.200602315

49. Gozem S, Luk HL, Schapiro I, Olivucci M (2017) Theory and simulation of the ultrafast double-bond isomerization of biological chromophores. Chem Rev 117:13502–13565. https://doi.org/10.1021/acs.chemrev.7b00177

50. Warshel A (1976) Bicycle-pedal model for the first step in the vision process. Nature 260:679–683. https://doi.org/10.1038/260679a0
51. Polyakov IV, Grigorenko BL, Mironov VA, Nemukhin AV (2018) Modeling structure and excitation of biliverdin-binding domains in infrared fluorescent proteins. Chem Phys Lett 710:59–63. https://doi.org/10.1016/j.cplett.2018.08.068
52. Wiebeler C, Rao AG, Gärtner W, Schapiro I (2019) The effective conjugation length is responsible for the red/green spectral tuning in the cyanobacteriochrome Slr1393g3. Angew Chemie Int Ed 58:1934–1938. https://doi.org/10.1002/anie.201810266
53. Zhuang X, Wang J, Lan Z (2013) Tracking of the molecular motion in the primary event of photoinduced reactions of a phytochromobilin model. J Phys Chem B 117:15976–15986. https://doi.org/10.1021/jp408799b
54. Altoè P, Climent T, De Fusco GC et al (2009) Deciphering intrinsic deactivation/isomerization routes in a phytochrome chromophore model. J Phys Chem B 113:15067–15073. https://doi.org/10.1021/jp904669x
55. Gouterman M (1961) Spectra of porphyrins. J Mol Spectrosc 6:138–163. https://doi.org/10.1016/0022-2852(61)90236-3
56. Soret J-L (1883) Analyse spectrale: Sur le spectre d'absorption du sang dans la partie violette et ultra-violette. Comptes rendus l' Académie des Sci 97:1269-
57. Gouterman M (1959) Study of the effects of substitution on the absorption spectra of porphin. J Chem Phys 30:1139–1161. https://doi.org/10.1063/1.1730148
58. Platt JR (1956) Radiation biology. In: A Hollaender (ed) Radiation biology, vol. III. Visible and near-visible light. McGraw-Hill Book Company, Inc., New York
59. Gouterman M (1978) The Porphyrins V3: Physical Chemistry Elsevier. In: Dolphin D (ed) The Porphyrins V3: Physical Chemistry Elsevier. Elsevier, pp 1–165
60. Slavov C, Fischer T, Barnoy A, et al (2020) The interplay between chromophore and protein determines the extended excited state dynamics in a single-domain phytochrome. Proc Natl Acad Sci 201921706. https://doi.org/10.1073/pnas.1921706117
61. Spillane KM, Dasgupta J, Mathies RA (2012) Conformational homogeneity and excited-state isomerization dynamics of the bilin chromophore in phytochrome cph1 from resonance raman intensities. Biophys J 102:709–717. https://doi.org/10.1016/j.bpj.2011.11.4019
62. Spillane KM, Dasgupta J, Lagarias JC, Mathies RA (2009) Homogeneity of phytochrome Cph1 vibronic absorption revealed by resonance raman intensity analysis. J Am Chem Soc 131:13946–13948. https://doi.org/10.1021/ja905822m
63. Dasgupta J, Frontiera RR, Taylor KC et al (2009) Ultrafast excited-state isomerization in phytochrome revealed by femtosecond stimulated Raman spectroscopy. Proc Natl Acad Sci USA 106:1784–1789. https://doi.org/10.1073/pnas.0812056106
64. Bizimana LA, Epstein J, Brazard J, Turner DB (2017) Conformational homogeneity in the Pr isomer of phytochrome Cph1. J Phys Chem B 121:2622–2630. https://doi.org/10.1021/acs.jpcb.7b02180
65. Song C, Essen LO, Gärtner W et al (2012) Solid-state NMR spectroscopic study of chromophore-protein interactions in the Pr ground state of plant phytochrome A. Mol Plant 5:698–715. https://doi.org/10.1093/mp/sss017
66. Song C, Lang C, Kopycki J et al (2015) NMR chemical shift pattern changed by ammonium sulfate precipitation in cyanobacterial phytochrome Cph1. Front Mol Biosci 2. https://doi.org/10.3389/fmolb.2015.00042
67. Song C, Lang C, Mailliet J et al (2012) Exploring chromophore-binding pocket: high-resolution solid-state 1H–13C interfacial correlation NMR spectra with windowed PMLG scheme. Appl Magn Reson 42:79–88. https://doi.org/10.1007/s00723-011-0196-6
68. Escobar FV, Lang C, Takiden A et al (2017) Protonation-dependent structural heterogeneity in the chromophore binding site of cyanobacterial phytochrome cph1. J Phys Chem B 121:47–57. https://doi.org/10.1021/acs.jpcb.6b09600
69. Macaluso V, Cupellini L, Salvadori G, et al (2020) Elucidating the role of structural fluctuations, and intermolecular and vibronic interactions in the spectroscopic response of a bacteriophytochrome. Phys Chem Chem Phys 22:8585–8594. https://doi.org/10.1039/d0cp00372g

70. Gouterman M, Wagnière GH, Snyder LC (1963) Spectra of porphyrins: part II. four O rbital model. J Mol Spectrosc 11:108–127. https://doi.org/10.1016/0022-2852(63)90011-0
71. Mardirossian N, Head-Gordon M (2017) Thirty years of density functional theory in computational chemistry: an overview and extensive assessment of 200 density functionals. Mol Phys 115:2315–2372. https://doi.org/10.1080/00268976.2017.1333644
72. Runge E, Gross EKU (1984) Density-functional theory for time-dependent systems. Phys Rev Lett 52:997–1000. https://doi.org/10.1103/PhysRevLett.52.997
73. Van Thor JJ, Mackeen M, Kuprov I et al (2006) Chromophore structure in the photocycle of the cyanobacterial phytochrome Cph1. Biophys J 91:1811–1822. https://doi.org/10.1529/biophysj.106.084335
74. Osoegawa S, Miyoshi R, Watanabe K et al (2019) Identification of the deprotonated pyrrole nitrogen of the Bilin-based photoreceptor by raman spectroscopy with an advanced computational analysis. 123:3242–3247. https://doi.org/10.1021/acs.jpcb.9b00965
75. Matute RA, Contreras R, González L (2010a) Time-dependent DFT on phytochrome chromophores: a way to the right conformer. J Phys Chem Lett 1:796–801. https://doi.org/10.1021/jz900432m
76. Wiebeler C, Schapiro I (2019) QM/MM benchmarking of cyanobacteriochrome Slr1393g3 absorption spectra. Molecules 24:1720. https://doi.org/10.3390/molecules24091720
77. Maitra NT, Zhang F, Cave RJ, Burke K (2004) Double excitations within time-dependent density functional theory linear response. J Chem Phys 120:5932–5937
78. Elliott P, Goldson S, Canahui C, Maitra NT (2011) Perspectives on double-excitations in TDDFT. Chem Phys 391:110–119. https://doi.org/10.1016/j.chemphys.2011.03.020
79. Maitra NT (2017). Charge-transfer in time-dependent density functional theory. https://doi.org/10.1088/1361-648X/aa836e
80. Dreuw A, Head-Gordon M (2004) Failure of time-dependent density functional theory for long-range charge-transfer excited states: the zincbacteriochlorin-bacteriochlorin and bacteriochlorophyll-spheroidene complexes. J Am Chem Soc 126:4007–4016. https://doi.org/10.1021/ja039556n
81. Baer R, Livshits E, Salzner U (2010) Tuned range-separated hybrids in density functional theory. Annu Rev Phys Chem 61:85–109. https://doi.org/10.1146/annurev.physchem.012809.103321
82. Christiansen O, Koch H, Jørgensen P (1995) The second-order approximate coupled cluster singles and doubles model CC2. Chem Phys Lett 243:409–418. https://doi.org/10.1016/0009-2614(95)00841-Q
83. Schirmer J (1982) Beyond the random-phase approximation: a new approximation scheme for the polarization propagator. Phys Rev A 26:2395–2416. https://doi.org/10.1103/PhysRevA.26.2395
84. Dreuw A, Wormit M (2015) The algebraic diagrammatic construction scheme for the polarization propagator for the calculation of excited states. Wiley Interdiscip Rev Comput Mol Sci 5:82–95. https://doi.org/10.1002/wcms.1206
85. Schreiber M, Silva-Junior MR, Sauer SPA, Thiel W (2008) Benchmarks for electronically excited states: CASPT2, CC2, CCSD, and CC3. J Chem Phys 128. https://doi.org/10.1063/1.2889385
86. Harbach PHP, Wormit M, Dreuw A (2014) The third-order algebraic diagrammatic construction method (ADC(3)) for the polarization propagator for closed-shell molecules: efficient implementation and benchmarking. J Chem Phys 141. https://doi.org/10.1063/1.4892418
87. Rhee YM, Head-Gordon M (2007) Scaled second-order perturbation corrections to configuration interaction singles: efficient and reliable excitation energy methods. J Phys Chem A 111:5314–5326. https://doi.org/10.1021/jp068409j
88. Lischka H, Nachtigallová D, Aquino AJA et al (2018) Multireference approaches for excited states of molecules. Chem Rev 118:7293–7361. https://doi.org/10.1021/acs.chemrev.8b00244
89. Schapiro I, Melaccio F, Laricheva EN, Olivucci M (2011) Using the computer to understand the chemistry of conical intersections. Photochem Photobiol Sci 10:867–886

90. Polyakov I, Grigorenko B, Nemukhin A (2019) Modeling structure and absorption spectra of the bacteriophytochrome-based fluorescent protein IFP1.4. https://doi.org/10.26434/CHE MRXIV.10297925.V1

91. Rockwell NC, Njuguna SL, Roberts L et al (2008) A second conserved GAF domain cysteine is required for the blue/green photoreversibility of cyanobacteriochrome Tlr0924 from *Thermosynechococcus elongatus* [†]. Biochemistry 47:7304–7316. https://doi.org/10. 1021/bi800088t

92. Tomasi J, Mennucci B, Cammi R (2005) Quantum mechanical continuum solvation models. Chem Rev 105:2999–3093. https://doi.org/10.1021/cr9904009

93. Matute RA, Contreras R, Pérez-Hernández G, González L (2008) The chromophore structure of the cyanobacterial phytochrome Cph1 as predicted by time-dependent density functional theory. J Phys Chem B 112:16253–16256. https://doi.org/10.1021/jp807471e

94. Chung LW, Sameera WMC, Ramozzi R et al (2015) The ONIOM method and its applications. Chem Rev 115:5678–5796. https://doi.org/10.1021/cr5004419

95. Mroginski MA, Mark F, Thiel W, Hildebrandt P (2007) Quantum mechanics/molecular mechanics calculation of the Raman spectra of the phycocyanobilin chromophore in α-C-phycocyanin. Biophys J 93:1885–1894. https://doi.org/10.1529/biophysj.107.108878

96. Kästner J, Thiel S, Senn HM et al (2007) Exploiting QM/MM capabilities in geometry optimization: a microiterative approach using electrostatic embedding. J Chem Theory Comput 3:1064–1072. https://doi.org/10.1021/ct600346p

97. Pezeshki S, Lin H (2015) Recent developments in QM/MM methods towards open-boundary multi-scale simulations. Mol Simul 41:168–189. https://doi.org/10.1080/08927022.2014. 911870

98. Lin H, Truhlar DG (2007) QM/MM: what have we learned, where are we, and where do we go from here? In: Theoretical Chemistry Accounts. pp 185–199

99. Olsen JMH, Steinmann C, Ruud K, Kongsted J (2015) Polarizable density embedding: a new QM/QM/MM-based computational strategy. J Phys Chem A 119:5344–5355. https://doi.org/ 10.1021/jp510138k

100. Senn HM, Thiel W (2009) QM/MM methods for biomolecular systems. Angew Chemie Int Ed 48:1198–1229. https://doi.org/10.1002/anie.200802019

101. Benighaus T, Thiel W (2009) A general boundary potential for hybrid QM/MM simulations of solvated biomolecular systems. J Chem Theory Comput 5:3114–3128. https://doi.org/10. 1021/ct900437b

102. Filippi C, Buda F, Guidoni L, Sinicropi A (2012) Bathochromic shift in green fluorescent protein: a puzzle for QM/MM approaches. J Chem Theory Comput 8:112–124. https://doi. org/10.1021/ct200704k

103. Steinbrecher T, Elstner M (2013) QM and QM/MM simulations of proteins. pp 91–124

104. Wanko M, Hoffmann M, Strodel P et al (2005) Calculating absorption shifts for retinal proteins: computational challenges. J Phys Chem B 109:3606–3615. https://doi.org/10.1021/ jp0463060

105. Falklöf O, Durbeej B (2013a) Modeling of phytochrome absorption spectra. J Comput Chem 34:1363–1374. https://doi.org/10.1002/jcc.23265

106. PW Atkins, J De Paula JK (2018) Atkins' physical chemistry. Oxford University Press

107. Hollas J (2004) Modern Spectroscopy. John Wiley & Son's

108. Palmer CW (1989) Theory and methods of calculation of molecular spectra. J Mod Opt 36:1545–1545. https://doi.org/10.1080/09500348914551591

109. Franzen S (2003) Use of periodic boundary conditions to calculate accurate β-sheet frequencies using density functional theory. J Phys Chem A 107:9898–9902. https://doi.org/10.1021/ jp035215k

110. Rüger R, van Lenthe E, Lu Y et al (2015) Efficient calculation of electronic absorption spectra by means of intensity-selected time-dependent density functional tight binding. J Chem Theory Comput 11:157–167. https://doi.org/10.1021/ct500838h

111. Alata I, Bert J, Broquier M et al (2013) Electronic spectra of the protonated indole chromophore in the gas phase. J Phys Chem A 117:4420–4427. https://doi.org/10.1021/jp4 02298y

112. Kaczor A, Reva I, Fausto R (2013) Influence of cage confinement on the photochemistry of matrix-isolated e -β-ionone: FT-IR and DFT study. J Phys Chem A 117:888–897. https://doi.org/10.1021/jp310764u

113. Uppsten M, Durbeej B (2012) Quantum chemical comparison of vertical, adiabatic, and 0–0 excitation energies: the PYP and GFP chromophores. J Comput Chem 33:1892–1901. https://doi.org/10.1002/jcc.23027

114. Send R, Sundholm D (2007) Coupled-cluster studies of the lowest excited states of the 11-cis-retinal chromophore. Phys Chem Chem Phys 9:2862–2867. https://doi.org/10.1039/b616137e

115. Sodt AJ, Mei Y, König G et al (2015) Multiple environment single system quantum mechanical/molecular mechanical (MESS-QM/MM) calculations. 1. estimation of polarization energies. J Phys Chem A 119:1511–1523. https://doi.org/10.1021/jp5072296

116. Groenhof G (2013) Introduction to QM/MM simulations, pp 43–66

117. Lei M, Zavodszky MI, Kuhn LA, Thorpe MF (2004) Sampling protein conformations and pathways. J Comput Chem 25:1133–1148. https://doi.org/10.1002/jcc.20041

118. Modi V, Donnini S, Groenhof G, Morozov D (2019) Protonation of the biliverdin IXα chromophore in the red and far-red photoactive states of a bacteriophytochrome. J Phys Chem B 123:2325–2334. https://doi.org/10.1021/acs.jpcb.9b01117

119. Duan M, Liu N, Zhou W et al (2016) Structural diversity of ligand-binding androgen receptors revealed by microsecond long molecular dynamics simulations and enhanced sampling. J Chem Theory Comput 12:4611–4619. https://doi.org/10.1021/acs.jctc.6b00424

120. Boltzmann L (1898) Vorlesungen über Gastheorie: Th. Theorie van der Waals'; Gase mit zusammengesetzten Molekülen; Gasdissociation; Schlussbemerkungen

121. Mauro JC, Gupta PK, Loucks RJ (2007) Continuously broken ergodicity. J Chem Phys 126. https://doi.org/10.1063/1.2731774

122. Warshel A, Levitt M (1976) Theoretical studies of enzymic reactions: Dielectric, electrostatic and steric stabilization of the carbonium ion in the reaction of lysozyme. J Mol Biol 103:227–249. https://doi.org/10.1016/0022-2836(76)90311-9

123. Battocchio G, González R, Rao AG, et al (2020) Dynamic Properties of the Photosensory Domain of Deinococcus radiodurans Bacteriophytochrome. J Phys Chem B 124:1740–1750. https://doi.org/10.1021/acs.jpcb.0c00612

124. Freddolino PL, Liu F, Gruebele M, Schulten K (2008) Ten-microsecond molecular dynamics simulation of a fast-folding WW domain. Biophys J 94. https://doi.org/10.1529/biophysj.108.131565

125. Ugur I, Marion A, Aviyente V, Monard G (2015) Why does Asn71 deamidate faster than Asn15 in the enzyme triosephosphate isomerase? Answers from microsecond molecular dynamics simulation and QM/MM free energy calculations. Biochemistry 54:1429–1439. https://doi.org/10.1021/bi5008047

126. Paulikat M, Mata RA, Gelabert R (2019) A high-throughput computational approach to UV-Vis spectra in protein mutants. Phys Chem Chem Phys 21:20678–20692. https://doi.org/10.1039/c9cp03908b

127. Li Z, Hirst JD (2017) Quantitative first principles calculations of protein circular dichroism in the near-ultraviolet. Chem Sci 8:4318–4333. https://doi.org/10.1039/c7sc00586e

128. Durbeej B, Borg OA, Eriksson LA (2004) Phytochromobilin C15-Z, syn → C15-E, anti isomerization: concerted or stepwise? Phys Chem Chem Phys 6:5066–5073. https://doi.org/10.1039/b411005f

129. Durbeej B, Eriksson LA (2006) Protein-bound chromophores astaxanthin and phytochromobilin: excited state quantum chemical studies. Phys Chem Chem Phys 8:4053–4071

130. Borg OA, Durbeej B (2007a) Relative ground and excited-state p K a values of phytochromobilin in the photoactivation of phytochrome: a computational study. J Phys Chem B 111:11554–11565. https://doi.org/10.1021/jp0727953

131. Strambi A, Durbeej B (2011) Initial excited-state relaxation of the bilin chromophores of phytochromes: a computational study. Photochem Photobiol Sci 10:569–579. https://doi.org/10.1039/c0pp00307g

132. Falklöf O, Durbeej B (2013b) Red-light absorption and fluorescence of phytochrome chromophores: a comparative theoretical study. Chem Phys 425:19–28. https://doi.org/10.1016/j.chemphys.2013.07.018
133. Elstner M, Porezag D, Jungnickel G et al (1998) Self-consistent-charge density-functional tight-binding method for simulations of complex materials properties. Phys Rev B 58:7260–7268. https://doi.org/10.1103/PhysRevB.58.7260
134. Niehaus TA, Elstner M, Frauenheim T, Suhai S (2001) Application of an approximate density-functional method to sulfur containing compounds. J Mol Struct Theochem 541:185–194. https://doi.org/10.1016/S0166-1280(00)00762-4
135. Elstner M, Hobza P, Frauenheim T et al (2001) Hydrogen bonding and stacking interactions of nucleic acid base pairs: a density-functional-theory based treatment. J Chem Phys 114:5149–5155. https://doi.org/10.1063/1.1329889
136. Yang Y, Linke M, von Haimberger T et al (2014) Active and silent chromophore isoforms for phytochrome Pr photoisomerization: an alternative evolutionary strategy to optimize photoreaction quantum yields. Struct Dyn 1:014701. https://doi.org/10.1063/1.4865233
137. Rockwell NC, Shang L, Martin SS, Lagarias JC (2009) Distinct classes of red/far-red photochemistry within the phytochrome superfamily. Proc Natl Acad Sci USA 106:6123–6127. https://doi.org/10.1073/pnas.0902370106
138. Feliks M, Lafaye C, Shu X et al (2016) Structural determinants of improved fluorescence in a family of bacteriophytochrome-based infrared fluorescent proteins: insights from continuum electrostatic calculations and molecular dynamics simulations. Biochemistry 55:4263–4274. https://doi.org/10.1021/acs.biochem.6b00295
139. Maximowitsch E, Domratcheva T (2020) A hydrogen bond between linear tetrapyrrole and conserved aspartate causes the far-red shifted absorption of phytochrome photoreceptors. https://doi.org/10.26434/CHEMRXIV.12278780.V1
140. Durbeej B (2009) On the primary event of phytochrome: quantum chemical comparison of photoreactions at C4, C10 and C15. Phys Chem Chem Phys 11:1354–1361. https://doi.org/10.1039/b811813b
141. De Vetta M, Baig O, Steen D et al (2018) Assessing configurational sampling in the quantum mechanics/molecular mechanics calculation of Temoporfin absorption spectrum and triplet density of states†. Molecules 23:2932. https://doi.org/10.3390/molecules23112932
142. Kjellgren ER, Haugaard Olsen JM, Kongsted J (2018) Importance of accurate structures for quantum chemistry embedding methods: which strategy is better? J Chem Theory Comput 14:4309–4319. https://doi.org/10.1021/acs.jctc.8b00202
143. De La Lande A, Alvarez-Ibarra A, Hasnaoui K et al (2019) Molecular simulations with indeMon2k QM/MM, a tutorial-review. Molecules 24:1653. https://doi.org/10.3390/molecules24091653

QM/MM Study of Bioluminescent Systems

Isabelle Navizet

Abstract Bioluminescence is a chemical reaction of (usually multi-steps) oxidation of a substrate (luciferin) in the presence of an enzyme (luciferase). The oxidation leads to a product in its electronic excited state, which releases the chemical energy in the form of light. Quantum mechanic/molecular mechanic (QM/MM) methods are approaches of choice to study bioluminescent reactions as they allow the study of electronic transitions taking into account the protein environment. In this chapter, we will present the QM/MM methods that are used for studying bioluminescent reactions. How the enzymatic environment can be taken into account? What are the difficulties to describe the excited states? Which experimental data are needed to be able to perform studies of bioluminescent systems? How simulations can help interpreting and predicting experimental observation? Tentative answers to these questions and some examples of studies of firefly's bioluminescent system have been reported in this chapter.

Keywords Bioluminescence · QM/MM · Luciferin · Oxyluciferin · Luciferase · Firefly

1 What Is Bioluminescence?

Bioluminescence is a phenomenon that produces and emits light by a living organism through a chemical reaction. The word "bioluminescence" comes from the Greek term *bios*, that means life and Latin term *lumen* for light. Fireflies are the oldest known bioluminescent systems and Aristotle was already mentioning them in his book *De Anima*. Other species, such as land animals and aquatic animals, are also known to be able to emit light. Among them, we can cite railroad worm, click beetle, firefly, jellyfish, protozoa, sponges, marine bacteria, dinoflagellates, crustaceans, and

I. Navizet (✉)
Laboratoire Modélisation et Simulation Multiéchelle, MSME UMR 8208, Université Gustave Eiffel, UPEC, CNRS, 77454 Marne-la-Vallée, France
e-mail: isabelle.navizet@univ-eiffel.fr; isabelle.navizet@u-pem.fr

© Springer Nature Switzerland AG 2021
T. Andruniów and M. Olivucci (eds.), *QM/MM Studies of Light-responsive Biological Systems*, Challenges and Advances in Computational Chemistry and Physics 31, https://doi.org/10.1007/978-3-030-57721-6_5

molluscs. The vegetal world has also examples of bioluminescent systems, notably amongst mushrooms. In the living species, bioluminescent can be used to communicate (the fireflies choose their mates by recognizing the frequency of the light flashes), to defend themselves (light flashes or cloud of bioluminescent substance sent to distract or blind a predator), to hide (bioluminescent blue light emitted from the belly of fishes or medusa that mimics the blue light of the sky and prevents the predators that are below the animal to see it), to attract as a lure (used by certain fishes with bacterial bioluminescent lantern). More information on all uses of bioluminescence in the sea can be found in the 2010 review from Haddock [28].

The scientific research field of bioluminescence has seen a lot of progress in the last half century. Since the first crystal structures were resolved, new technologies have been developed not only in the experimental laboratories but also in the modeling field. A better understanding of chemical mechanism of the system has also helped to develop new applications, especially in the domain of sensors. The chemistry Nobel Prize of 2008 was award to Shimomura, Chalfie and Tsien for the discovery and the development of the Green Fluorescent Protein (GFP) of the *Aequorea victoria* jellyfish, protein that is activated by a bioluminescent reaction.

2 The Bioluminescent Reaction: An Overview of the Reaction Scheme

Chemically, all bioluminescent reactions involve the oxidation of a substrate (a small organic molecule called luciferin) in the presence of an enzyme (luciferase). This reaction produces an intermediate of reaction that usually shows a cyclic peroxide moiety that breaks down to produce a molecule in an electronic excited state (the light emitter, called oxyluciferin in firefly's system), that is able to release energy as light. In this section, we will describe first the reaction steps in the firefly's bioluminescence system. Then we will present the other bioluminescent systems and highlight the similarities and the differences with the firefly's system.

In fireflies, light comes from the electronic de-excitation of the oxyluciferin molecule. The oxyluciferin comes from the oxidation of the luciferin substrate in the cavity of the luciferase. The firefly luciferase is composed of about 550 amino acids. In the protein cavity, the D-luciferin reacts with adenosine triphosphate (ATP) to form a D-luciferyl adenylate intermediate. This reaction intermediate reacts with a dioxygen molecule to lead to a firefly dioxetanone intermediate. Decarboxylation of the dioxetanone intermediate conducts to a carbonyl function on the oxyluciferin molecule, excited in its first singlet excited state. The relaxation of oxyluciferin toward the ground state is accompanied by the emission of a photon in the visible spectrum (see Fig. 1).

From the previous description of the firefly bioluminescent reaction, we can discern three main steps in the mechanisms that are common with all chemi- and bioluminescent reactions: first the generation of a high-energy intermediate (HEI) (the

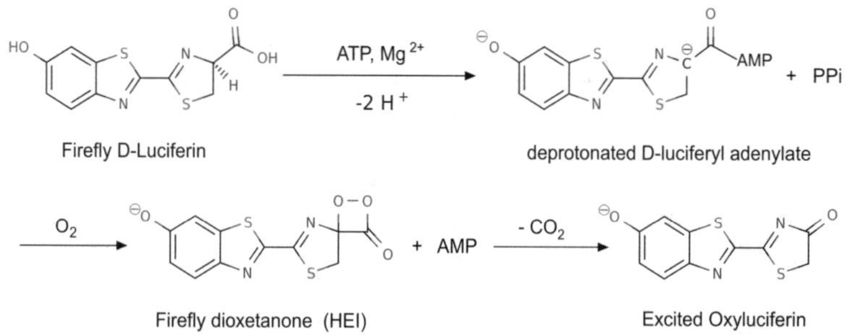

Fig. 1 Reaction scheme of firefly's bioluminescent reaction. The firefly D-luciferin reacts inside the protein cavity with ATP leading to a reaction intermediate D-luciferyl adenylate that is oxidized to a high-energy intermediate (HEI), the firefly dioxetanone. The decomposition of HEI leads to the oxyluciferin product in its first singlet excited state. Reprinted (adapted) with permission from (Vacher et al. Chemical Reviews, 2018, 118(15):6927–6974.). Copyright (2018) American Chemical Society

firefly dioxetanone in firefly bioluminescence) (see Sect. 5), second the decomposition of the HEI that produce an excited light emitter (here the oxyluciferin in its first excited state) (see Sect. 6), and third the physical electronic phenomena of light emission (see Sect. 7). All of these steps occur in the cavity of the protein in bioluminescent reactions. Therefore, the contribution of computational chemistry to better understand or predict each step is more complete if including the protein environment. This requires, for example, the use of QM/MM simulations.

Before discussing the contribution of the QM/MM studies on each of the three steps (see Sects. 5, 6 and 7), it is worth highlighting the similarities in different bioluminescent systems. These similarities were first identified in a 2011 review [60] and classified depending on the nature of the HEI in a 2018 review [88]. The HEI is the intermediate that is not stable and dissociates leading to a system in its excited state. It involves usually a peroxy bond and, in firefly, this peroxy bond is part of a dioxetanone cycle. We can classify the bioluminescent systems by the ones with a dioxetanone HEI like the firefly HEI and the ones that have only a peroxy bond (see Figs. 2 and 3).

The formation of the HEI takes place in the ground state and is so-called thermal reaction. The theoretical study of this reaction step consists of understanding the different paths leading or not to the HEI. In Firefly bioluminescent reaction, the reaction step involving the molecular oxygen is a big challenge to model as the molecular oxygen is in its triplet state and the resulting system is in a singlet state, forcing to explore intersystem crossings.

The decomposition of the HEI is also very challenging for the theoretical chemists as it involves crossing among different electronic states. Therefore, the use of very accurate methods able to describe the crossings (like the time-consuming multiconfigurational methods) is required. The simulations aim to understand what, in bioluminescent systems, favors the formation of the light emitter in an excited state over the decomposition of the HEI without light emission. This usually involves the

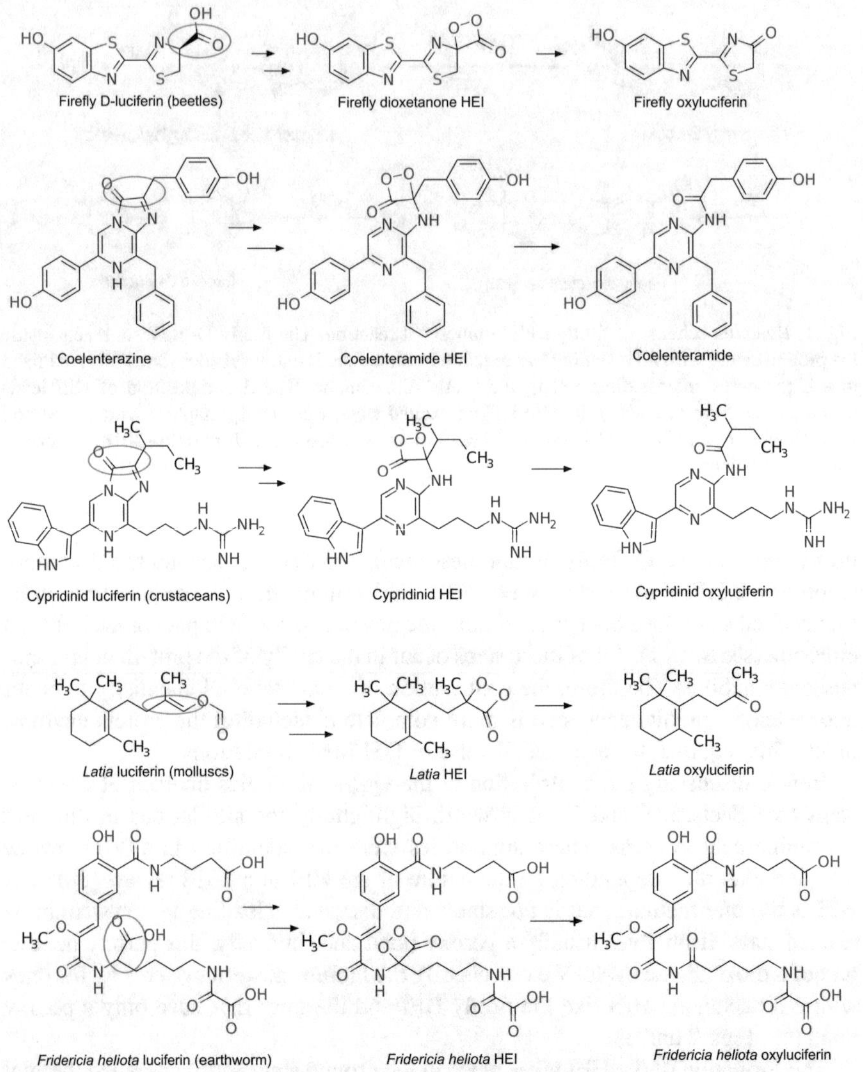

Fig. 2 Commonly proposed reactions of known luciferins through a 1,2-dioxetanone high-energy intermediate (HEI). The oxidation process takes place in the molecular part circled in the luciferin. All structures are shown in their protonated neutral states. The protonation state of the different species is for most cases an unsolved question. The first steps leading to the HEI are not explicitly drawn and represented by two arrows. Reprinted (adapted) with permission from (Vacher et al. Chemical Reviews, 2018, 118(15):6927–6974.). Copyright (2018) American Chemical Society

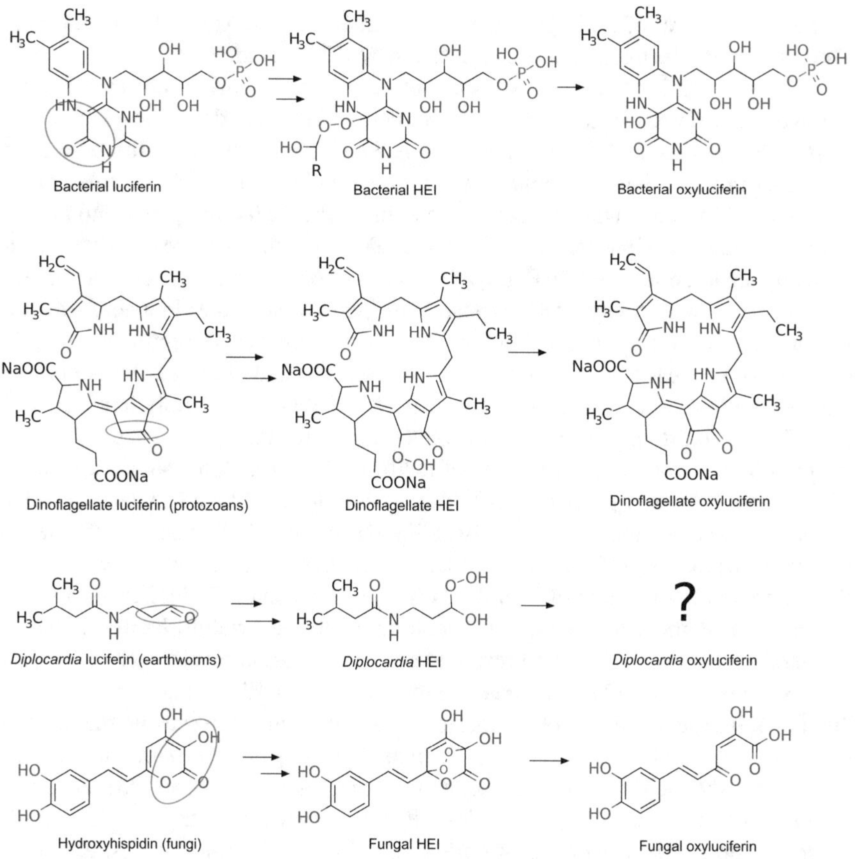

Fig. 3 Commonly proposed reactions of known luciferins through a non-dioxetanone high-energy intermediate (HEI). The oxidation process takes place in the molecular part circled in the luciferin. All structures are shown in their protonated neutral states. The protonation state of the different species is for most cases an unsolved question. The first steps leading to the HEI are not explicitly drawn and represented by two arrows. Reprinted (adapted) with permission from (Vacher et al. Chemical Reviews, 2018, 118(15):6927–6974.). Copyright (2018) American Chemical Society

presence of a so-called activator: a part of the molecule or external catalyst rich in electrons that promotes the formation of the product in the excited state. The protein surrounding can modulate this activator.

The final part of the reaction, the electronic transition that leads to the emission of light, is the most studied part of the reaction with theoretical tools. Computational chemistry is helping to understand the origin for the experimental emission spectra and the reasons for color modulation. Questions about the chemical form of the light emitter, the electronic nature of transition or the charge-transfer character can be investigated. Experimentally, the color modulation can be induced by mutation in the protein [9, 38, 54, 58, 84], chemical modifications of the light emitter or changes of the temperature [53], pH [3, 4], or concentration of cations in the solvent [92]. These can partly be explained or predicted by QM/MM studies.

In Figs. 2 and 3 of reaction schemes of different bioluminescent systems, we see that the different systems have in common an HEI that decomposes to a light emitter, as discussed for the firefly's system. Some similarities and differences not presented in these figures should be highlighted, notably the different roles of the protein and cofactors in the reaction scheme. In firefly, the luciferin covalently bonds with an adenosine triphosphate (ATP) molecule inside the enzyme cavity to create an adenylate intermediate (see Fig. 1). Without ATP or enzyme, the reaction does not occur. The enzyme cavity probably stabilizes the two molecules in a conformation which allows the reaction. The first function of the luciferase is therefore to bring the reactants next to each other in a confined environment and to force them adopting a reactive conformation that leads to the HEI. Without enzyme, firefly luciferin with ATP, Mg^{2+} and dioxygen does not produce light. To mimic the reaction without the protein, a system of condensation of luciferin with propylphosphonic anhydride (T3P) and a base was proposed [39]. This system leads to an adenylate-like intermediate that is decomposed to an HIE alike the firefly's HIE. The use of T3P for studying the reaction without the enzyme, i.e., the chemiluminescence, is very attractive compared to the fastidious synthesis and purification of the luciferyl adenylate intermediate without an enzyme. Coelenterazine and cypridinid luciferin share a common core moiety. They are present in numerous marine species (jellyfishes, sea pens, anemones, etc.) [74]. The reaction leading to the HEI does not need any ATP but the luciferase (aequorin, obelin,...) requires Ca^{2+} cations to activate the reaction of the coelenterazine with dioxygen to form the HEI [90]. For cyprodinil bioluminescence, only the luciferin, the enzyme, and the dioxygen are needed [75]. *Federicia heliota*'s bioluminescent system needs ATP and dioxygen, like the firefly's bioluminescent system [20]. The mucus of *Latia* limpet-like snails, from the New Zealand's freshwaters, emits a blue-green light. Unlike most luciferins that have π-conjugated moieties, *Latia* luciferin is not fluorescent and does not absorb visible light. This supposes that a very different bioluminescence mechanism may involve several partners like a chromophore bounded to *Latia* luciferase [62]. In summary, the primary role of the luciferase is to catalyze the reaction leading to HEI. It can control the reaction rate of bioluminescence by binding to either ions or other cofactors to produce light. This gives to some organisms the ability to control their light emission [74, 95].

3 Modeling a Bioluminescent Reaction

The reaction of bioluminescence is a chemiluminescent reaction catalyzed by an enzyme, the luciferase. Chemiluminescent reactions are reactions of oxidation of a reactant, which produces a species in an excited state, the light emitter. The decay of the light emitter to its ground state produces light. In order to better apprehend the chemiluminescent reaction, we propose here to compare it on one hand with thermal reaction and on the other hand with other cold light phenomena that are able to emit light by electronic transition.

3.1 Modeling a Thermal Reaction

The thermal chemical reaction involved breakage and/or formation of bonds starting from one or several reactants to produce one or several products. Kinetic models help to better understand the chemical reactions. The reaction can be elementary, when it involves only one step reaction, or complex, described by multiple elementary reaction steps and forming intermediates.

In elementary reactions (see Fig. 4a), the so-called reaction coordinate (RC) is the coordinate that permits to follow the energy of the system from the reactant to the product. It could be, for example, the length of the bond that is broken or created or a more complicated coordinate like the combination of angles or/and bonds values. For example, in a nucleophilic substitution of second-order S_N2, the RC is a combination of the lengths of a breaking bond and a forming bond. From the reactant to the product, the energy of the system rises to reach a maximum so-called transition state (TS) and then decreases along the RC. Reactant and product correspond to states that are in a minimum of the potential energy surface of the electronic ground state (GS). The energy barrier, corresponding to the difference of energy of the TS and the one of the reactants, is crossed to form the product thanks to enough thermal energy.

Fig. 4 Energetic profiles (cut of the electronic potential energy surface along the reaction coordinate) of two thermal reactions. **a** Example of an elementary reaction. **b** Example of a complex reaction. Intermediates of reaction (IR) correspond to structures at a minimum in energy. Transition states (TS) correspond to structures at a maximum of energy

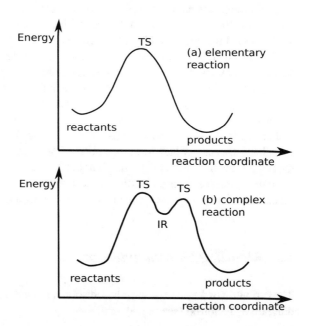

In complex reaction, intermediates of reactions corresponding to minimum of energy are subsequently formed separated by TS, corresponding to maximum of energy (see Fig. 4b). For example, in the nucleophilic substitution of first-order S_N1, the first step corresponds to the breakage of the bond between the substrate and the leaving group leading to a carbocation as an intermediate of reaction. Then, formation of the bond between the carbocation and the nucleophile during the second step leads to the product of the reaction. The RC corresponds to the bond between the carbon and the leaving group in the first step and, in the second step, to the bond between the carbon and the nucleophile. The energetic profile corresponds to a cut of the electronic potential energy surface of the system in the electronic ground state (GS). In the previous description, no excited state (ES) is involved. Therefore, this type of reaction cannot lead to products in an electronic excited state (ES).

3.2 Cold Light Phenomena

Two phenomena can lead to light: The incandescence or more commonly called thermal glow is produced by hot objects that emit thermal radiation in the visible spectra and follows the law of the black-body (examples are the incandescent lights bulbs and the sun). The other phenomena are called "cold light phenomena" (in contrast to incandescence of hot body). Light comes from the relaxation of a light emitter, a molecule or system that is in an electronic excited state (ES), to a lower excited state and in particular to the electronic ground state (GS) of the system.

Bioluminescence is an example of the cold light phenomena. Before presenting the modeling of bioluminescence and photoluminescence in details in the next section, we herein list the other cold light phenomena.

The different types of cold light luminescence can be classified by the excitation mode, i.e., how the excited state of the light emitter is populated. In electroluminescence, the species were excited by an electric field, as it is used in the light-emitting diode (LED) [52]. The modification of a crystal structure can lead to crystalloluminescence, the modification of temperature to a thermoluminescence, and a mechanical action to the triboluminescence (when bonds of the material are broken) [55]. Piezoluminescence comes from the action of pressure and sonoluminescence from the imploding of bubbles [73]. The photoluminescence is the result of absorption of photons. Fluorescence is a photoluminescence involving a singlet–singlet electronic relaxation while phosphorescence involves a triplet–singlet electronic relaxation.

3.3 Modeling Photoluminescence

The evolution of a system after it has absorbed a photon can be explained using the diagram of Jablonski (see Fig. 5).

Fig. 5 Jablonski's diagram with the different possible photophysical processes after absorption (abs) of a photon. Radiative processes include fluorescence (fluo) and phosphorescence (phos). Non-radiative processes include vibrational relaxation (VR), internal conversion (IC), and intersystem crossing (ISC)

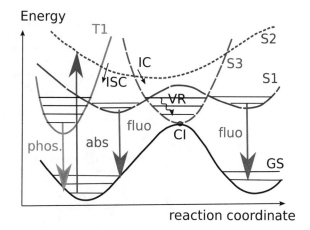

3.3.1 Absorption

First, the system is considered to be in the most stable conformation of its electronic GS. Finding the minimum of the electronic GS is a routine algorithm in most quantum chemistry packages. After photon absorption in the range of ultraviolet (UV) to infrared (IR), the system reaches an electronic ES. The wavelength of the absorbed photon is calculated at the first approximation from the difference of the energy of the electronic ES and the GS at the GS optimized structure (GSoptStr).

Here, it can be taken into account of the vibrational and rotational states in both ground and excited state. The system energy depends on the electronic state and the vibrational and rotational quantum numbers, denoted ν and J respectively for diatomic molecules. Indeed, the system at 0 K in its electronic GS is populating at least the first level of its vibronic states. In the Born–Oppenheimer approximation, the energy of the vibronic states can be taken as the sum of electronic, vibrational, and rotational energy. In the Fig. 6, the vibronic levels are represented for the different electronic states. The system starts in the electronic ground state in its vibrational ground state $\nu'' = 0$. The vibrational states $\nu' = i, i + 1, i + 2, ..., j$ reached in the excited state are the ones following the Franck–Condon principle, i.e., with the same nuclear coordinates. This is called a vertical transition. Therefore the absorption spectrum comes from the contribution of the electronic transitions and the vibronic couplings.

In practice, experiments give the shape of the absorption spectra, i.e., the intensity for a range of wavelengths, and the wavelength λ_{max}^{exp} corresponding to the maximum of the intensity of the absorption spectra. Simulations try to estimate, predict, or explain the shape on one hand and the relative values of λ_{max}^{exp} on the other hand when comparing similar systems. As a first approximation, and for simplicity, λ_{max}^{exp} is compared to the calculated wavelength corresponding to the vertical electronic transition from the ground state to the excited state that has the highest oscillator strength value f. The calculations can be refined, taken into account the energy of the

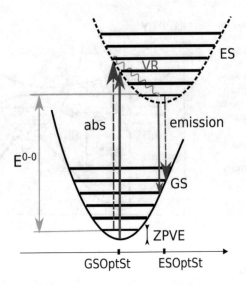

Fig. 6 Electronic transitions. Absorption is calculated as the vertical transition at the ground-state optimized structure GSOptSt (red plain arrow), or the vertical transition from the first vibrational state taking into account the zero point vibrational energy level (ZPVE) (blue dotted arrow). Emission is calculated as the vertical transition at the excited state optimized structure ESOptSt (red plain arrow) or the transition corresponding to the blue dotted arrow when taking into account the ZPVE of the ES. E_{0-0} is the difference of energy of the ES at the ESOptStr taking into account the ZPVE of the ES and the energy of the GS at the GSOptStr taking into account the ZPVE of the GS

zero point vibrational energy (ZPVE) obtained by harmonic approximation at the GSoptStr.

3.3.2 Fluorescence

Once the photon is absorbed, the system is in an excited state. The vibrational relaxation leads to the optimized structure of the excited state (ESOptStr) (see Fig. 6). Quantum chemistry packages can optimize the structure in excited state and calculate the vertical transition energy at this structure. The wavelength corresponding to the vertical energy can be compared to the wavelength of the most intensity of emission spectra. Again, this comparison neglects the vibrational states and ZPVE in the excited state. To take them into account, it is needed to perform the calculation of the matrix of second energy-derivatives (the Hessian) of the excited state energy that yields harmonic vibrational frequencies upon diagonalization.

The approximation of comparing the calculated vertical transition energies to the energies corresponding to wavelengths of the maxima of experimental absorption or emission spectra gives good correlation for rigid systems. For more flexible systems, the energy corresponding to the crossing of the absorption and emission spectra, AFCP (Absorption-Fluorescence Crossing Point) can be compared to the E_{0-0}, i.e., the difference of energy of the ES at the ESOptStr taking into account the ZPVE of the ES and the energy of the GS at the GSOptStr taking into account the ZPVE of the GS [36]. Vibrational resolved absorption and emission spectra can be computed (see references [16, 25, 30] for firefly oxyluciferin systems) but are computational time consuming. The shape of the spectra can be refined (instead of a rough Gaussian-like shape band, having a vibronic band shape) but for many systems the resolution of the experimental spectra does not show the vibrational transitions. Therefore, theoretical studies on absorption and fluorescence spectra of the light emitter of bioluminescent systems are mostly ignoring this aspect.

3.3.3   Non-radiative Processes

As shown in the Jablonski's diagram (see Fig. 5), other phenomena can occur once the photon is absorbed. If two states of the same spin multiplicity cross, an internal conversion (IC) can occur where the system goes from one excited state to the other excited state throw a conical intersection (CI). If the second state (or the last step after few CI) is the GS then no radiative phenomena (light emissions) are observed. The structure relaxation can lead to the same structure of the system before radiation or to another arrangement of the nuclei, leading to a photoproduct of the reaction. If the second state is an excited state, the fluorescence emission can then occurs after vibrational relaxation in this excited state.

   If the state reached by the absorption crosses a state with a different multiplicity, i.e., a singlet state crossing a triplet state, an intersystem crossing can occur. As the transition is spin-forbidden, it will be allowed if the spin–orbit coupling at the crossing structure is high enough. After vibronic relaxation in the reached state, for example, triplet state, the radiative transition to the singlet GS is called phosphorescence. As the transition is spin-forbidden, the phosphorescence is much slower than the fluorescence phenomena.

   The phenomena of non-radiative processes involved conical intersections. Avoided crossing between states will lead to barrier in the lower state, as can be seen on the S1 state in Fig. 5, due to an avoided crossing between state S1 and state S2.

3.4   Back to the Chemiluminescent and Bioluminescent Reactions

The main difference between the chemiluminescent reaction and a fluorescent experiment is the way the system reaches the excited state. All radiative and non-radiative processes that can occur in the fluorescent experiment can also occur in the chemiluminescent reaction.

Fig. 7 Potential energy curves involved in a bioluminescent reaction for the decomposition of the HEI to light emitter. The presence of a conical intersection (CI) near the transition state of the GS is required to allow the system to thermally go from the GS to the ES

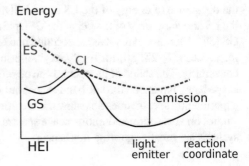

Let consider a simple case involving only the GS and the first singlet ES of the reaction step that permits the system to reach the ES (Fig. 7). In this figure, we only look at the last step of the chemi- or bioluminescent scheme, i.e., the decomposition of HEI intermediate. The two states should cross in order the system to pass from one state to the other state without radiation phenomena. The conical intersection is nearby the transition state of the reaction and the energetic barrier to reach it should be low enough for the system to be able to reach it. Once in the ES, the system undergoes vibrational relaxation before light emission process.

A fluorescence experiment on the product of the chemiluminescence can lead or not to the same light emission as obtained by the chemiluminescence experiment. Indeed, once the product system relaxed in the GS, the conformation of the system has changed from the ES conformation obtained after chemiluminescence reaction. The fluorescence experiment on this product can lead either to an relaxed excited state with an other conformation as the one of the chemiluminescent ES conformation or to the same conformation. This leads to different or same transition energy as the chemiluminescence reaction. A lot of experimental studies of chemiluminescent systems are performed by exploring the fluorescent properties of the molecules produced by chemiluminescent reaction. The validation of this practice is therefore important. We will go back to this point in Sect. 7.1.

Bioluminescent reactions are chemiluminescent reactions that occur in the cavity of a protein. The proteinic environment plays a role in the reaction steps leading to the HEI (thermal reaction steps not shows in Fig. 7 leading to the HEI in its GS) by lowering energetic barrier and constraining reactants to reactive conformations. It can play a role on the height of the energetic barrier to reach the CI of the decomposition of the HEI to the light emitter. It plays definitely a role on the color modulation of the light emission. Studying the complete scheme of a bioluminescent reaction means then to study thermal reaction, reactions involving excited states and the physical phenomena of light emission. Taking into account the environment, and especially proteinic environment that involved thousands of atoms require techniques that combine quantum description to properly calculate electronic transitions or bond formations and breakages and classical molecular mechanic to model the environment. These hybrid methods (QM/MM) involving quantum mechanic (QM) and molecular mechanic (MM) will be presented in the next section.

4 How to Perform the Theoretical Studies: The Choice of the QM/MM Approaches

In order to compare the molecular property obtained with quantum tools and the experimental data, the effect of the proteinic surrounding in case of bioluminescence and solvent surrounding in case of chemiluminescence has to be taken into account. We will present here the different choices to make in order to perform QM/MM calculations. In QM/MM calculations, part or whole of the surrounding is explicitly described. This corresponds to a case of "explicit solvation" if we extend the notion of solvation into any environment. On another hand, to avoid the cost of QM/MM calculation, implicit solvation or implicit description of the protein cavity can be used. This two schemes are therefore presented here before introducing the choices for QM/MM, i.e., the QM/MM frontier, the MM level, the QM level, the type of embedding and the process to perform a calculation.

4.1 Explicit or Implicit Solvation?

Two ways exist to model the environment: explicitly or implicitly.

Implicit solvation. In implicit solvation, the solute is considered in a cavity of the solvent. This latter is considered as a continuum. The specific interactions between a solvent molecule and the solute are not taken into account but the global contribution of the solvent is considered in the calculation of the energy of the system. In classical molecular mechanics, implicit solvents are known as General Born (GB) model and, taking into account the hydrophobic solvent accessible surface area (SA) of the solute, as GBSA [41]. In quantum mechanics, the most popular implicit solvation models are the ones from Pise, Italy: polarizable continuum model (PCM) [10, 48], from Minnesota, USA: solvent model based on density (SDM) [47], or from Regensburg, Germany: Conductor-like Screening Model (COSMO) [98]. Implicit solvent models have the advantage to be fast but the drawback of not describing specific interactions, especially the H-bonding between solvent and solute in protic solvent. The dielectric constants of the PCM solvent used to mimic the protein are taken from the range from 1 to 20 [27]. To mimic the influence of a hydrophobic enzymatic cavity for bioluminescent, the parameters of the chloroform solvent ($\varepsilon = 4.7$) [7], dibutyl ether ($\varepsilon = 3.05$) [79, 101], or dimethylsulfoxide DMSO ($\varepsilon = 46.8$) [17] were used. Experimentally, DMSO was also used to mimic the protein environment [94].

Explicit solvation. To take into account the influence of the environment, a better modeling is to explicitly describe the solvent molecules or the atoms of the protein in the case of bioluminescence. The bioluminescent reaction occurs in the cavity of a protein, named luciferase. Even if the computer powers allow now including hundreds or even thousands of atoms in quantum calculation, it is not necessary to calculate the whole protein at the quantum level. Moreover, a high accuracy in the results is needed to be able to compare calculated electronic transitions that are sometime only different from 0.01 eV. To have this accuracy, the level of theory used for the calculation cannot be performed on molecules of more than few dozens of atoms. Therefore QM/MM calculations are the most adapted methods to investigate bioluminescent properties.

4.2 Choice of the QM/MM Frontier

When performing QM/MM calculations, choices have to be done: which part of the system will be treated at the QM level (high layer) and which part of the system will be treated at the MM level (low layer). Contrary to the Green Fluorescence Protein where the light emitter or chromophore is covalently bound to the protein, most bioluminescent systems have emitter that is not covalently bond to the protein. Therefore, when the light emitter molecule is small enough like the one in the firefly bioluminescence, the choice is usually done to take the light emitter in the high layer and the rest of the molecules: protein, water molecules, and other molecules in the cavity like the Adenosine monophosphate (AMP) in the low layer.

When the light emitter, the reactant or the reaction intermediate that the researcher want to study is too big to be treated at the quantum level of theory, it has to be cut to select the important part of the molecule and put it in the high layer and put the rest of the system in the low layer. The boundaries of the layers are therefore cutting some bonds. The atoms at the boundary of the layers have to be treated in order to take into account the broken bonds. Many models exist and the most common one is the one of "link-atom". Atoms (usually hydrogen) are added to saturate the dangling bonds of the high layer. These atoms are not seen in the calculation of the lower layer and "follow" the broken bond.

Finally, it can be important to include in the high layer part of the residues of the cavity or water molecules that can, for example, be involved in protonation or deprotonation of the substrate. Again, this can imply to cut covalent bond between the two layers.

The choice of the frontier is motivated by the size of the high layer and the type of bond to be broken: the best is to cut bonds involving two carbon atoms connected by a single bond in the Lewis scheme. It is also important not to disrupt the π-conjugation extend of the QM part.

Before going further in the choices of QM/MM description, let go briefly to the choices of the QM and the MM level independently.

4.3 Choice of the MM Level

When describing a molecular system at molecular mechanic (MM) level, the electronic degrees of freedom are neglected and the system is described as a charged spheres' ensemble, linked by springs. The energy of the system is given by the equation:

$$E = \underbrace{V_{\text{bonds}} + V_{\text{angles}} + V_{\text{dihedrals}} + V_{\text{improper angles}}}_{V_{\text{bonded}}} + \underbrace{V_{vdW} + V_{el}}_{V_{\text{non-bonded}}} \quad (1)$$

with

$$V_{\text{bonds}} = \sum_{\text{bond } i} \frac{1}{2} k_i (l_i - l_{0i})^2 \quad (2)$$

$$V_{\text{angles}} = \sum_{\text{angle } \theta} \frac{1}{2} k_\theta (\theta - \theta_0)^2 \quad (3)$$

$$V_{\text{improper angles}} = \sum_{\text{improper angle } \omega} \frac{1}{2} k_\omega (\omega - \omega_0)^2 \quad (4)$$

$$V_{\text{dihedrals}} = \sum_{\text{dihedral } \phi} \sum_n \frac{V_{n,\phi}}{2} [1 + \cos(n\phi - \gamma_0)] \quad (5)$$

$$V_{el} = \sum_{i<j} f_{ij}^{el} \frac{q_i q_j}{\epsilon_0 r_{ij}} \quad (6)$$

$$V_{vdW} = \sum_{i<j} f_{ij}^{LJ} e_{ij}^* \left[\left(\frac{r_{ij}^*}{r_{ij}}\right)^{12} - 2 \left(\frac{r_{ij}^*}{r_{ij}}\right)^6 \right] \quad (7)$$

Equation (1) gives the energy of the system by the sum of the energy of the bonds (Eq. 2), angles (Eq. 3), improper angles (Eq. 4), and dihedral angles (Eq. 5), as well as the electrostatic energies (Eq. 6) and van der Waals energies (Eq. 7). Terms in Eq. (1) can have different forms (here is the form in the AMBER force field) [31] and depend on parameters like the stiffness (k_i, k_θ, k_ω constants), relaxed parameters of the springs (l_{0i}, θ_0, ω_0,...), the charges (q_i), equilibrium distances (r_{ij}^*), well depths (e_{ij}^*), and corrected factors (f_{ij}^{el}, f_{ij}^{LJ}). The ensemble of parameters and forms of the energy equation is called a classical force field. To allow the evaluation of the MM energy and gradient to be fast, the force field equations are simple analytical functions that depend on atom coordinates and fitted parameters.

The researchers choose the force field following different criteria: the availability of the parameters of the force field for the studied system and in the chosen software, as well as the habit of the researcher. When studying bioluminescent systems, many popular force fields (like Amber [31], CHARMM [32], GROMOS [64], universal force field (UFF) [67] to cite only the most commonly used) were developed for the protein. For the substrates, parameters are available like General Amber Force field (GAFF) [91] or CHARMM General Force Field (CChenFF) [89]. These force fields were parametrized for molecules in their electronic ground state. When dealing with excited states like in the study of bioluminescent systems, the parameters of the light emitter have to be set using QM and QM/MM simulations. Iterative simulation (alternating QM/MM calculations and molecular dynamics (MD) simulations) can be used for this parametrization until convergence (see for example supporting information in Ref. [24]).

4.4 Choice of the QM Level

The choice of the description of the system at the quantum level is important and many criteria have to be taken into account. In bioluminescent reactions, several electronic states are involved. If the reactivity is to be studied, the crossings between the states have to be described properly. The challenges of describing conical intersections, avoided crossings, intersystem crossings, and/or multi-states systems with a large number of electrons are responsible for the fact that theoretical methods have been used for studying bioluminescent systems since only few decades. The experimental scientist community is now convinced that the theoretical studies can contribute toward a better understanding on the bioluminescent systems. Depending on the questions to answer, multiconfigurational wave function methods, density functional theory and/or semi-empirical methods are employed to investigate these systems.

Electronic structure methods are based on the time-independent Schrödinger equation (8):

$$\hat{H}|\Psi\rangle = E|\Psi\rangle, \tag{8}$$

where $\hat{H}$ is the molecular Hamiltonian operator, E is the total energy of the system, and $|\Psi\rangle$ is the wave function. The wave function depends on the coordinates of all particles (nucleus and electrons) of the system. The Born–Oppenheimer approximation allows separating the electronic and nuclear degrees of freedom, leading to an expression of $|\Psi\rangle$ as the solution of the Schrödinger equation of the system of the electronic particles for a fixed nuclear geometry.

Hartree–Fock method. A first way of resolving the Schrödinger equation is to consider that each electron sees the average distribution of the other electrons, this is a mean-field approximation. This resolution of the equation is known as the Hartree–Fock method (HF), where the wave function $|\Psi\rangle$ is described as a single Slater determinant (an antisymmetrized product of one-electron functions or orbitals, named basis functions). The solution is obtained by minimizing the calculated energy. As the interactions between electrons are averaged, the chemical interactions are only approximate.

Basis sets are the basis functions that are chosen for the calculations. There are several basis sets and they will not be reviewed here. However, the reader of this chapter will meet the following basis sets: ANO-RCC-VDZP [70] and 6-31G(d,p) [29] in the examples.

Electron correlated methods. Methods are used to include the static electron correlation (or non-dynamical correlation) energy to improve the description of the electron–electron interactions neglected in the HF method. To go beyond HF, where the wave function is expressed as a single Slater determinant, one expresses the wave function as a linear combination of multiple Slater determinants. The method is known as multiconfigurational method, as each Slater determinant corresponds to a particular electronic configuration, i.e., a specific set of orbitals occupied by one or two electrons. As the number of configurations grows exponentially with the number of electrons and basis functions, the computational resources limit the number of electrons that can be studied. For systems involving more than fifteen or so electrons, a selection of the most important orbitals involved in the description of the reactivity is made. All possible configurations generated with these orbitals and a chosen number of electrons in these orbitals form the so-called complete active space (CAS) of configurations. The selected orbitals are iteratively optimized and the method is called CAS self-consistent field (CASSCF) [69]. Using CASSCF method requires choosing the active space, i.e., the number of electrons and the orbitals that will be optimized. In articles, the number of electrons and orbitals are described as, for example, (16e,14o) or (16,14) or 16-in-14 for 16 electrons in 14 orbitals and the description of the orbitals selected is given (see Fig. 8). The choice of the orbitals selected depends on the reaction to investigate.

In bioluminescence system, two main questions are raised. For the reactivity when breaking or forming a bond, the bonding/antibonding orbitals of the bond will be included in the active space. For the electronic transition, the valence orbitals of π character will be included. One has to play with the limited size of the active space and the need for the correct investigation. For more technical information on the use of these methods, I recommend the book of B. O. Ross et al. [71].

Dynamical correlation. To get a more accurate description of the electron–electron interactions, further methods include the dynamical correlation, i.e., the correlation due to the movement of electrons. These methods are performed by adding a second-order perturbation theory (PT2) to multiconfigurational methods. For example, CASPT2 stands for complete active space PT2 [2, 22, 76], MCQDPT2 for multiconfigational quasi-degenerate PT2 [57] and NEVPT2 for n-electrons valence state PT2 [5].

Density functional theory. Another way of determining the electronic structure of a system is the density functional theory (DFT) based on the Hohenberg–Kohn theorems that describe systems of electrons moving under the influence of an external potential. The ground-state energy of a system depends on the electronic density ρ. The energy is a functional of the density ρ, itself function of the only three special coordinates (x, y, z). The density ρ is obtained by minimizing the energy, which is written as a sum of external energy due to the nuclei, an electron–electron Coulomb repulsion, and an unknown functional that includes kinetic, exchange and correlation energy. The method developed by Kohn and Sham introduces a fictitious non-interacting system and leads to Kohn–Sham (KS) equations. This puts all non-

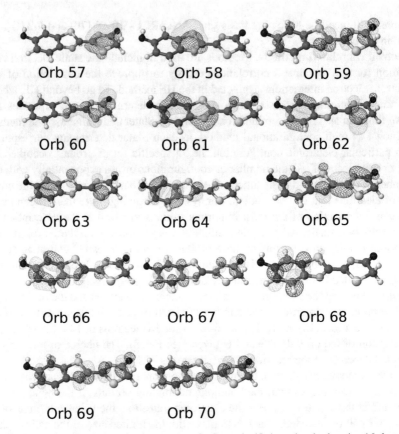

Fig. 8 Illustration of a choice of active space on firefly oxyluciferin molecule choosing 16 electrons in 14 orbitals (16-in-14)

determined terms of the equation to be solved in an exchange–correlation functional. Researchers using DFT have to choose the most appropriate exchange–correlation functional between a vast number of choices. The simplest functionals are based on the local density approximation (LDA) where the functional depends only on the density at each point in space $\rho(r)$ [6, 43]. Introducing a dependency on the density gradient $(\nabla \rho(r))$ leads to generalized gradient approximation (GGA) functionals and a dependency of the second derivative $(\nabla^2 \rho(r))$ to the meta-GGA approximation [65, 86]. Hybrid exchange functionals include some components of the exchange energy that are not anymore only depending on the density but on KS orbitals treated as HF wave functions. The most popular hybrid functional is the B3LYP [43] for Becke 3-parameters combined with Lee–Yang–Parr (LYP) correlation functional. Double-hybrid functionals include some of the correlation energy by applying PT2 to the KS orbitals. Every year, new functionals are proposed and tested on benchmarks, set of molecules, and compared to existing functionals.

TD-DFT. In order to study bioluminescent systems, there is a need to describe not only the ground state (GS) but also the excited states. The time-dependent DFT (TDDFT or TD-DFT) [87] is based on the Runge–Gross theorem, which establishes a direct relationship between the time-dependent electron density and the time-dependent potential; hence the potential and the wave function are dependent on the density. If a very small disturbance (δt) is applied to the wave function of the GS, the orbitals will respond to this disturbance and the TD-DFT equations are obtained via the approximation of the linear response (first-order response) of the orbitals and density to this perturbation. The vertical excitation energies are obtained as the response of the system to a frequency-dependent perturbation. The adiabatic approximation to TD-DFT from Casida [12] is the one most implemented in the simulation packages.

DFT and TD-DFT calculation can be applied for system much larger than the ones possible with multiconfigurational methods. However researchers have to choose the functional and should justify this choice by comparing with experimental results and/or multiconfigurational calculations. The functional used in TD-DFT are the functional optimized for the ground state (the optimization of a functional for ES is very costly and there is lack of reliable experimental data). A benchmark is usually done at the beginning of the study of a new system in order to choose the "best" or "suitable" functional and basis set to be selected for the calculation. Readers interested in the benchmarks on TD-DFT can read the review from A. Laurent and D. Jacquemin [42]. As the functionals optimized for the DFT calculation underestimate charge-transfer excitation energies, some range-separated hybrid functionals have been designed by combining functionals that treat differently the short and long interelectronic distances. One example used in bioluminescent systems is the Coulomb-attenuated B3LYP (CAM-B3LYP) method [97].

4.5   Choices for the QM/MM

The researcher has to choose the high layer (the part of the system to be treated at the QM level) and the low layer (the rest of the system to be treated at the MM level). Also, the researcher has to choose the level of the method (*ab initio* methods or TD-DFT) and the force field for the MM description. The researcher has eventually cut some bonds and chosen how to treat the cuts (for example, with the link-atom). There are still some choices to be made.

Subtractive and additive schemes. Once the high and low layers were chosen, the researcher has to choose between different schemes that are available in QM/MM program packages. There are two classes of schemes: the additive scheme and the subtractive scheme. In the additive scheme, the energy of the system is calculated as the sum of three terms: the energy of the high layer at the QM level, the energy of the low layer at the MM level and the energy due to the interactions between the two layers. In the subtractive scheme, the energy of the system is calculated as the sum of the energy of the whole system at the MM level and the energy of the high layer at the

Subtractive scheme:

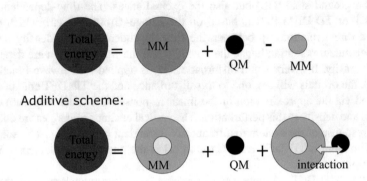

Additive scheme:

Fig. 9 Illustration of the two QM/MM schemes

QM level minus the energy of the high layer at the MM level (see Fig. 9). This is the case in the very popular Own N-layered Integrated Molecular Orbital and molecular Mechanics (ONIOM) [18] method. In the later scheme, the interactions between the two layers are described at the MM level when the whole system is calculated. This implies that the force field is described for the whole system.

Mechanical embedding. The geometrical constraint due to the van der Waals interactions and the location of the QM/MM boundary, when it exists, implies steric effects on the high layer. This is called the mechanical embedding (ME) and is the first level of the QM/MM coupling. The steric repulsions are described at the MM level, with the van der Waals parameters of the force field.

Electrostatic embedding. The electrostatic interaction between the high layer and the low layer system are included when the electrostatic embedding or electronic embedding (EE) is described. The potential created by the point charges put on the nucleus of the low layer in the classical force field polarizes the electronic wave function or density of the high layer. The electrostatic potential fitted (ESPF) [21] method implemented in the coupling between the softwares MOLCAS and TINKER or GAUSSIAN and TINKER is one example of implementation of such embedding. The wave function and QM energy of the high layer is optimized by taking into account the electrostatic potential due to multipoles locate on a grid around the high layer and imitating the potential created by the low layer's charged points. To gain time, when optimizing the geometry of the whole system, at each optimization step of the optimization algorithm of the wave functions, the low layer part is moved in order to optimize the MM part of the energy, taking into account the point charges artificially calculate on the high layer nucleus (for example, Mulliken charges). With classical force field, the charges parameters for the low layer are fixed by the force field and not modified by the presence of the high layer. In order to avoid the over-polarization of the QM calculated density at a shared bond between high and low layer (when frontier between the layers has broken a bond), the MM charge of the low layer atom "linked" to the high layer should be set to zero. The corresponding charge is then distributed to the adjacent atoms in order to have a whole number for the high layer's charge.

Polarizable embedding. In order to see the influence of the electrostatic change of the high layer on the low layer, one should use polarizable force fields. Examples of polarizable force fields are Atomic Multipole Optimized Energies for Biomolecular Applications (AMOEBA) [45], and Sum of Interactions Between Fragments Ab initio computed (SIBFA) [66]. The polarizable embedding (PE) is very time consuming and used only if ME and EE failed to describe correctly the system. With the increase of the computer capacity and the implementation in the software, the PE is and will be more and more used.

4.6 Protocol to Prepare a QM/MM Calculation

Each researcher has its own protocol to perform QM/MM studies. Here are the things to be aware and usually done by researcher in order to perform QM/MM calculations for studying bioluminescence.

- *Perform the calculation on the high level system in gas phase*. Before performing QM/MM, the system that will be chosen in the high layer is studied in gas phase. This step is usually also done in an implicit solvent (PCM). This allows having a first idea on the system. For example, here can be chosen how many excited states will be taken in future calculations of the spectra. This allows choosing the QM level of theory. Does the functional of the TD-DFT gives energy transitions that are close to the more accurate level like CASSCF/CASPT2? Which smaller basis set gives values of energy transitions close to the one obtained with larger (but more expensive) basis set? In case of CASSCF calculation, does the chosen active space gives results close to the one with a bigger (but more expensive) active space? In summary, this step allows choosing the compromises that can be done for performing the calculations. Later, the preliminary results obtained in gas phase and with implicit solvation will be compared to the ones obtained with explicit environment (QM/MM calculation) to better define the influence of the presence of the protein in the studied properties. If no many changes are observed, it will allow eventually validating the use of the much faster implicit approaches for further investigation.

  For example, in the first QM/MM study of the oxyluciferin emission in the protein done in 2010 [59], the transition involved in the emission has a $\pi/\pi*$ character. The selection of the active space was 16-in-14, corresponding to all $\pi/\pi*$ orbitals except the one on the thiazole-ring sulfur, because using a larger active space including the one on the thiazole-ring sulfur (i.e., a 18-in-15 active space) gives nearly the same calculated emission energies in gas phase and is much more time consuming. The choice of the basis set was also checked. 6-31G(d) and ANO-L-VTZP basis sets give similar transitions in gas phase leading to the less computer time basis set 6-31G(d) for later calculations. The transition energies have been computed at the CASPT2 level of theory using 2-roots SA-CASSCF wave functions, as the third root calculated in a gas phase model was high enough not to be taken it into account.

It is to be noted that criteria evolve and can change depending on the focus of the study. In bioluminescent, the reliable calculations of energy transitions (wavelengths and oscillator strengths) are usually searched.

- *Add missing residues, atoms, and/or loops on the structure of the protein.* Most of X-ray crystal structures are not completed. The positions of the most fluctuated residues are often not determined. Therefore, the missing atoms have to be modeled. Usually these residues belong to external loops and are not in the cavity where the bioluminescent reaction occurs.

  There exist different programs available to model the missing residues, Disgro [85], I-TASSER [72] or Modeler [93] to cite some of them. The Leap software of Amber suite of program is also able to built missing residues when the length of the loop is not too long.

  For example, in the PDB file 4G37 [83] of the luciferase of Photinus Pyralis Firefly, the missing residues are the 3 first residues, the residues 200 to 2004, 527 to 528, and 544 to 550. In Ref. [7], the missing residues were added with Disgro program. Once the missing residues are added, minimisation and relaxation with molecular dynamic are usually needed to complete the model. The missing residues are far away from the enzymatic pocket where the substrate is bound and therefore the accuracy of the missing loops construction should not influence too much the results. Furthermore, it should not influence the next step that consists of the docking of the substrate in the cavity.

- *Dock the substrate in the cavity and other molecules if needed.* This step can be done with docking programs or "by hand" if the position of the substrate is known. For example, in firefly bioluminescent systems, crystallographic structure of the protein with an analogue of the reaction intermediate, DLSA, is known (PDB code 4G36) [83]. It is "easy" to replace by hand the DLSA with the oxyluciferin and AMP molecule before relaxation and equilibration of the system. In this particular case, using a docking program like AutoDock [56] leads to some poses where the substrate is in the cavity but with a position where the benzothiazole moiety is positioned on the location of the thiazole cycle of the DLSA. These poses are not correct as they do not correspond to experimental data. A good understanding of the system (from experimental data and available crystallographic structures) can avoid choosing them. In case of firefly luciferase–luciferin systems, the luciferin is a quite rigid structure therefore many conformations of the molecule are found by the docking. Larger bioluminescent systems like coelenterazine can be more problematic to model with many different conformations of similar free binding energy. Again, experimental structural data help to decipher which structure(s) to take for further investigation.

- *Protonate the residues of the protein and add ions to perform the study at a certain pH.* The protonation state of the residues and substrates depends on the pH. Therefore it is important to first check the local pKa of the residues (there are programs like MolProbity [19], PROpKa [63] or H++ [1] that are doing this task), and to change the protonation state depending on the pH. The histidine residues are usually to be double checked by hand as their protonation state also depends on the H-bonding in which they can be involved. Ions are usually added to neutralize the whole system before performing molecular dynamics.

- *Calculate the force field parameters for the substrate if not available.* This part is certainly the most time consuming. When the purpose is the study of the substrate in its excited state, conventional parameters are not adapted anymore. The protocol to get the parameters for the substrate is usually done in an iterative process (alternating QM/MM calculations and MD simulations). One description can be seen in Ref. [24]. The accuracy of the parameters is important to a good sampling of the conformation for the calculation of the emission spectra of the firefly oxyluciferin in an explicit water solvent. The impact is negligible on the emission spectra of the firefly oxyluciferin inside the luciferase cavity. This comes from the fact that constraints inside the cavity are stronger and the modification of the residues and water molecule in the cavity are less sensitive to the parametrization than the modification of the water shell in water solvent. Therefore, when 3 iterations are needed for the parametrization of the oxyluciferin charges for emission in water solvent, only one is needed for the study of oxyluciferin in protein.
- *Perform classical minimization following by classical MD simulation.* This step is done to equilibrate the system and to get samples of the system.
- *Extract some snapshots.* The number of snapshots taken from the MD depends on the type of method used for QM/MM: for very time consuming like CASSCF/CASPT2 QM description, only few snapshots will be chosen, for less consuming methods like TD-DFT, 100 snapshots can be chosen. The choice of number of snapshots chosen is a compromise between the time of the calculation and the accuracy of the results. We have shown that 100 were sufficient to sample correctly the system to get the absorption and emission spectra of the firefly bioluminescent system [24]. Indeed, taking $N = 100$ snapshots or $N = 1000$ leads to the same spectra.
- *Create the input files for the QM/MM calculations.* This step converts output files from the MD simulation to input files for the QM/MM calculation.
- *Run QM/MM calculations.* QM/MM calculations are done on each snapshot to extract the property investigated.

In the case of simulation of the absorption and emission spectra of oxyluciferin, the light emitter of the firefly bioluminescent system, the vertical energy transition and corresponding oscillator strengths are extracted. Then the absorption or emission spectrum is simulated as a convolution of the vertical transition energies of N snapshots, using a gaussian function with a full-width at half-maximum of (for example) 0.2 eV [24]. The obtained spectra are used to predict and/or explain the experimental spectra.

5 Formation of the HEI in Firefly Bioluminescent System

In firefly, the bioluminescent reaction catalyzed by the protein luciferase involves first the reaction of the adenosine triphosphate (ATP) with the D-luciferin in presence of magnesium ions leading to an intermediate of reaction where the adenosine monophosphate (AMP) is covalently bound to the luciferin moiety (see first step in Fig. 1). The first step involving decomposition of the ATP, to my knowledge, has not been studied yet with theoretical tools. Therefore, it will not be discussed specifically here. This step has similarity to the decomposition of the adenosine triphosphate in adenosine diphosphate and free phosphate ion responsible, in ATPase, of the release of energy used for other transformations. The difficulty to do simulation of this specific step remains in a good description of the protein environment and the role of the magnesium cation. More experimental information are needed to build a model that could be used for the study of this step. Indeed no crystal structures of luciferase with ATP, magnesium ions and luciferin are available. Modeling the conformation of the reactive molecules in the protein is still a hazardous task. Moreover, as seen in the study of ATPases, the difficulty to study the reaction of ATP hydrolysis inside the protein with QM/MM methods comes from the highly charged ATP that polarize the surrounding protein residues in the cavity and necessity to treat some of the residues involved in the reaction at the QM level [40].

The D-luciferyl adenylate intermediate reacts then with the molecular oxygen to create the HEI. Some QM/MM studies were done on the HEI formation starting with D-luciferyl adenylate substrate with the deprotonated hydroxybenzothiazole moiety, inside the protein [8]. They will be presented here.

The first step of the formation of the dioxetanone corresponds to the deprotonation at the α-position of the carbonyl group. Two hypotheses can be explored leading in three proposed mechanisms (see Fig. 10). In mechanism (a) exposed in Fig. 10, a basic amino acid of the cavity deprotonates the D-luciferyl adenylate, leading to a carbanion substrate. The carbanion shares its two electrons to create a bond with the triplet dioxygen. In mechanism (b) the first step is the same as for hypothesis (a) then a single electron is transferred from the negative charged carbon of the substrate to molecular oxygen, leading to the creation of a radical superoxide anion and a radical substrate. The homolytic bond formation of C–O follows. In mechanism (c) (see Fig. 10) the transfer of the proton is directly from the D-luciferyl adenylate to the molecular oxygen, rising to singlet ground-state hydroperoxide molecule that forms back a homolytic bond with the resulting radical substrate.

The hypothesis (c) was experimentally studied [83] and the (a) and (b) were investigated with the aim of computational simulation studies [8, 50]. QM/MM studies show that the single electron transfer is instantaneous when molecular oxygen is inside the protein cavity (this would favor the pathway (b)). The system initially in its triplet ground state undergoes an ISC as the two lower singlet and two lower triplet states are degenerated until the formation of the first C–O bond. Once in the cavity, the molecular oxygen is stabilized by interaction with water molecules and, in the case of the theoretically studied hypothesis (a) and (b), with the amino acid

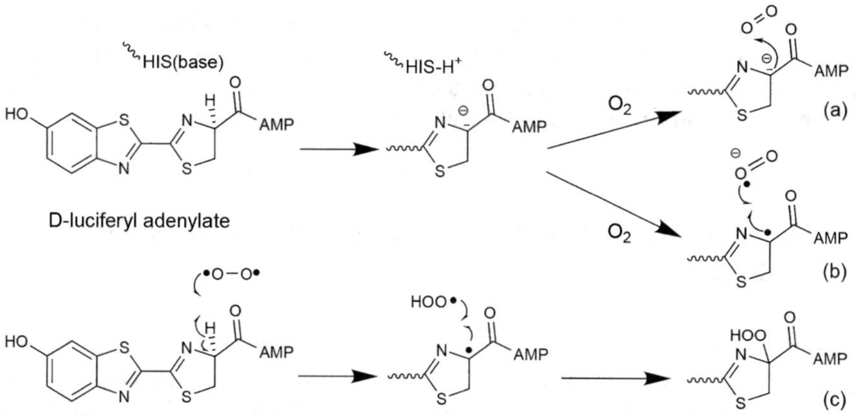

Fig. 10 Suggested pathways for the reaction of triplet molecular oxygen and the D-luciferyl adeny-late intermediate leading to singlet dioxetanone. **a** Deprotonation mechanism from histidine in the cavity and direct reaction with molecular oxygen. **b** Deprotonation mechanism from histidine in the cavity and formation of a superoxide anion before creating a homolytic bond. **c** Hydrogen abstraction mechanism from molecular oxygen followed by homolytic formation of C–O bond. Reprinted (adapted) with permission from (Vacher et al. Chemical Reviews, 2018, 118(15):6927–6974.). Copyright (2018) American Chemical Society

that has deprotonated the luciferin. An easy proton transfer to the formed superoxide anion in case (a) can be observed forming the hydroperoxide anion. The reaction would then finish by following mechanism (c). Therefore, the mechanisms (b) and (c) may coexist or be part of a more complex mechanism. Furthermore, the QM/MM studies show that the formation of the dioxetanone ring is only possible when the bond between the AMP moiety and the D-luciferyl moiety is firstly broken [8].

To be able to perform a QM/MM study of the reactivity of the molecular oxy-gen on a deprotonated D-luciferyl adenylate in the luciferase cavity, the first step is to start with a proper model. In the previous study [8], the crystallographic struc-ture of the firefly luciferase with a substrate that looks like the D-luciferyl adeny-late was used. The missing residues were completed, the deprotonated D-luciferyl adenylate and molecular oxygen were added and after relaxation by molecular dynamics, the approach of O_2 was performed by successive QM/MM optimiza-tions under constraint. CASSCF calculations were not possible to be done includ-ing the whole D-luciferyl adenylate molecule. Therefore, this substrate was partly described at the high level and partly at the low level with the frontier modeled with link-atoms (see Sect. 4.2). The calculations were performed at the CASSCF/ANO-RCC-VTZP/AMBER99ff with a (8,10) active space. Single points were done at the CASPT2 level. We see here that the protocol to use is not a black box and that the computational researcher has to make discerning choices to answer the questions.

A lot of questions remain to understand the formation of the dioxetanone systems. Some can be answered by studying part of the system outside of the protein on chemiluminescent systems that mimic the reaction steps, other will need to take into account the protein environment with all of the difficulties that this drives.

6 Decomposition of the HEI in Firefly Leading to the Light Emitter

The HEI of firefly's bioluminescent system is firefly's dioxetanone. The HEI decomposes into carbon dioxide and oxyluciferin, the light emitter in its first singlet excited state. To achieve this, the dioxetanone moiety is bonded to an extended π-conjugated system, very rich in electrons, that plays the role of activator for the reaction. Without this activator moiety, the activation barriers of 1,2-dioxetane and 1,2-dioxetanone dissociation are too high and lead to a highly populated triplet state, inducing a low efficiency of light emission [88].

How firefly HEI decomposition deciphers from the chemiluminescence of 1,2-dioxetane and 1,2-dioxetanone? Does the protein environment play a role? What is the protonation state of the HEI? Why the bioluminescent efficiency of firefly bioluminescence is around 40%, which makes firefly one of the most efficient systems? Here are some questions that future QM/MM calculations will have to answer. The study of the HEI decomposition inside the cavity is quite challenging. Until nowadays, few QM/MM studies were performed on the decomposition of firefly HEI in the protein, and the theoretical studies of this step were rather focused on the corresponding "chemiluminescent reaction", i.e., in gas phase or in an implicit solvent. Reader has to have in mind that these studies are only "theoretical" as there is no observed chemiluminescence of firefly luciferin without enzyme or ATP. In this section, some results in solvent are first presented as they show the difficulties to study this step of the reaction. A study done with QM/MM will be then presented [101] and the difficulties to overcome with QM/MM studies will be discussed.

Computational studies on the firefly HEI decomposition were first performed in solvent ("chemiluminescent" reaction) to a better understanding of the influence of the protonation state of the hydroxybenzothiazole moiety [61, 99], as the protonation form of hydroxybenzothiazole moiety (protonate or deprotonated) is still in debate.

The study of the neutral form of HEI (protonated hydroxybenzothiazole form) shows that this form leads to low chemiluminescent efficiency as the barrier to reach the excited state is high (around 26 kcal/mol) and the first deformation, that is, the elongation of the O–O bond, as in 1,2-dioxetanone, leads to a biradical system where GS, S1, and T1 states are degenerated, favoring an ISC and the population of the triplet state.

The study of the anionic form (deprotonated hydroxybenzothiazole form) shows a completely different pathway. The reaction starts with the elongation of the O–O bond until reaching the TS. A charge transfer from the phenolate to the peroxide bond

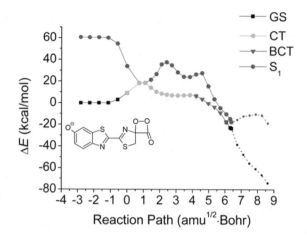

Fig. 11 Calculated CASPT2//CAM-B3LYP potential energy plot of GS and S1 along the dissociation path of the deprotonated firefly's dioxetanone in benzene solvent (PCM). The path is colored in the reference depending on the character of the transition: originally green (here light gray balls in the GS): CT, partial electronic charge transfer from oxyluciferin to carbon dioxide; originally magenta (here dark triangles in the GS): BCT, back chargetransfer of partial electronic charge from carbon dioxide to oxyluciferin moiety. Reprinted (adapted) with permission from (Yue et al. Journal of the American Chemical Society, 2012, vol. 134, no 28, pp. 11632–11639.). Copyright (2012) American Chemical Society

destabilizes the O–O bond and opens it to a biradical state. The chemiluminescent reaction is therefore feasible with a barrier of only 15 kcal/mol [60, 61, 99]. The following is the breakage of the C–C bond leading to a CI, from where the system can decay to the GS (non-radiative deactivation) or lead to the S1 state of the light emitter.

As an illustration, Fig. 11 shows the energy of the GS and S1 of the decomposition of the firefly HEI calculated on an intrinsic reaction coordinate (IRC) at the CAM-B3LYP/6-31G(d,p) level followed by CASPT2(18-in-15)/ANO-RCC-VDZP in benzene solvent (PCM). The partial charges on the oxyluciferin moiety and on the CO_2 moiety are evaluated. First a partial electronic charge transfer (CT) is observed from oxyluciferin to carbon dioxide then the electron gradually goes back to the carbonyl substrate, the back charge transfer (BCT). No full electron is transferred during the dissociation and the gradually charge transfer suggests the description of a Charge Transfer Induced Luminescence (CTIL) mechanism [35].

To take into account the protein environment, QM/MM single energy point was performed on firefly dioxetanone (HEI) inside the protein [61, 101] and the thermolysis transition state leading to the light emitter [101]. Zhou et al. have performed a QM/MM (ONIOM UCAM-B3LYP/6- 311+G(d,p)/UFF with an electrostatic embedding scheme) study on HEI (resp. the transition state) surrounding by 20 residues of the cavity and AMP described at MM level and implicit solvent (dibutyl ether) with PCM. The authors has show that the protein surrounding lower the energy barrier of the reaction comparing to gas phase due to the change of the local electrostatic field, and therefore has a catalytic effect on the reaction.

QM/MM modeling of the decomposition of the HEI inside the protein cavity is certainly one of the most challenging simulations to set up for computational researchers. First, taking into account the surrounding raises the question of choices of a good description of the protein conformation, as no crystallographic structure exists with the HEI inside the protein because the HEI is unstable. Furthermore, the same questions as for the study of the light emitter (see next Sect. 7 on the protonation state of the substrate, the protonation state of the surrounding residues and AMP, and on the conformation of the protein) have to be addressed. Moreover, the studies of the reaction path in vacuum and with implicit solvation are already very challenging concerning the QM level of theory to be used. Conical intersections are not well defined by TD-DFT methods. The size of the active space to use CASSCF approach is one of the drawbacks to perform the calculations. Development of methods to increase the active space like the density matrix renormalization group (DMRG) [80] will allow making such studies more feasible.

7 Light Emission in Bioluminescent Systems

The present section focuses on the last step of the bioluminescent reaction, i.e., the electronic transition that leads to photon emission.

From a theoretical point of view, the study of the light emitter implies to find the structure of the most stable local minimum in first singlet ES (S1) via an optimization algorithm. When the system is in the excited state optimized structure (ESOptSt), the difference of energy between the GS and S1 gives an estimation of the energy of the emitted light (and therefore the emitted wavelength and color). Moreover, an estimation of the relative expected intensity of the light is given from the calculations by the oscillator strength (f) of the transition. It should be mention here that oscillator strength gives roughly the radiation probability. In order to estimate the emission quantum yield, it should be also taking into account the yield of the reaction leading to the light emitter, and also all possible relaxation channels, i.e., all non-radiative pathways that prevent the emission. Assessing all of these pathways is above what we can do in the study of bioluminescent systems including the protein environment.

Experimentally, studying the bioluminescence is a challenge and, often, fluorescent experiments are performed on the product of the reaction. Before interpreting the fluorescent experiments to explain the bioluminescent or chemiluminescent reaction, the researcher has to make sure that the light emitter is the same in the two phenomena. The chemical form of the light emitter, and especially its protonation state is also a question that theoretical simulation can investigate. Indeed, some interexchange reaction (protonation, tautomerization) can occur in the ground and excited state of the light emitter. Which one(s) is(are) responsible for the light emission is under debate. Finally, a better understanding of the nature of the light emitter allows proposing mutations in the protein or chemical modifications of the light emitter. The structural modifications of the light emitter designed to tune the emitted color can be explained and predicted by computational studies. The effect of the environment, especially the protein surrounding, and other conditions like pH modifications start to be elucidated by QM/MM investigations.

7.1 Bioluminescent Versus Fluorescent Reactions

It is a habit to overcome the experimental difficulty of studying the bioluminescent reaction by recording the fluorescent properties of the product obtained after the bioluminescent reaction. It is taken for granted that the conformation of the bioluminescent species and the one obtained by photoexcitation (in a fluorescent experiment) are the same. From the simulation point of view, this means that the conformation of the bioluminescent product state (BS), corresponding to the local energy minimum on the first singlet ES S1 reached after the reaction, is alike the conformation of the fluorescent experiment state (FS) obtained from minimization of the S1 energy after excitation from the minimum GS conformation.

Depending on the studied system, this assumption may not always be true. Indeed, if the energy barrier in S1 is significantly high, then the BS and FS correspond to two different chemical species (Fig. 12a), leading the photoexcitation and bioluminescent experiments to different colors of the emitted light. If the energy barrier connecting the BS and FS is small, the system can reach both conformations (BS and FS) before emitting light (Fig. 12b). Finally, if there is no energy barrier connecting the BS and FS, then BS and FS correspond to the same species (Fig. 12c). Computational studies provide the geometries and electronic structures of both BS and FS, give insights into the energetic barrier and help to discern between the different scenarios.

The question of the difference of light emitter nature in the case of fluorescence study and in the bioluminescent reaction has risen while studying the bioluminescence of the obelin system [68]. A small model of the coelenterazine and cypridinid luciferin was designed and the decomposition of the HEI and departure of CO_2 was modeled at the CASSCF/CASPT2 level of theory. The obtained light emitter was compared to the one obtained by exciting the model of the product of the reaction. The comparison was done between the chemiluminescent emitter (CS) (as the study was done in gas phase), that still has a tetrahedral carbonyl, and the fluorescent emitter, that already has a trigonal carbonyl group. Hence, the electronic nature of the excited species responsible for the chemiluminescence and the fluorescence is different, as well as the corresponding geometrical structures. The color of the light emission in the two experiments can be different and the conclusion from the fluorescence experiment cannot be transferred to the chemiluminescent reaction. The systematic investigation of the fluorescent and chemiluminescent states of the coelenteramide, the light emitter of obelin bioluminescent system at the TD-CAM-B3LYP/6-31+G(d,p) level of theory shows that the not planar CS is a dark state, a state that has a small f value and therefore that should produce low/no intense emission [14]. The explanation of intense light emission is proposed by a low energy

Fig. 12 Different scenarios for the relative position of the bioluminescent state (obtained from HEI decomposition, arrow starting from the left) and the fluorescent state (arrow starting from the right minimum of the GS). **a** The barrier between the bioluminescent state (BS) and the fluorescent state (FS) is high, the two experiments lead to different color emissions. **b** The barrier is low, there could be a mixing of the two emissions in both experiments. **c** There is no barrier; the two conformations are the same

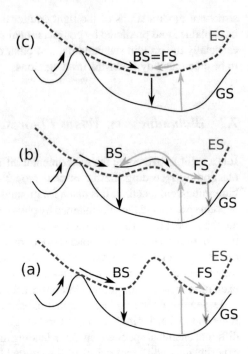

barrier in the first excited singlet state from the CS state to the FS that allows the system to proceed to the FS brighter (high oscillator strength f value) state. Taking into account the protein environment was not performed on this system that is a quite large and flexible system.

The nature of the fluorescence and bioluminescent states of firefly oxyluciferin in the luciferase protein of the firefly bioluminescent system was investigated [61]. The thermal decomposition path of the firefly dioxetanone HEI and the light emission states of the firefly oxyluciferin responsible for the bioluminescence and the fluorescence experiment of firefly were investigated at CAM-B3LYP/6-31G(d,p) on the intrinsic reaction coordinate (IRC) path followed by single points energy calculations at the CASPT2(16in13)/ANO-RCC-VDZP level of theory [61]. The chemiluminescent and fluorescent states were found to be the same. Putting the chemiluminescent structure in the protein and performing QM/MM optimization in the S1 state, lead to the same state at the one obtained by fluorescence simulation in the protein, without showing any energy barrier in the protein environment. These studies were done on one of the tautomer forms of the light emitter and further investigation has to be performed on the nature of the light emitter.

7.2 Nature of the Light Emitter

In the last subsection, we have discussed the possibility of the light emitter to be in different chemical forms or conformations, depending how it was created: from a photoexcitation of the light emitter initially in its ground state or as the product of a reaction leading to the light emitter directly in its ES. We discuss here a little further of the modifications that can take place at the ES before light is emitted. Indeed, looking back to Figs. 2 and 3, we can see phenol groups or amino groups in most structures of the light emitters. These groups can be deprotonated in the cavity of the protein and/or while the system is in an ES (when the system shows an excited state proton transfer (ESPT)). Moreover, some light emitters also show keto moiety that can be tautomerized to enol group. We will present here some results of QM and QM/MM studies on firefly oxyluciferin.

The Firefly oxyluciferin molecule has a phenol group, which may or may not be protonated, and a ketone group with a protonated α-carbon, which allows a keto–enol tautomerization. This induces six possible forms of photo emitter (Fig. 13). The nature of the light emitter during the bioluminescence process is still an open question. It is due to the fact that the bioluminescent emission spectrum can be complex (with more than a single band emission) and experimental investigations of bioluminescent systems are not easy task.

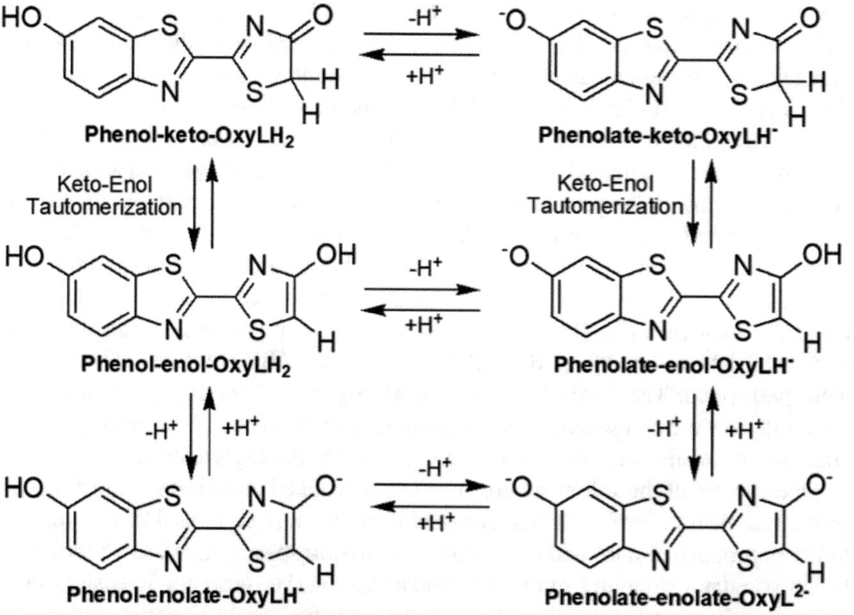

Fig. 13 The six forms of firefly oxyluciferin due to protonation/deprotonation state and keto/enol tautomerization. Reprinted (adapted) with permission from (García-Iriepa et al. JCTC, 2018, 14, 2117–2126). Copyright (2018) American Chemical Society

Experimentally, the color of the emission can be modulated from green-yellow to red by changing the pH. Combined experimental and theoretical studies have been performed in order to give some clues to understand this phenomena and to try to find which chemical forms are responsible for the color modulation [13, 49, 77].

Experimentally, fluorescent spectra of oxyluciferin in water in different pH were performed [26]. However, if acid-base equilibrium or keto–enol tautomerization equilibrium occurs, they can be difficult to control. Playing with the pH can drive the equilibrium but in the case of oxyluciferin, too many forms still can be present in many pH values. Therefore, oxyluciferin analogues whose protons are replaced by methyl groups were synthetized to prevent target equilibrium. Six analogues were proposed to mimic the six different forms of oxyluciferin (see Fig. 14). It remains to be seen whether the analogues mimic correctly the natural forms. The experimental and theoretical studies of the absorption and fluorescence of the analogues were done in water [24, 25] and are still in progress in the protein [23].

The computational simulations of the emission spectra are very useful to do such study. Indeed, while modeling fluorescence spectra of the oxyluciferin, the theoretical researcher can choose to freeze the protonation or tautomer form of the molecule, which is not possible experimentally.

First computational simulations of the emission spectra were done in vacuum or modeling a solvent with implicit model. The advantage of using implicit model is that these methods are very fast. Modeling explicitly the water solvent with QM/MM approaches (with the light emitter in the high layer and the water molecules in the low layer) improved the comparison between simulation and experimental results and gave insight into the H-bond network formed around the studied molecule. The drawback is that these methods are much time consuming, as they need to equilibrate the system by molecular dynamics (MD) and to perform many QM/MM calculations. The spectra in the protein by mean of QM/MM approaches were also performed.

It should be noted that taking into account the solvent for comparison with fluorescent experiments or the protein for comparison with bioluminescent and fluorescent in protein experiments are essential [25]. There are quite some shifts due to the presence or not of the surrounding. It has been shown that absorption spectra and emission spectra simulated in explicit water solvent could reproduce the experimental spectra of the analogues [24, 25]. For this, 10 ns molecular dynamics of the analogues in a water box following by B3LYP/6-311g(2d,p)/AMBERff99 on 100 snapshots were performed. The QM/MM studies of analogues and the forms of the natural oxyluciferin were compared and show promising applications of the analogues. The simulations are also important to interpret and guide the experiments.

One can recall the following conclusions: calculated emission spectra from the protonated neutral forms (phenol–keto and phenol–enol) are blue-shifted compared to the deprotonated ones and experimental emission spectra. Enolate–phenol form has its oscillator strength f of the emission transition (i.e., emission intensity) much lower than the ones computed for the other deprotonated forms. Experimental studies and theoretical calculations show that the benzophenone should be deprotonated [44]. Studies comparing simulated and experimental emission spectra of the different chemical forms of oxyluciferin and some synthetic analogues have demonstrated that

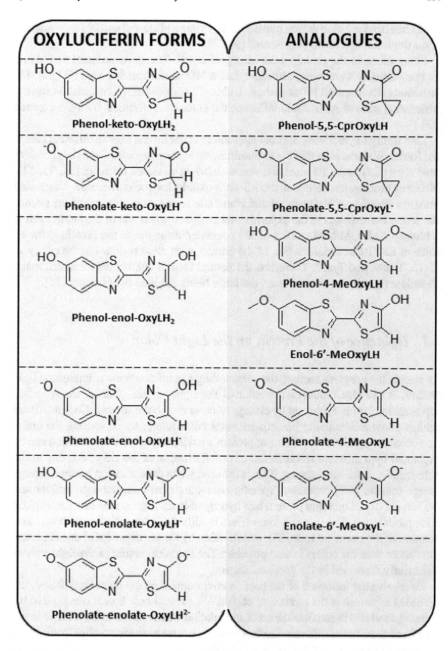

Fig. 14 The six forms of firefly oxyluciferin and the synthetic analogues that are studied to better understand the protonation/deprotonation states and keto/enol tautomerization. Reprinted (adapted) with permission from (García-Iriepa et al. JCTC, 2018, 14, 2117–2126). Copyright (2018) American Chemical Society

an efficient excited state proton transfer (ESPT) takes place and so, only the emission of the deprotonated forms is observed [24, 78].

Simulations in the protein were also preformed. The same protocol as the one for simulation in water was performed, i.e., a MD simulation followed by QM/MM transition calculation on 100 snapshots. It shows that the protein conformation and the protonation state of AMP could influence the emission of fireflies' bioluminescence [23].

Like firefly oxyluciferin, the coelenteramide in obelin can exist in different chemical forms. We have seen that the bioluminescent systems evolved from the chemiexcited state (CS) to the FS state, the one with bright emission intensity [14, 15, 51]. Different protonation forms of the FS are possible for coelenteramide. Computed emission energies of both neutral and phenolate forms show that one of the phenolate forms corresponds to the experimental emission when one of the neutral forms is blue shifted. QM/MM studies of the coelenteramide inside the protein show an efficient ESPT due to interaction of the emitter with close residues in the cavity as His16, Trp86, and Tyr82. Therefore, the neutral form is quickly deprotonated to the phenolate form in the excited state, the latter being the blue-light emitter [15].

7.3   Influence of the Protein on the Light Color

As seen in the previous section, the surrounding can influence the nature of the light emitter. It can also influence the color of the light emitted. Indeed, the so-called solvatochromism is explained by change of color due to the solvent. Computational studies allow rationalizing the experimental color tuning by computing the emission considering different solvent or protein's environments. The modulation can be explained by a different stabilization or destabilization of the GS or S1, due to the polarization of the surrounding. This is the case when the excitation presents a large charge-transfer (CT) character. Specific interactions between the light emitter and the solvent or surrounding protein like hydrogen-bond interactions can also explain color modulation. The two previous effects modify the color emission without modification of the chemical form. The stabilization by the environment of one chemical form more than the others is also possible. The chemical nature of the light emitter was briefly discussed in the previous section.

To resolve the influence of the protein environment on the color modulation, the QM/MM approach is the method of choice. We present here how it was used to try to understand and rationalize the color modulation when one mutation is made in the protein or to compare different firefly's species, whose proteins are slightly different and therefore emit in a different range of color.

The resolution of the wild type *Luciola cruciata* and a red mutant (S288N) luciferase structures published by Nakatsu et al. [58] in 2006 at different stages of the reaction suggested that modulation of the light emitted by fireflies depends on the open or closed nature of the cavity. According to the authors, the cavity structure of the single mutated S286N remains open, allowing the light emitter to relax before emitting a red light. The native protein would close on the molecule preventing it from relaxing and leading to a wavelength emitted in the green. To test this hypothesis, QM/MM studies of the phenolate-keto form of the oxyluciferin inside the cavity were performed [59]. After placing the molecule in the cavity, performing a small classical molecular dynamics and extracting a snapshot in the two cases: an open cavity or a closed cavity, the oxyluciferin was chosen as high layer and the rest of the protein as low layer (see Sect. 4.2). QM/MM optimization of the system in its first singlet excited state at the CASSCF(16,14)/6-31G*//AMBERParm99 level, allowing the water molecules at less than 5 Å from the oxyluciferin and side chains of ARG220, PHE249, THR253, ILE288, SER349, and ALA350 to move was followed by CASPT2//AMBERParm99 calculations of the vertical electronic transition. The results showed that the open or closed nature of the cavity did not influence the emitted wavelength. However the hydrogen-bond network involving the residues, water molecules, and the oxyluciferin's benzothiazole moiety was responsible for the color modulation of the emitted light. The mutation S286N, present at the door of the cavity, changes the H-bonding network and then the polarity of the microenvironment at the phenolate moiety of oxyluciferin molecule, destabilizing the HOMO orbitals, and then a red-shift of the emitted light. The influence of the presence of the water molecule or H-bonding on the color modulation was latter showed experimentally by spectroscopy results on a single monohydrated molecule of oxyluciferin in gas phase showing the red-shifting due to the presence of the water molecule [81].

A more recent application of QM/MM study of color modulation by the firefly luciferase has been published in 2018 [11]. Two new crystallographic structures were determined: one structure from the *Amydetes Vivianii* firefly (GBAv) whose light emission is green-blue shifted from the Japanese *Luciola cruciatia* yellow-green bioluminescent system (YGLc), and one from the glow-worm *Phrixothrix hirtus*, whose light emission is red. Thanks to MD followed by QM/MM calculations, it has been possible to study and understand the impact of new luciferases on the wavelength of bioluminescence. The two new crystallographic structures were determined without any substrate (neither luciferin or oxyluciferin nor AMP in the cavity). The protocol used followed the step of (1) docking the oxyluciferin and AMP, (2) performing 10 ns classical MD, (3) calculating of the vertical transition by TD-DFT/MM.

For GBAv, the conformation obtained by X-ray with the open C-terminal domain is not the conformation allowing bioluminescence. Thus the folding of this domain (closing the cavity) is a necessary prerequisite for obtaining the emission. Moreover, the superposition of the resulting GBAv model and YGLc shows a displacement on the substrate in GBAv, exposing the emitting molecule to other amino acids than in YGLc and affecting the color emission. The conclusion of the QM/MM investigations explained the blue-shift of the wavelength in *Amydetes Vivianii* by a change of environment inside the cavity. In a second step, the study of several *in silico* mutations makes it possible to better understand the influence of the environment on the substrate and the emission. The difficulty resides on a systematic protocol to be able to measure the influence of each residue and taking into account the dynamics of the protein due to mutations. The ultimate goal would be to be able to predict the influence of each residue and to be able to point out the relevant mutations.

This study raises the difficulty and the limit of the QM/MM simulations. The modeling of the color modulation of the emitted light relies on the starting crystallographic structures and on the sampling of different conformations of the protein. This means that it still relies on the experimental data. Mutations can cause huge change of conformation and this should be taken into account in the modeling of the phenomena. Experimental-theoretical collaboration remains the driving force of scientific progress in the field of bioluminescence.

7.4 Toward New Light Emitters

In order to design red-shift emitted light, crucial for biological application, where low in energy light is required to ensure a deep tissue penetration, numerous analogues of firefly luciferin molecule have been proposed [38]. Computational studies are useful

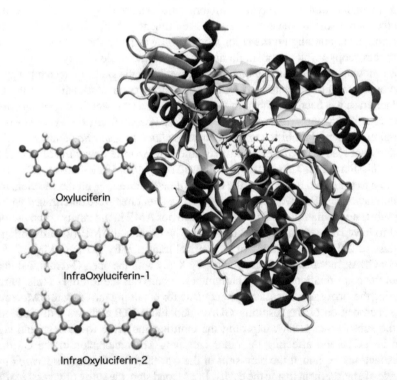

Fig. 15 Infrared analogues of oxyluciferin studied by QM/MM simulations in the cavity of firefly oxyluciferase. Two conformations of the InfraOxyluciferin were explored (InfraOxyluciferin-1 and InfraOxyluciferin-2 conformations shown at the left of the figure) and compared to the native phenolate-keto oxyluciferin (at the top left of the figure). The protein is drawn as a ribbon structure. Reproduced from Ref. R. Berraud-Pache and I. Navizet, Phys. Chem. Chem. Phys., 2016, 18, 27460 with permission from the PCCP Owner Societies

tools to understand the modification of the bioluminescent properties (efficiency, color, and intensity) of the synthetic derivatives.

For example, QM/MM studies of electronic transitions of two possible conformations of a synthetic derivative of firefly oxyluciferin were performed [7] to understand the experimental results of Ref. [37]. Figure 15 shows the two conformers of the studied analogue which possesses an extra double C=C between the two cycles of the natural oxyluciferin. QM/MM calculations (B3LYP/6-311G(2d,p)/AMBER and MS-CASPT2/ANO-RCC-VTZP/AMBER) of the electronic transition from the optimized first ES show that the protein cavity constraints the molecule differently in the two conformers and that the electrostatic effect of the protein lowers the transition energy. Comparing QM/MM calculations and experimental data shows that the conformer InfraOxyluciferin-2 is probably the light emitter. A recent experimental study [82] has confirmed the conformation predicted by QM/MM studies.

Few QM/MM studies have been performed to simulate the absorption and fluorescence spectra of some synthetic derivatives of firefly oxyluciferin (see Fig. 16), in solvent and inside the protein [100]. The experimental shifts observed in fluores-

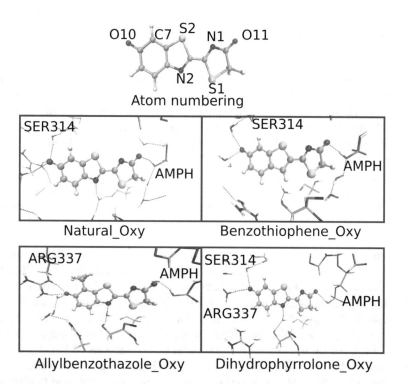

Fig. 16 H-bond network in the protein cavity around the analogues of oxyluciferin studied by QM/MM simulations. Adapted from Ref. M. Zemmouche, C. García-Iriepa and I. Navizet, Phys. Chem. Chem. Phys., 2020, 22(1), 82–91, DOI: https://doi.org/10.1039/C9CP04687A with permission from the PCCP Owner Societies

cent spectra [33, 34, 96] were well reproduced by QM/MM (B3LYP/6-311G(2d,p)/ AMBER) calculations [100]. The calculated spectra correspond to the experimental ones. Moreover an analyse of the type of transition and the H-bonding surrounding inside the cavity is possible (see Fig. 16 with the H-bonding surrounding of the analogues). The comparison of QM/MM calculations with the ones considering an implicit solvent shows that the (much faster) last calculations can give a qualitative trend of the emission wavelength whereas, considering explicitly the surrounding (QM/MM methods) is needed for quantitative results.

In the two exposed examples, we can conclude that QM/MM calculations can be good predictive tools to design new analogues. Moreover, increasing the π-conjugated system is one of research leads that can be investigated. The addition of groups that attract or donate electrons in the system is to be exploited.

8 Conclusion

In this chapter, some examples of how QM/MM studies were performed to give better understanding of bioluminescent systems were presented. We have presented the different bioluminescent systems and the specificity of bioluminescent reactions that are complex reactions involving at least one electronic excited state. QM/MM methods were also presented and the possible protocols used to study bioluminescent systems were exposed. The idea was not to be exhaustive on all theoretical studies done on bioluminescent systems but some examples, taken essentially from my group on the firefly bioluminescence system, were exposed. The presented results illustrate the difficulties of such studies but also how QM/MM is a nice tool to the research community to better understand embedded systems.

There are still many questions to solve in the bioluminescent systems and researchers have to be inventive to propose smart protocols to answer the questions. The modulation of light in bioluminescent systems still keeps some secrets. Only a joint experimental–theoretical studies will be able to reach the knowledge on these systems.

What if we have infinite computer capacities (memory and processor time) as well infinite human time to interpret the data? We could consider exploring QM/MM dynamics of the dioxetanone intermediate's decomposition inside the protein cavity, as it is done in gas phase for smaller models [88] leading to the excited state of the oxyluciferin molecule. We could investigate further the pH effect on the light emission, taking into account the protonation equilibrium at different pH, as it is done in Ref. [46], but for the excited state and with highly accurate QM/MM methods. We could explore the whole reaction with QM/MM methods from the entrance of the substrates to the light emission (QM/MM dynamics would be used to study the different reaction steps); we could perform QM/MM dynamics in the excited states to better understand what is the form of the light emitter (keto or enol form). We could perform QM/MM dynamics in the excited state to have a better understanding on the quantum yield. Hopefully, a not so far future will make able all these investigations.

Acknowledgements Isabelle Navizet acknowledges support from the ANR BIOLUM project (ANR-16-CE29-0013) for QM/MM studies on firefly systems. Isabelle Navizet acknowledges M. Zemmouche and M. Sahihi and the reviewers of this chapter for their advice.

References

1. Anandakrishnan R, Aguilar B, Onufriev AV (2012) H++ 3.0: automating pK prediction and the preparation of biomolecular structures for atomistic molecular modeling and simulations. Nucleic Acids Res 40(1):537–541. https://doi.org/10.1093/nar/gks375
2. Andersson K, Malmqvist PÅ, Roos BO (1992) Second-order perturbation theory with a complete active space self-consistent field reference function. J Chem Phys 96(2):1218–1226. https://doi.org/10.1063/1.462209
3. Ando Y, Akiyama H (2010) pH-dependent fluorescence spectra, lifetimes, and quantum yields of firefly-luciferin aqueous solutions studied by selective-excitation fluorescence spectroscopy. Jpn J Appl Phys 49(11R):117002. https://doi.org/10.1143/jjap.49.117002
4. Ando Y, Niwa K, Yamada N, Enomoto T, Irie T, Kubota H, Ohmiya Y, Akiyama H (2008) Firefly bioluminescence quantum yield and colour change by pH-sensitive green emission. Nat Photonics 2(1):44–47. https://doi.org/10.1038/nphoton.2007.251
5. Angeli C, Cimiraglia R, Evangelisti S, Leininger T, Malrieu JP (2001) Introduction of *n*-electron valence states for multireference perturbation theory. J Chem Phys 114(23):10252–10264. https://doi.org/10.1063/1.1361246
6. Becke AD (1988) Density-functional exchange-energy approximation with correct asymptotic behavior. Phys Rev A 38(6):3098–3100. https://doi.org/10.1103/physreva.38.3098
7. Berraud-Pache R, Navizet I (2016) QM/MM calculations on a newly synthesised oxyluciferin substrate: new insights into the conformational effect. Phys Chem Chem Phys 18(39):27460–27467. https://doi.org/10.1039/c6cp02585d
8. Berraud-Pache R, Lindh R, Navizet I (2018) QM/MM study of the formation of the dioxetanone ring in fireflies through a superoxide ion. J Phys Chem B 122(20):5173–5182. https://doi.org/10.1021/acs.jpcb.8b00642, pMID: 29659277
9. Branchini BR, Ablamsky DM, Murtiashaw MH, Uzasci L, Fraga H, Southworth TL (2007) Thermostable red and green light-producing firefly luciferase mutants for bioluminescent reporter applications. Anal Biochem 361(2):253–262. https://doi.org/10.1016/j.ab.2006.10.043
10. Cances E, Mennucci B, Tomasi J (1997) A new integral equation formalism for the polarizable continuum model: theoretical background and applications to isotropic and anisotropic dielectrics. J Chem Phys 107(8):3032–3041. https://doi.org/10.1063/1.474659
11. Carrasco-López C, Ferreira JC, Lui NM, Schramm S, Berraud-Pache R, Navizet I, Panjikar S, Naumov P, Rabeh WM (2018) Beetle luciferases with naturally red- and blue-shifted emission. Life Sci Alliance 1(4). https://doi.org/10.26508/lsa.201800072
12. Casida ME (1995) Time-dependent density functional response theory for molecules. World Scientific, Singapore, chap 5:155–192. https://doi.org/10.1142/9789812830586_0005
13. Chen SF, Liu YJ, Navizet I, Ferré N, Fang WH, Lindh R (2011) Systematic theoretical investigation on the light emitter of firefly. J Chem Theory Comput 7(3):798–803. https://doi.org/10.1021/ct200045q
14. Chen SF, Navizet I, Roca-Sanjuán D, Lindh R, Liu YJ, Ferré N (2012) Chemiluminescence of coelenterazine and fluorescence of coelenteramide: a systematic theoretical study. J Chem Theory Comput 8(8):2796–2807. https://doi.org/10.1021/ct300356j
15. Chen SF, Ferré N, Liu YJ (2013) QM/MM study on the light emitters of aequorin chemiluminescence, bioluminescence, and fluorescence: a general understanding of the bioluminescence of several marine organisms. Chem-Eur J 19(26):8466–8472. https://doi.org/10.1002/chem.201300678

16. Cheng YY, Liu YJ (2016) Vibrationally resolved absorption and fluorescence spectra of firefly luciferin: a theoretical simulation in the gas phase and in solution. Photochem Photobiol 92(4):552–560. https://doi.org/10.1111/php.12601

17. Cheng YY, Liu YJ (2019) Luciferin regeneration in firefly bioluminescence via proton-transfer-facilitated hydrolysis, condensation and chiral inversion. ChemPhysChem 20(13):1719–1727. https://doi.org/10.1002/cphc.201900306

18. Dapprich S, Komáromi I, Byun K, Morokuma K, Frisch MJ (1999) A new ONIOM implementation in Gaussian98. Part I. The calculation of energies, gradients, vibrational frequencies and electric field derivatives. J Mol Struct: THEOCHEM 461–462:1–21. https://doi.org/10.1016/S0166-1280(98)00475-8

19. Davis IW, Leaver-Fay A, Chen VB, Block JN, Kapral GJ, Wang X, Murray LW, Bryan AIW, Snoeyink J, Richardson JS, Richardson DC (2007) Molprobity: all-atom contacts and structure validation for proteins and nucleic acids. Nucleic Acids Res 35(2):375–383. https://doi.org/10.1093/nar/gkm216

20. Dubinnyi MA, Kaskova ZM, Rodionova NS, Baranov MS, Gorokhovatsky AY, Kotlobay A, Solntsev KM, Tsarkova AS, Petushkov VN, Yampolsky IV (2015) Novel mechanism of bioluminescence: oxidative decarboxylation of a moiety adjacent to the light emitter of Fridericia luciferin. Angew Chem Int Ed 54(24):7065–7067. https://doi.org/10.1002/anie.201501668

21. Ferré N, Ángyán JG (2002) Approximate electrostatic interaction operator for QM/MM calculations. Chem Phys Lett 356(3–4):331–339. https://doi.org/10.1016/s0009-2614(02)00343-3

22. Finley J, Malmqvist PÅ, Roos BO, Serrano-Andrés L (1998) The multi-state CASPT2 method. Chem Phys Lett 288(2–4):299–306. https://doi.org/10.1016/s0009-2614(98)00252-8

23. García-Iriepa C, Navizet I (2019) Effect of protein conformation and amp protonation state on fireflies' bioluminescent emission. Molecules 24(8). https://doi.org/10.3390/molecules24081565

24. García-Iriepa C, Gosset P, Berraud-Pache R, Zemmouche M, Taupier G, Dorkenoo KD, Didier P, Léonard J, Ferré N, Navizet I (2018) Simulation and analysis of the spectroscopic properties of oxyluciferin and its analogues in water. J Chem Theory Comput 14(4):2117–2126. https://doi.org/10.1021/acs.jctc.7b01240

25. García-Iriepa C, Zemmouche M, Ponce-Vargas M, Navizet I (2019) The role of solvation models on the computed absorption and emission spectra: the case of fireflies oxyluciferin. Phys Chem Chem Phys 21:4613–4623. https://doi.org/10.1039/C8CP07352J

26. Ghose A, Rebarz M, Maltsev OV, Hintermann L, Ruckebusch C, Fron E, Hofkens J, Mély Y, Naumov P, Sliwa M, Didier P (2015) Emission properties of oxyluciferin and its derivatives in water: revealing the nature of the emissive species in firefly bioluminescence. J Phys Chem B 119(6):2638–2649. https://doi.org/10.1021/jp508905m, pMID: 25364813

27. Gnandt D, Na S, Koslowski T (2018) Simulating biological charge transfer: continuum dielectric theory or molecular dynamics? Biophys Chem 241:1–7. https://doi.org/10.1016/j.bpc.2018.07.001

28. Haddock SHD, Moline MA, Case JF (2010) Bioluminescence in the sea. Annu Rev Mar Sci 2(1):443–493. https://doi.org/10.1146/annurev-marine-120308-081028

29. Hehre WJ, Stewart RF, Pople JA (1969) Self-consistent molecular-orbital methods. I. Use of Gaussian expansions of slater-type atomic orbitals. J Chem Phys 51(6):2657–2664. https://doi.org/10.1063/1.1672392

30. Hiyama M, Noguchi Y, Akiyama H, Yamada K, Koga N (2015) Vibronic structures in absorption and fluorescence spectra of firefly oxyluciferin in aqueous solutions. Photochem Photobiol 91(4):819–827. https://doi.org/10.1111/php.12463

31. Hornak V, Abel R, Okur A, Strockbine B, Roitberg A, Simmerling C (2006) Comparison of multiple amber force fields and development of improved protein backbone parameters. Proteins: Struct Funct Bioinform 65(3):712–725. https://doi.org/10.1002/prot.21123

32. Huang J, Rauscher S, Nawrocki G, Ran T, Feig M, de Groot BL, Grubmüller H, MacKerell AD (2017) CHARMM36m: an improved force field for folded and intrinsically disordered proteins. Nat Methods 14(1):71–73. https://doi.org/10.1038/nmeth.4067

33. Ikeda Y, Saitoh T, Niwa K, Nakajima T, Kitada N, Maki SA, Sato M, Citterio D, Nishiyama S, Suzuki K (2018) An allylated firefly luciferin analogue with luciferase specific response in living cells. Chem Commun 54:1774–1777. https://doi.org/10.1039/C7CC09720D

34. Ioka S, Saitoh T, Iwano S, Suzuki K, Maki SA, Miyawaki A, Imoto M, Nishiyama S (2016) Synthesis of firefly luciferin analogues and evaluation of the luminescent properties. Chem Eur J 22(27):9330–9337. https://doi.org/10.1002/chem.201600278

35. Isobe H, Takano Y, Okumura M, Kuramitsu S, Yamaguchi K (2005) Mechanistic insights in charge-transfer-induced luminescence of 1,2-dioxetanones with a substituent of low oxidation potential. J Am Chem Soc 127(24):8667–8679. https://doi.org/10.1021/ja043295f

36. Jacquemin D, Planchat A, Adamo C, Mennucci B (2012) 5TD-DFT assessment of functionals for optical 0–0 transitions in solvated dyes. J Chem Theory Comput 8(7):2359–2372. https://doi.org/10.1021/ct300326f, pMID: 26588969

37. Jathoul AP, Grounds H, Anderson JC, Pule MA (2014) A dual-color far-red to near-infrared firefly luciferin analogue designed for multiparametric bioluminescence imaging. Angew Chem Int Ed 53(48):13059–13063. https://doi.org/10.1002/anie.201405955

38. Kaskova ZM, Tsarkova AS, Yampolsky IV (2016) 1001 lights: luciferins, luciferases, their mechanisms of action and applications in chemical analysis, biology and medicine. Chem Soc Rev 45(21):6048–6077. https://doi.org/10.1039/c6cs00296j

39. Kato D, Shirakawa D, Polz R, Maenaka M, Takeo M, Negoro S, Niwa K (2014) A firefly inspired one-pot chemiluminescence system using n-propylphosphonic anhydride (T3P). Photochem Photobiol Sci 13:1640–1645. https://doi.org/10.1039/C4PP00250D

40. Kiani FA, Fischer S (2015) Advances in quantum simulations of ATPase catalysis in the myosin motor. Curr Opin Struct Biol 31:115 – 123. https://doi.org/10.1016/j.sbi.2015.04.006 (Theory and simulation/macromolecular machines and assemblies)

41. Kollman PA, Massova I, Reyes C, Kuhn B, Huo S, Chong L, Lee M, Lee T, Duan Y, Wang W, Donini O, Cieplak P, Srinivasan J, Case DA, Cheatham TE (2000) Calculating structures and free energies of complex molecules: combining molecular mechanics and continuum models. Acc Chem Res 33(12):889–897. https://doi.org/10.1021/ar000033j, pMID: 11123888

42. Laurent AD, Jacquemin D (2013) TD-DFT benchmarks: a review. Int J Quantum Chem 113(17):2019–2039. https://doi.org/10.1002/qua.24438

43. Lee C, Yang W, Parr RG (1988) Development of the Colle-Salvetti correlation-energy formula into a functional of the electron density. Phys Rev B 37(2):785–789. https://doi.org/10.1103/physrevb.37.785

44. Liu F, Liu Y, De Vico L, Lindh R (2009) A CASSCF/CASPT2 approach to the decomposition of thiazole-substituted dioxetanone: substitution effects and charge-transfer induced electron excitation. Chem Phys Lett 484(1–3):69–75. https://doi.org/10.1016/j.cplett.2009.11.009

45. Loco D, Polack É, Caprasecca S, Lagardère L, Lipparini F, Piquemal JP, Mennucci B (2016) A QM/MM approach using the AMOEBA polarizable embedding: from ground state energies to electronic excitations. J Chem Theory Comput 12(8):3654–3661. https://doi.org/10.1021/acs.jctc.6b00385

46. Manuel de Almeida Barbosa N, Zemmouche M, Gosset P, García-Iriepa C, Ledentu V, Navizet I, Didier P, Ferré N (2019) pH-dependent absorption spectrum of oxyluciferin analogues in the presence of adenosine monophosphate. ChemPhotoChem 3(12):1219–1230. https://doi.org/10.1002/cptc.201900150

47. Marenich AV, Cramer CJ, Truhlar DG (2009) Universal solvation model based on solute electron density and on a continuum model of the solvent defined by the bulk dielectric constant and atomic surface tensions. J Phys Chem B 113(18):6378–6396. https://doi.org/10.1021/jp810292n, pMID: 19366259

48. Mennucci B, Cances E, Tomasi J (1997) Evaluation of solvent effects in isotropic and anisotropic dielectrics and in ionic solutions with a unified integral equation method: theoretical bases, computational implementation, and numerical applications. J Phys Chem B 101(49):10506–10517. https://doi.org/10.1021/jp971959k

49. Min CG, Ren AM, Guo JF, Li ZW, Zou LY, Goddard JD, Feng JK (2010) A time-dependent density functional theory investigation on the origin of red chemiluminescence. ChemPhysChem 11(1):251–259. https://doi.org/10.1002/cphc.200900607

50. Min CG, Ren AM, Li XN, Guo JF, Zou LY, Sun Y, Goddard JD, Sun CC (2011) The formation and decomposition of firefly dioxetanone. Chem Phys Lett 506(4):269–275. https://doi.org/10.1016/j.cplett.2011.01.064
51. Min CG, Pinto da Silva L, Esteves da Silva JC, Yang XK, Huang SJ, Ren AM, Zhu YQ (2017) A computational investigation of the equilibrium constants for the fluorescent and chemiluminescent states of coelenteramide. ChemPhysChem 18(1):117–123. https://doi.org/10.1002/cphc.201600850
52. Mitschke U, Bauerle P (2000) The electroluminescence of organic materials. J Mater Chem 10:1471–1507. https://doi.org/10.1039/A908713C
53. Mochizuki T, Wang Y, Hiyama M, Akiyama H (2014) Robust red-emission spectra and yields in firefly bioluminescence against temperature changes. Appl Phys Lett 104(21):213704. https://doi.org/10.1063/1.4880578
54. Modestova Y, Ugarova NN (2016) Color-shifting mutations in the C-domain of *L. mingrelica* firefly luciferase provide new information about the domain alternation mechanism. Biochim Biophys Acta Proteins Proteomics 1864(12):1818–1826. https://doi.org/10.1016/j.bbapap.2016.09.007
55. Monette Z, Kasar AK, Menezes PL (2019) Advances in triboluminescence and mechanoluminescence. J Mater Sci: Mater Electron 30(22):19675–19690. https://doi.org/10.1007/s10854-019-02369-8
56. Morris GM, Huey R, Lindstrom W, Sanner MF, Belew RK, Goodsell DS, Olson AJ (2009) AutoDock4 and AutoDockTools4: automated docking with selective receptor flexibility. J Comput Chem 30(16):2785–2791. https://doi.org/10.1002/jcc.21256
57. Nakano H (1993) Quasidegenerate perturbation theory with multiconfigurational self-consistent-field reference functions. J Chem Phys 99(10):7983–7992. https://doi.org/10.1063/1.465674
58. Nakatsu T, Ichiyama S, Hiratake J, Saldanha A, Kobashi N, Sakata K, Kato H (2006) Structural basis for the spectral difference in luciferase bioluminescence. Nature 440(7082):372–376. https://doi.org/10.1038/nature04542
59. Navizet I, Liu YJ, Ferré N, Xiao HY, Fang WH, Lindh R (2010) Color-tuning mechanism of firefly investigated by multi-configurational perturbation method. J Am Chem Soc 132(2):706–712. https://doi.org/10.1021/ja908051h
60. Navizet I, Liu YJ, Ferré N, Roca-Sanjuán D, Lindh R (2011) The chemistry of bioluminescence: an analysis of chemical functionalities. ChemPhysChem 12(17):3064–3076. https://doi.org/10.1002/cphc.201100504
61. Navizet I, Roca-Sanjuán D, Yue L, Liu YJ, Ferré N, Lindh R (2013) Are the bio- and chemiluminescence states of the firefly oxyluciferin the same as the fluorescence state? Photochem Photobiol 89(2):319–325. https://doi.org/10.1111/php.12007
62. Ohmiya Y, Kojima S, Nakamura M, Niwa H (2005) Bioluminescence in the limpet-like snail. Latia neritoides. Bull Chem Soc Jpn 78(7):1197–1205. https://doi.org/10.1246/bcsj.78.1197
63. Olsson MHM, Søndergaard CR, Rostkowski M, Jensen JH (2011) PROPKA3: consistent treatment of internal and surface residues in empirical pKa predictions. J Chem Theory Comput 7(2):525–537. https://doi.org/10.1021/ct100578z
64. Oostenbrink C, Villa A, Mark AE, Van Gunsteren WF (2004) A biomolecular force field based on the free enthalpy of hydration and solvation: the GROMOS force-field parameter sets 53A5 and 53A6. J Comput Chem 25(13):1656–1676. https://doi.org/10.1002/jcc.20090
65. Perdew JP, Ruzsinszky A, Tao J, Staroverov VN, Scuseria GE, Csonka GI (2005) Prescription for the design and selection of density functional approximations: more constraint satisfaction with fewer fits. J Chem Phys 123(6):062201. https://doi.org/10.1063/1.1904565
66. Piquemal JP, Gresh N, Giessner-Prettre C (2003) Improved formulas for the calculation of the electrostatic contribution to the intermolecular interaction energy from multipolar expansion of the electronic distribution. J Phys Chem A 107(48):10353–10359. https://doi.org/10.1021/jp035748t
67. Rappe AK, Casewit CJ, Colwell KS, Goddard WA, Skiff WM (1992) UFF, a full periodic table force field for molecular mechanics and molecular dynamics simulations. J Am Chem Soc 114(25):10024–10035. https://doi.org/10.1021/ja00051a040

68. Roca-Sanjuán D, Delcey MG, Navizet I, Ferré N, Liu YJ, Lindh R (2011) Chemiluminescence and fluorescence states of a small model for coelenteramide and cypridina oxyluciferin: a CASSCF/CASPT2 study. J Chem Theory Comput 7(12):4060–4069. https://doi.org/10.1021/ct2004758

69. Roos BO, Taylor PR, Siegbahn PEM (1980) A complete active space SCF method (CASSCF) using a density matrix formulated super-CI approach. Chem Phys 48(2):157–173. https://doi.org/10.1016/0301-0104(80)80045-0

70. Roos BO, Lindh R, Malmqvist PÅ, Veryazov V, Widmark PO (2004) Main group atoms and dimers studied with a new relativistic ANO basis set. J Phys Chem A 108(15):2851–2858. https://doi.org/10.1021/jp031064+

71. Roos BO, Lindh R, Malmqvist PÅ, Veryazov V, Widmark PO (2016) Multiconfigurational quantum chemistry. Wiley, Ltd. https://doi.org/10.1002/9781119126171

72. Roy A, Kucukural A, Zhang Y (2010) I-TASSER: a unified platform for automated protein structure and function prediction. Nat Protoc 5(4):725–738. https://doi.org/10.1038/nprot.2010.5

73. Sharipov GL, Abdrakhmanov AM, Gareev BM (2019) Visualization of luminescence of two types in an acoustic field in a liquid. Tech Phys Lett 45(12):1175–1177. https://doi.org/10.1134/S1063785019120137

74. Shimomura O (2006) Bioluminescence. World Scientific. https://doi.org/10.1142/6102

75. Shimomura O, Johnson FH, Masugi T (1969) Cypridina bioluminescence: light-emitting oxyluciferin-luciferase complex. Science 164(3885):1299–1300. https://doi.org/10.1126/science.164.3885.1299

76. Shiozaki T, Győrffy W, Celani P, Werner HJ (2011) Extended multi-state complete active space second-order perturbation theory: energy and nuclear gradients. J Chem Phys 135(8):081106. https://doi.org/10.1063/1.3633329

77. da Silva LP, Esteves da Silva JC (2011) Computational studies of the luciferase light-emitting product: oxyluciferin. J Chem Theory Comput 7(4):809–817. https://doi.org/10.1021/ct200003u

78. da Silva LP, Simkovitch R, Huppert D, Esteves da Silva JC (2013) Oxyluciferin photoacidity: the missing element for solving the keto-enol mystery? ChemPhysChem 14(15):3441–3446

79. da Silva LP, da Silva JCE (2014) Quantum/molecular mechanics study of firefly bioluminescence on luciferase oxidative conformation. Chem Phys Lett 608:45–49. https://doi.org/10.1016/j.cplett.2014.05.061

80. Stein CJ, Reiher M (2016) Automated selection of active orbital spaces. J Chem Theory Comput 12(4):1760–1771. https://doi.org/10.1021/acs.jctc.6b00156, pMID: 26959891

81. Støchkel K, Hansen CN, Houmøller J, Nielsen LM, Anggara K, Linares M, Norman P, Nogueira F, Maltsev OV, Hintermann L, Nielsen SB, Naumov P, Milne BF (2013) On the influence of water on the electronic structure of firefly oxyluciferin anions from absorption spectroscopy of bare and monohydrated ions in vacuo. J Am Chem Soc 135(17):6485–6493. https://doi.org/10.1021/ja311400t

82. Stowe CL, Burley TA, Allan H, Vinci M, Kramer-Marek G, Ciobota DM, Parkinson GN, Southworth TL, Agliardi G, Hotblack A, Lythgoe MF, Branchini BR, Kalber TL, Anderson JC, Pule MA (2019) Near-infrared dual bioluminescence imaging in mouse models of cancer using infraluciferin. eLife 8:e45801. https://doi.org/10.7554/eLife.45801

83. Sundlov JA, Fontaine DM, Southworth TL, Branchini BR, Gulick AM (2012) Crystal structure of firefly luciferase in a second catalytic conformation supports a domain alternation mechanism. Biochemistry 51(33):6493–6495. https://doi.org/10.1021/bi300934s

84. Tafreshi NK, Hosseinkhani S, Sadeghizadeh M, Sadeghi M, Ranjbar B, Naderi-Manesh H (2007) The influence of insertion of a critical residue (Arg^{356}) in structure and bioluminescence spectra of firefly luciferase. J Biol Chem 282(12):8641–8647. https://doi.org/10.1074/jbc.m609271200

85. Tang K, Zhang J, Liang J (2014) Fast protein loop sampling and structure prediction using distance-guided sequential chain-growth Monte Carlo method. PLoS Comput Biol 10(4):e1003539. https://doi.org/10.1371/journal.pcbi.1003539

86. Tao J, Perdew JP, Staroverov VN, Scuseria GE (2003) Climbing the density functional ladder: nonempirical meta-generalized gradient approximation designed for molecules and solids. Phys Rev Lett 91(14):146401. https://doi.org/10.1103/physrevlett.91.146401

87. Ullrich CA, Yang Z (2014) A brief compendium of time-dependent density functional theory. Braz J Phys 44(1):154–188. https://doi.org/10.1007/s13538-013-0141-2

88. Vacher M, Fdez Galván I, Ding BW, Schramm S, Berraud-Pache R, Naumov P, Ferré N, Liu YJ, Navizet I, Roca-Sanjuán D, Baader WJ, Lindh R (2018) Chemi- and bioluminescence of cyclic peroxides. Chem Rev 118(15):6927–6974. https://doi.org/10.1021/acs.chemrev.7b00649, pMID: 29493234

89. Vanommeslaeghe K, Hatcher E, Acharya C, Kundu S, Zhong S, Shim J, Darian E, Guvench O, Lopes P, Vorobyov I, Mackerell AD Jr (2010) CHARMM general force field: a force field for drug-like molecules compatible with the CHARMM all-atom additive biological force fields. J Comput Chem 31(4):671–690. https://doi.org/10.1002/jcc.21367

90. Vysotski ES, Markova SV, Frank LA (2006) Calcium-regulated photoproteins of marine coelenterates. Mol Biol 40(3):355–367. https://doi.org/10.1134/s0026893306030022

91. Wang J, Wolf RM, Caldwell JW, Kollman PA, Case DA (2004) Development and testing of a general Amber force field. J Comput Chem 25(9):1157–1174. https://doi.org/10.1002/jcc.20035

92. Wang Y, Kubota H, Yamada N, Irie T, Akiyama H (2011) Quantum yields and quantitative spectra of firefly bioluminescence with various bivalent metal ions. Photochem Photobiol 87(4):846–852. https://doi.org/10.1111/j.1751-1097.2011.00931.x

93. Webb B, Sali A (2016) Comparative protein structure modeling using modeller. Curr Protoc Protein Sci 86(1):2.9.1–2.9.37. https://doi.org/10.1002/cpps.20

94. White EH, Rapaport E, Hopkins TA, Seliger HH (1969) Chemi- and bioluminescence of firefly luciferin. J Am Chem Soc 91(8):2178–2180. https://doi.org/10.1021/ja01036a093

95. Wilson T, Hastings JW (1998) Bioluminescence. Annu Rev Cell Dev Biol 14(1):197–230. https://doi.org/10.1146/annurev.cellbio.14.1.197, pMID: 9891783

96. Woodroofe CC, Meisenheimer PL, Klaubert DH, Kovic Y, Rosenberg JC, Behney CE, Southworth TL, Branchini BR (2012) Novel heterocyclic analogues of firefly luciferin. Biochemistry 51(49):9807–9813. https://doi.org/10.1021/bi301411d, pMID: 23164087

97. Yanai T, Tew DP, Handy NC (2004) A new hybrid exchange-correlation functional using the Coulomb-attenuating method (CAM-B3LYP). Chem Phys Lett 393(1–3):51–57. https://doi.org/10.1016/j.cplett.2004.06.011

98. York DM, Karplus M (1999) A smooth solvation potential based on the conductor-like screening model. J Phys Chem A 103(50):11060–11079. https://doi.org/10.1021/jp992097l

99. Yue L, Liu YJ, Fang WH (2012) Mechanistic insight into the chemiluminescent decomposition of firefly dioxetanone. J Am Chem Soc 134(28):11632–11639. https://doi.org/10.1021/ja302979t

100. Zemmouche M, García-Iriepa C, Navizet I (2020) Light emission colour modulation study of oxyluciferin synthetic analogues via QM and QM/MM approaches. Phys Chem Chem Phys 22:82–91. https://doi.org/10.1039/C9CP04687A

101. Zhou JG, Yang S, Deng ZY (2017) Electrostatic catalysis induced by luciferases in the decomposition of the firefly dioxetanone and its analogue. J Phys Chem B 121(49):11053–11061. https://doi.org/10.1021/acs.jpcb.7b08000, pMID: 29168632

QM/MM Approaches Shed Light on GFP Puzzles

Alexander V. Nemukhin and Bella L. Grigorenko

Abstract The green fluorescent protein (GFP) is one of the most important light-responsive biological systems. We describe applications of a particular QM/MM version based on the flexible effective fragment potential theory to characterize the properties of this protein including transformations between different forms of GFP. We also show how the chemical reactions of chromophore maturation in GFP and the photochemical reactions leading to GFP malfunction are characterized in QM/MM simulations.

Keywords Green fluorescent protein · Photochemistry · QM/MM · Protein maturation · Photobleaching

1 Introduction

We limit materials of this chapter to QM/MM studies of only one light-responsive biological system, namely, the green fluorescent protein (GFP). We dare to assume that development of QM/MM would be fully justified even if this method were applied to characterize only GFP, because of the enormous significance of this protein as a parent species of multiple fluorescent biomarkers in living cells [1]. When describing GFP by QM/MM approaches, it is hard to avoid using popular templates, like "the tale of two Nobel Prizes". Indeed, the 2008 award was for the discovery and development of GFP and the 2013 award for the development of multiscale models (QM/MM) for complex chemical systems. Importantly, the Nobel prize from 2014 for the development of super-resolved fluorescence microscopy is also directly related to GFP-like proteins.

A. V. Nemukhin (✉) · B. L. Grigorenko
Department of Chemistry, Lomonosov Moscow State University, Leninskie Gory, 1/3, Moscow 119991, Russia
e-mail: anem@lcc.chem.msu.ru

Emanuel Institute of Biochemical Physics, Russian Academy of Sciences, Kosygina 4, Moscow 119333, Russia

© Springer Nature Switzerland AG 2021
T. Andruniów and M. Olivucci (eds.), *QM/MM Studies of Light-responsive Biological Systems*, Challenges and Advances in Computational Chemistry and Physics 31, https://doi.org/10.1007/978-3-030-57721-6_6

Being held captive by templates, we borrow the "shed light" expression for the title of our chapter. Probably, one of the first examples of using this idiom with a reference to GFP was a title of the paper by Weber et al. "Shedding light on the dark and weakly fluorescent states of green fluorescent protein" [2] devoted to quantum chemical simulations of GFP chromophore in the gas phase. The title of the important paper by Filippi et al. "Bathochromic shift in green fluorescent protein: A puzzle for QM/MM approaches" [3], which describes applications of the advanced electronic structure methods for GFP modeling prompted us to borrow another "lego" for our title.

We demonstrate in this chapter that applications of QM/MM approaches allows us to describe GFP "from birth to death". More specifically, we describe chemical reactions of chromophore maturation in GFP, changes in the states of this fluorescent protein in its photo-cycle, and photochemical reactions leading to protein malfunction.

A wide part of this program is covered by the works carried out with the use of a particular version of the QM/MM method, namely, the flexible effective fragment approach developed by us [4, 5]. We present here a short description of this QM/MM variant followed by its applications to model the GFP properties.

2 GFP and the Efforts to Model Its Properties

Several excellent reviews [6–10] exhaustively cover results of multiple experimental studies of GFP, including structures of this protein and its mutants, their optical and vibrational spectra and other photophysical properties. Initially these works aimed at development of efficient fluorescence markers based on GFP and GFP-like proteins for in vivo imaging [1]. Recently, studies of GFP-like proteins in reversible photo-switching reactions, which plays a key role in the fluorescence microscopy, gain an increasing attention [11, 12]. In this section, we briefly summarize the GFP features, which are essential for computational simulations.

The GFP chromophore, the 4-(p-hydroxybenzylidene)-5-imidazolinone molecule, which is covalently bound to the peptide chain in the protein matrix, is depicted in Scheme 1. Here, the molecule is shown in the neutral state, i.e. protonated on the phenolic part; the corresponding state of GFP with the neutral chromophore is traditionally called the A-form. The anionic state of the chromophore (deprotonated

Scheme 1 The chemical structure of the GFP chromophore

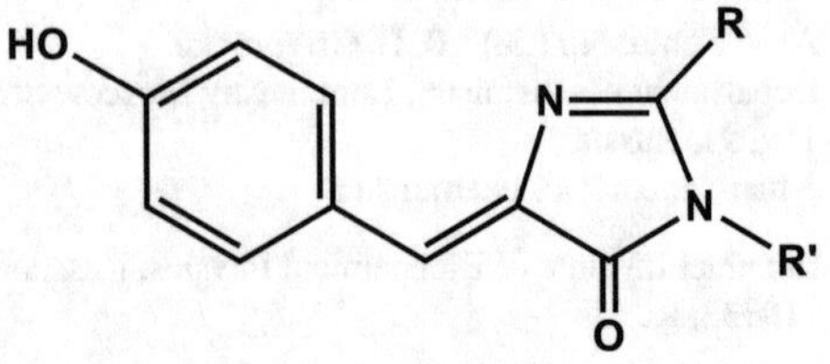

on the phenolic part) accounts for another, the B-form, of GFP. The A and B terminology reflects labelling of the observed spectral bands in GFP in the first experimental studies [13].

From the theoretical side, the GFP chromophore molecule has been widely characterized in the gas phase and in solutions using different approaches of quantum chemistry. We already mentioned one of the pioneering papers by Weber et al. [2], in which semi-empirical methods were utilized. In subsequent publications, the level of theory applied for this molecule was dramatically enhanced, e.g., [14]. In one of the most recent papers by Send et al. [15], the coupled-cluster approach CC2/def2-TZVP was used to describe the neutral and anionic forms of the chromophore in the gas phase. Since this book is devoted to QM/MM simulations, we do not discuss here the works on the isolated GFP chromophore; the focus of this chapter is on proteins containing this chromophore.

The chromophore in the protein is surrounded by the side chains of amino acid residues, which are important for GFP function. First of all, we should point out those in the immediate vicinity of the chromophore, Arg96, Glu222, Ser205, His148, Thr203, the role of which is emphasized in GFP studies [6, 7, 10, 13]. Atomic coordinates of the wild-type GFP and its mutants are known from multiple crystallography experiments, including those, that report high resolution structures (e.g., [16]). Coordinates from the crystal structures deposited to the Protein Data Bank usually serve as initial data in simulations of protein properties as well as data for the validation of computational predictions. However, it should be noted that proteins are rather flexible objects in solution, and a crystal structure by itself is not enough to define protein's features. The recent work [17] shows a fascinating example how packing of fluorescent proteins into crystals changes photo-reaction pathways. With respect to the latter issue, we acknowledge that a huge number of dynamical studies of GFP are performed in both ground and excited state (e.g., [8, 9]). It is unrealistic to describe in this chapter even principal findings in this research area; therefore, we focus here on QM/MM based simulations related to structures on potential energy surfaces and optical spectra of GFP.

The picture in Fig. 1 is a typical cartoon presented in multiple papers on QM/MM simulations showing a general view of the entire system and molecular groups assigned to the QM subsystem in insets. Here, we illustrate the approach for the wild-type GFP and draw in the insets a proposal for the QM subsystem, which includes the chromophore, the nearest amino acid side chains Arg96, His148, Ser205, Glu222 and water molecules. This QM composition was applied in Grigorenko et al. [18] to model proton transfer routes in GFP.

In general, partition of a model protein system into QM and MM parts for QM/MM simulations is primarily guided by intuition and researcher's experience, although attempts to apply automated strategies are known [19]. In the case of GFP, an instructive approach would be to include into QM the entire chromophore and the surrounding molecular groups hydrogen bonded to the chromophore. Studies of the role of the GFP chromophore environment in computing spectra of model systems have been carried out already in the first QM-based simulations [20–24]. In particular, presence of the charged residue Arg96 in QM is strongly advised for correct

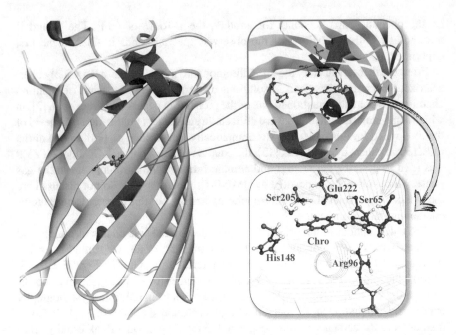

Fig. 1 The shape of GFP (left) and a part of the chromophore-containing pocket (right). Here and in other figures, carbon atoms are colored green, oxygen—red, nitrogen—blue, hydrogen—light grey

modeling of electronic excitations in GFP. The hydrogen-bond network connecting the chromophore, a water molecule, Ser205, Glu222 is critical for interpretation of the basic photophysical properties of GFP, i.e. switching between the A and B forms [13]. Importance of His148 and Thr203 for an adequate description of structural features of GFP was underlined in Ref. [10]. Therefore, at least the amino acid side chains listed in this paragraph should be considered as likely candidates for the QM subsystem.

Clearly, the smaller the size of QM is the higher level of quantum chemistry can be used for the system and vice versa. To this end, it would be desirable to leave only the chromophore molecule in QM and to apply, say, the coupled-cluster method with large basis sets as in Ref. [13] to describe at least ground state structures and vibrational spectra of GFP with a high accuracy. However, this strategy would require implementation of advanced electronic embedding schemes in the QM/MM theory, which are still under development (e.g., Refs. [25, 26]). Practical approaches when modeling ground state properties of GFP are based on the conventional density functional theory (DFT) for QM subsystems composed of the chromophore and the nearby residues, conventional MM force fields and simple electronic embedding treatment (say, inserting the MM charges to the QM Hamiltonian) in QM/MM. We mentioned above that to model switching between the neutral and anionic forms of the chromophore in GFP additional molecular groups besides the chromophore should be included into QM.

The same arguments hold when optical spectra of GFP and excited state dynamics in GFP are simulated. The very first applications of QM/MM to GFP [20, 22] revealed that multiconfigurational methods of quantum chemistry, based on the configuration interaction and complete active space SCF approximations perform better if molecular groups near the chromophore are included to QM along with the chromophore. The same refers to calculations of the ionized states of the GFP chromophore in the protein [24]. A rule of thumb is to use the same QM/MM partitioning both for structure optimization in the ground state and for simulations of absorption spectra. A more difficult issue is to optimize structures on the excited state potential energy surface to compute fluorescence spectra and conical intersection points. Usually, this requires reduction of the QM subsystem. Studies of electronic excitation in GFP with small QM subsystem but with the use of advanced coupling with MM are also described in the literature [3, 23, 25].

In the subsequent sections, we show how various GFP properties have been simulated with a particular version of the QM/MM method, namely, the flexible effective fragment QM/MM technique [4, 5].

3   Flexible Effective Fragment QM/MM Method

Application of the theory of effective fragment potentials (EFPs) in QM/MM simulations was described in the feature article by Gordon et al. [27] and extended in the subsequent papers [28, 29]. In this technique, the MM part is represented by a collection of molecular groups interacting with the QM subsystem via EFPs. Specifically, each group assigned to the MM subsystem creates potentials, which are added to the one-electron part of the QM Hamiltonian matrix. In general, the EFP includes the electrostatic, polarization, exchange-repulsion and dispersion contributions. One of the important findings in the EFP-based QM/MM method was an elegant idea to introduce a buffer region between the QM and MM parts, which helped to solve a long-standing problem of a practical treatment of the QM-MM boundary.

In the original version [27] of the EFP-based QM/MM method, effective fragments are supposed to be geometrically frozen during chemical transformations in the QM area. A proposal [4, 5] to overcome this restriction assumed a partitioning of the MM groups to flexible chains of relatively small effective fragments. Another important idea was to model fragment-fragment interactions with the help of conventional force field parameters of molecular mechanics. This proposal is especially valuable for applications to proteins: although each fragment is geometrically frozen, polypeptide chains are flexible enough to describe the protein matrix around the QM groups.

We illustrate the flexible effective fragment QM/MM approach with an application to GFP modeling in Fig. 2. The entire molecular model system shown in the left side in Fig. 2 is partitioned to the QM and MM subsystems with an assignment of the chromophore and several groups (nearest to Chro) to the QM part. The remaining

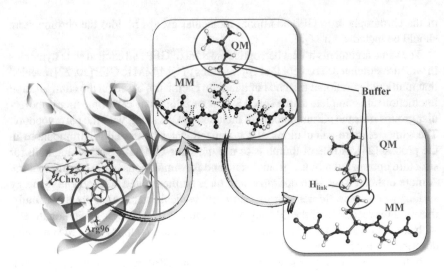

Fig. 2 Illustration of the flexible effective fragment QM/MM. The treatment of the QM-MM boundary with the help of the buffer fragment is explained in the insets

groups from the polypeptide chains in the protein constitute the MM part. Let us focus on the Arg96 amino acid residue, which must participate in transformations with GFP upon its photocycle. As shown in the upper inset in Fig. 2, the arginine side chain consisting of three carbon aliphatic chains ending in the guanidinium group is split into the QM- and MM-parts. The charged guanidinium group and the nearest CH_2 group are in QM, while other atoms including the rest of the Arg side chain and the backbone are in MM. As shown by red dotted lines in the inset, the latter are subdivided to small effective fragments, e.g., NH-CO, CH_3, etc. These fragments are described by the frozen internal coordinates, but conformations of the chains are flexible, when geometry optimization of the entire system is performed in QM/MM calculations.

The insets in Fig. 2 clarify a treatment of the QM-MM boundary across the covalent bond. The CH_2 group separating the carbon aliphatic chain and the guanidinium head in Arg96 is a buffer effective fragment contributing to both QM and MM subsystems. In the lower inset in Fig. 2 the arginine side chain is artificially separated in space between the QM and MM parts for better visibility. The broken valence in the QM subsystem is saturated by a conventional link hydrogen atom, H_{link}. Therefore, the corresponding CH_3 group is considered as a special fragment in QM. Exactly the same geometry configuration is assumed for the corresponding effective fragment CH_2 in MM. Atoms of the buffer fragment bear no charges to prevent artificial polarization of the QM part.

As described in detail in Refs. [4, 5], implementation of the flexible effective fragment QM/MM method for modeling proteins required the creation of the library of parameters of effective fragments. Partitioning of the backbone chain into fragments is illustrated here in Fig. 2. Partitioning of the amino acid side chains was carried out manually (see, e.g., an example for the arginine in Fig. 2), and this scheme was not changed in all applications. The electrostatic contributions in EFP are modeled by distributed multipoles, extending from charges up to octupoles; they are centered at each MM atom and each bond midpoint shown by small black circles in Fig. 2. The library of these parameters were computed in ab initio calculations using GAMESS(US) [30]. The exchange-repulsion interaction between each effective fragment and QM subsystem is modeled by one-electron potentials added to the Hamiltonian matrix. Such potential is represented [27] by a linear combination of Gaussian functions located at MM atomic centers, parameters of which should be fitted in preliminary calculations. The first experience with the flexible effective fragment QM/MM method for proteins [4, 5] showed that keeping polarization contributions in EFPs along with the fitted exchange-repulsion potentials as suggested in the original approach [27] for modeling phenomena in solvents led to unnecessary complication. The same results could be obtained by omitting polarization and readjusting parameters of repulsive potentials. Parameters in linear combinations of Gaussian functions representing the terms in EFP supplementary to electrostatic contributions for all fragment in the library were fitted as follows. Since description of hydrogen bonding is of the primary importance in modeling biomolecules, a water molecule served as a probing vehicle in the fitting procedure. A variety of directions along which a water molecule could reach an effective fragment were considered, and reference data were created by ab initio calculations. Then, the fitting of the EFP parameters supplementary to those from the electrostatic terms was carried out. The reliability of such parameterization demonstrating, in particular, the influence of the boundaries, sensitivity to the threshold values used in geometry optimization, and other details are discussed in the methodical papers [4, 5].

Practical steps in modeling a selected biomolecule using the flexible effective fragment QM/MM assume that, first of all, a model system is constructed. In the case of GFP, a suitable structure from the Protein Data Bank is selected, which serves as a source of initial coordinates of heavy atoms. Depending on the goals of modeling either the entire protein (with or without surrounding solvent shells) or a large region around the chromophore-containing pocket with the water molecules seen in the X-ray structure compose a molecular model system.

The QM subsystem is specified, and the corresponding buffer fragments at the QM-MM boundary are selected. The MM subsystem is partitioned into effective fragments each of which is represented in the library. The corresponding parameters of EFPs are included to the data in the input files compatible with the GAMESS(US) program. To model EFP-EFP interactions in the MM subsystem, conventional force field parameters (AMBER, CHARMM) are employed. Typical size of QM is about 150 atoms, whereas about 3000 atoms grouped into ~ 700 effective fragments constitute the MM subsystem. In the course of calculations, all contributions to the energy and the energy gradient are computed in the code implemented in the GAMESS(US) program and the algorithms of this program are used to model the entire QM/MM system. It should be recognized that these algorithms are not adjusted for an iterative optimization of QM and MM parts alternately, what in practice requires quite a number of cycles to reach convergent gradients.

Optimization of geometry structures on the ground state potential energy surface is an important step in modeling. Predominantly, the DFT methods were applied in QM for this goal using functionals available in GAMESS(US). A strong feature in GAMESS(US) is the use of efficient CASSCF-based algorithms for quantum chemical calculations. Correspondingly, this method was often applied in modeling excited state properties in GFP.

4 QM/MM Modeling of the Chromophore Maturation in GFP

The formation of the chromophore molecule in GFP occurs upon autocatalytic protein maturation in the presence of molecular oxygen. In the wild-type protein, the process requires a series of modifications of the tripeptide sequence Ser65-Tyr66-Gly67 finally resulting in the chromophore molecule shown in Scheme 1. The pioneers of GFP experimental studies have tentatively suggested [31] the basic steps of this interesting chemical reaction subdivided to the cyclization, dehydration, oxidation steps and measured the corresponding rate constants [32]. However, subsequent attempts to dissect the molecular mechanism and to describe a chain of elementary reaction steps did not result in a consistent knowledge.

A series of crystallography studies of possible intermediates in the GFP chromophore synthesis was carried out in the Getsoff's group [34–38]. The proposed mechanism, which is known presently as the Getsoff's mechanism of GFP chromophore maturation, assumes the reaction sequence cyclization → dehydration → oxidation. Several structures of tentative intermediates at the ends of cyclization and dehydration steps preceding oxidation are suggested. Spectroscopic and biochemical studies of GFP maturation including studies of kinetic isotope effects performed in the Wachter's group [39–41] led the authors to an alternative proposal of the reaction sequences, namely, cyclization → oxidation → dehydration. Both proposals agree on the first cyclized intermediate, but further steps are described differently. Experimental results from other groups [42, 43] basically agree with the sequence cyclization → dehydration → oxidation.

It would be beneficial for the theory to resolve the controversy between these two hypotheses on the base of QM/MM simulations. One such attempt is described in Refs. [33, 44]. The results of the flexible effective fragment QM/MM method applied to the wild-type GFP [33] provide support to the mechanism involving the cyclization-dehydration-oxidation sequence of the chromophore's maturation reactions as illustrated in Fig. 3. We comment instantly, that although the experimental evidences are partly explained by this QM/MM model, it is too early to claim a win. The model well describes structures along the reaction pathway and reasonably well energy barriers at different steps, which correlate with the experimental kinetic data. However, some arguments favoring the alternative sequence cyclization → oxidation → dehydration are neither confirmed, nor invalidated in simulations. Probably,

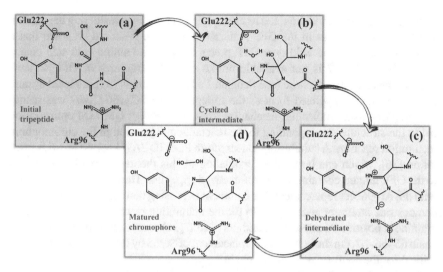

Fig. 3 The mechanism of chromophore maturation in GFP consistent with the cyclization →
dehydration → oxidation reaction pathway

a diversity of mechanisms of chromophore maturation in GFP-like proteins may
occur, e.g., in different GFP mutants as actually used in different experiments.

It is not instructive to duplicate in this chapter the published materials of QM/MM
simulations [33, 44]. Instead, we focus here on selected issues of the GFP maturation reaction; namely, we briefly describe the QM/MM computational protocol and
introduce the results for structures and energy barriers.

The cyclization and dehydration reactions were modeled starting from a structure of the protein with the non-cyclized Ser65-Tyr66-Gly67 tripeptide. The QM
subsystem included the reactants Ser65, Tyr66, Gly67, the side chains of Arg96,
His148, Ser205, Glu222, and four water molecules. The computed cyclization-
dehydration pathway led to the reaction intermediate prepared for the interaction
with molecular oxygen (Fig. 3c). At this point, the molecular oxygen was added
to the system and the oxidation reaction was modeled. The DFT method with the
PBE0 functional and 6-31G* basis set was used in QM, whereas fragment-fragment
interactions in MM were modeled with the AMBER force field. At the oxidation
step, initially the triplet state oxygen molecule was present in the system, and the
unrestricted version of DFT was applied to scan the corresponding section of the
potential energy surface. The triplet to singlet state switching was assumed after the
activation of dioxygen due to the electron transfer from the chromophore precursor
leading to the reactive O_2^- species. Next, the conventional restricted DFT method
was applied to describe the reaction pathway to the mature GFP and the hydrogen
peroxide molecule inside the protein. Energy reliefs computed in Ref. [33] showed
several reaction intermediates and the corresponding transition states.

Prediction of protein structures, i.e. geometry configurations of the minimum
energy points on the ground electronic state energy surfaces, is one of the strongest

features of QM/MM modeling. Let us exemplify this issue by considering the starting and terminal structures in the GFP maturation reaction. It should be noted that a crystal structure of the wild-type GFP precursor with the amino acid side chains Ser65, Tyr66, Gly67 is not available unlike multiple structures of GFP variants with the mature chromophore (see Scheme 1). In simulations [33], a model system containing the pre-cyclized form of the tripeptide Ser65-Tyr66-Gly67 was manually prepared using molecular mechanics tools. Subsequent QM/MM optimization led to the structure corresponding to the reactant configuration in the chromophore maturation reaction. A suitable crystal structure PDB ID 2AWJ solved at 1.6 Å resolution [36], which can be used for validation of this theoretical prediction, refers to the GFP mutant containing the non-cyclized tripeptide Thr65-Tyr66-Gly67 along with the Arg96Met replacement. We comment in passing that these experiments [36] demonstrate an importance of Arg96 for the maturation reaction. Only substituting arginine at position 96 allowed the authors of Ref. [36] to crystallize the non-cyclized chain 65–66–67. On the other hand, replacement of Ser65 by threonine does not seem critical for the conformation of this chain.

The left side in Fig. 4 illustrates a fragment of the model system mimicking the protein with the amino acid side chains Ser65, Tyr66, Gly67 in the pre-cyclized form as obtained in QM/MM optimization. The table in the right side in Fig. 4 lists the relevant geometry parameters. Clearly, distances between the heavy atoms and the respective valence angles in the model agree very well with those in the crystal structure, confirming that the model structure accurately represents the conformation of the pre-cyclized tripeptide. Of a special importance are the distances distinguished

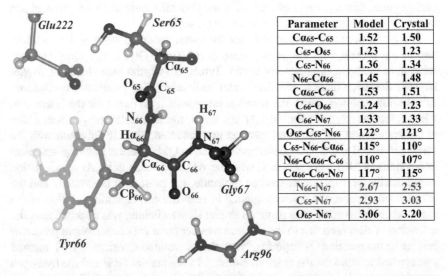

Parameter	Model	Crystal
$C\alpha_{65}$-C_{65}	1.52	1.50
C_{65}-O_{65}	1.23	1.23
C_{65}-N_{66}	1.36	1.34
N_{66}-$C\alpha_{66}$	1.45	1.48
$C\alpha_{66}$-C_{66}	1.53	1.51
C_{66}-O_{66}	1.24	1.23
C_{66}-N_{67}	1.33	1.33
O_{65}-C_{65}-N_{66}	122°	121°
C_{65}-N_{66}-$C\alpha_{66}$	115°	110°
N_{66}-$C\alpha_{66}$-C_{66}	110°	107°
$C\alpha_{66}$-C_{66}-N_{67}	117°	115°
N_{66}-N_{67}	2.67	2.53
C_{65}-N_{67}	2.93	3.03
O_{65}-N_{67}	3.06	3.20

Fig. 4 The structure with the pre-cyclized tripeptide Ser65-Tyr66-Gly67. Geometry parameters in the table in the column "Model" correspond to the results of QM(PBE0/6-31G*)/MM(AMBER) optimization; parameters in the column "Crystal" refer to the structure PDB ID 2AWJ. The atom–atom distances are given in Å

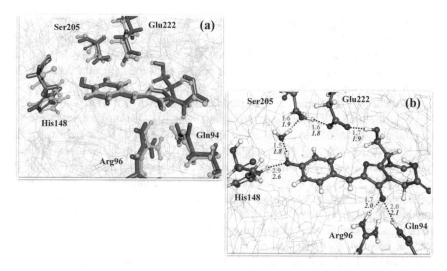

Fig. 5 Comparison of the computed structure of GFP with the matured chromophore and the crystal structure PDB ID 6JGH. Panel (a): superposition of two structures, blue sticks refer to the crystal, green balls and sticks refer to the computationally derived structure. Panel (b): comparison of the selected intermolecular distances in Å, blue and green characters refer to the crystal and the computational model, respectively

by the red color in the table, $N_{66}-N_{67}$ and $C_{65}-N_{67}$. The reaction pathway at the cyclization step (compare the panels (a) and (b) in Fig. 3) is sensitive to the initial values of these atom–atom distances.

Figure 5 illustrates an important comparison of the computed structure with the reaction products (the system with the mature chromophore) with the recent X-ray structure of the GFP (PDB ID 6JGH) [16] solved at the subatomic resolution. In this case, the coordinates of the hydrogen atoms in the protein are available.

Although the study [16] reports a structure of the GFP mutant F99S/M153T/V163A/T203I, its geometry parameters in the chromophore-containing pocket should be close to those in the wild-type protein. Comparison of the results of QM/MM simulations with this particular crystal structure PDB ID 6JGH is especially valuable because the starting coordinates of the model system with the chromophore precursor (see Fig. 4), i.e. those of the reactants, were obtained independently from the coordinates of PDB ID 6JGH, and the model structure of the products shown in Fig. 5 was created in a series of chemical reactions described at the QM/MM level without any correction at the intermediate steps. This comparison reveals that the computed with the QM/MM method intramolecular geometry parameters, i.e. distances between covalently bound atoms and valence angles in the amino acid side chains and in the mature chromophore are well consistent with the experimental data; see table in Fig. 4 as an example. The differences between geometry parameters of hydrogen-bonded species obtained in the simulations and observed in the crystal structure as shown in Fig. 5b are typical for the QM/MM method. We intentionally show here the most tricky parameters, namely, distances to hydrogens in the hydrogen-bonded pairs. Deviations between computed and observed intermolecular distances between the corresponding heavy atoms are typically less than 0.2 Å. We reiterate that protein structures are well described in QM(DFT)/MM calculations.

As for energy barriers on the computed reaction pathways, the QM/MM calculations described in Ref. [33] reported the highest barrier 17 kcal/mol at the cyclization-dehydration route and 21 kcal/mol at the oxidation route. The kinetic measurements have been performed [32] separately for these reactions (cyclization-dehydration and oxidation); the corresponding rate constants can be converted to the free energy barriers of 20.7 and 22.7 kcal/mol, respectively. The computational protocol used in [33] can be improved to obtain higher quality results for the direct comparison with the experimental kinetic data, namely, calculations of free energy profiles are required, a higher level of the theory than the PBE0/6-31G* approximation in QM can be used. Nevertheless, comparison of the computed and experimentally derived energy barriers shows that QM/MM modeling successfully determines that the oxidation step should be the rate-limiting one in the multistep maturation reaction.

5   QM/MM Modeling of the Transformations Between the A and B Forms in GFP

We remind that the GFP photocycle is associated with the transformations between the state of the protein with the neutral chromohore (the A-form) and the state with the anionic chromophore (the B-form) [11, 45]. To explain the observed photophysical properties of GFP, it was proposed to introduce one more state along the transformation route, an intermediate between A and B (the I-form of the protein). A tentative reaction scheme for the transitions between the protein forms A, B and I, as suggested in experimental studies [11, 45], is summarized, e.g., in Ref. [10]. The transformation from A to I assumes changes in the protonation states of the chromophore and Glu222, whereas the transition from I to B includes a slow conformational change of the protein.

The upper panels in Fig. 6 illustrate the computed and experimental transitions between the S_0 and S_1 electronic states of the wild-type GFP in these three forms. The bottom panels show the corresponding molecular groups.

In calculations described in [18], model molecular systems were designed using atomic coordinates taken from the crystal structure PDB ID 1EMA of the GFP mutant Ser65Thr. The side chain of threonine at position 65 was replaced by serine; hydrogen atoms were added manually using molecular mechanics tools. It was reasonable to assign the chromophore and the side chains of Arg96, Glu222, Ser205, His148, Thr203, as well as 2 water molecules to the QM subsystem. The most important molecular groups are shown in the bottom panels in Fig. 6.

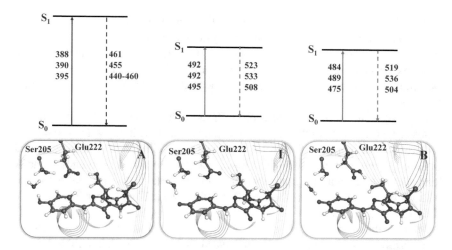

Fig. 6 The upper panels show the vertical transitions $S_0 \rightarrow S_1$ and $S_1 \rightarrow S_0$ in the A, I, and B states characterized by the corresponding wavelengths in nm. The corresponding values are arranged in the following order: the XMCQDPT2 results, the SOS-CIS(D) results, and the experimental data. The panels in the bottom row show the QM/MM optimized structures in the ground electronic state

The flexible effective fragment QM/MM program [4, 5] was applied in simulations. Geometry configurations on the ground state potential energy surface were optimized using the QM(PBE0/6-31G*)/MM(AMBER) approach. The choice of initial coordinates for optimization was guided by consideration of the proton transfer route along the hydrogen-bonded network connecting the phenolic hydroxyl of the chromophore and the carboxyl group of Glu222 via the water molecule and the Ser205 side chain (see the bottom panels in Fig. 6). First, the structures with the anionic chromophore (forms B and I) were designed, then the protons were transferred along the A-I-B route to prepare the structure with the neutral chromophore (form A). The energy profile connecting the ground state minimum energy points was estimated following reaction coordinates along the same proton transfer route. The located saddle points separating minimum energy structures were verified by computing multiple passages in the forward and backward directions.

Vertical excitation energies $S_0 \rightarrow S_1$ were computed using two different approaches. First, the extended multi-configuration quasi-degenerate perturbation theory (XMCQDPT2) [46] was applied for the molecular clusters composed of the chromophore, and the side chains of Arg96, Glu222, Ser205, and water molecules (i.e. for the part of the QM subsystem) embedded into the set of effective fragments representing the rest of the model system. The perturbation theory corrections were estimated on the base of the CASSCF(12/11)/cc-pVDZ wavefunctions. The second approach, SOS-CIS(D) [47, 48], is a version of the configuration interaction method. The SOS-CIS(D)/cc-pVDZ calculations were performed for the QM parts from the corresponding QM/MM model systems.

The QM/MM method with the CIS/6-31G* approximation in QM was used in Ref. [18] to estimate coordinates of stationary points on the excited-state S_1 potential energy surface. The vertical $S_1 \rightarrow S_0$ transition energies were calculated using the same two approaches as for the excitation energies. We expect that the accuracy of the computed emission energies must be somewhat lower than that of the excitation energies.

As a result, the QM/MM calculations [18] successfully describe the experimental evidences with respect to the A $\rightarrow$ I $\rightarrow$ B transformations in the wild-type protein and provide full support to the three-state model of the GFP photocycle summarized in the beginning of this section. The lowest energy structure corresponds to the A form. Consistently with the experimental findings on the relative populations of the A and B forms of the protein, the form B lies about 1 kcal/mol higher. The form I is almost isoenergetic with B.

We note that these QM/MM calculations contribute to the computational studies of the proton transfer along the transient hydrogen-bond network in GFP [49, 50].

6   QM/MM Modeling of Photo-Induced Irreversible Transformations in GFP

Finally, we describe QM/MM simulations of "GFP death", i.e. simulations of irreversible transformations in GFP (decarboxylation and photobleaching) induced by intense irradiation of the protein. The common feature of these processes is that they are initiated by an excitation of the system from the ground electronic state S_0 to highly excited states of the charge-transfer character, lying above the level of the first excited state S_1 responsible for basic optical properties of GFP. Needless to say that modeling such transitions requires substantial efforts.

Decarboxylation of the side chain Glu222 in the A-form of GFP (Fig. 7) was observed experimentally by van Thor et al. [51] (illumination by 254 nm) and by Bell et al. [52] (illumination by 254 and 280 nm).

The QM/MM simulations described in Ref. [53] modeled the excitation of the system corresponding to the electron transfer from deprotonated Glu222 to the chromophore (Chro) using the flexible effective fragment QM(PBE0/cc-pVDZ)/MM(AMBER) method. The QM part for geometry optimization included the following species: the chromophore, the side chains of Arg96, Ser205, Glu222, and the water molecule connecting Ser205 and the phenolic hydroxyl of the chromophore by hydrogen bonds. A slightly reduced QM subsystem was used for modeling electronic excitations. In all cases, contributions to the quantum Hamiltonian from the rest of the protein were modeled with the effective fragments in the MM part.

To bypass the problem of prohibitively huge sizes of active spaces that should comprise occupied and virtual orbitals from both Glu222 and Chro species in the conventional CASSCF-based approaches, the authors of Ref. [53] applied the occupation restricted multiple active space (ORMAS) approach [54]. This approximation

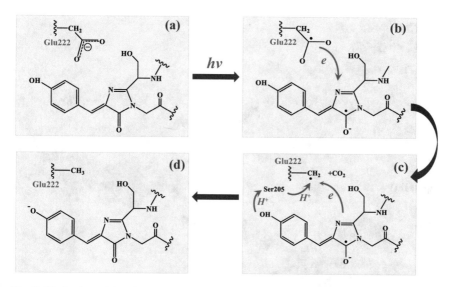

Fig. 7 Mechanism of decarboxylation of the side chain of Glu222 in GFP

allows one to re-arrange the multiconfigurational expansions and to recognize the charge transfer states associated with the electron transfer from Glu222 to Chro as relatively low-lying roots in the configuration interaction matrices. To this end, a careful analysis of the orbitals and the corresponding CI coefficients is required.

Figure 8 shows the most important orbitals involved in the transitions to the charge-transfer states of the system. As such, we consider the lp1 and lp2 lone pair orbitals of the glutamine 222 side chain and the π^* orbital of the chromophore. Two excitations $S_0 \rightarrow S_{CT1}$ and $S_0 \rightarrow S_{CT2}$ (either from lp1 or from lp2) are characterized by close values of oscillator strengths. Experimentally, the charge transfer states can be obtained in one-photon or two-photon excitations [51, 52].

A similar problem, namely, identification of the excited electronic states of the charge-transfer character in GFP was examined by QM/MM modeling of the protein photobleaching. Under the relevant experimental conditions [55], photoexcitation occurs from the protein structure with the anionic chromophore and the protonated Glu222 in the presence of molecular oxygen near the chromophore. These conditions imposed additional difficulties to construct model systems and to carry out calculations [56]. First, the oxygen molecule was inserted to the chromophore-containing pocket, and its position was optimized in QM/MM calculations. Second, the initial steps in the photochemical reaction were described for the total triplet state of the system. Third, the point of the transition from the triplet to the singlet states was located, from which the evolution of the system on the singlet potential energy surface led to the reaction products with the decomposed chromophore [56].

Figure 9 clarifies the key step of identification of the charge-transfer states corresponding to the electron transfer from the anionic chromophore to the oxygen molecule.

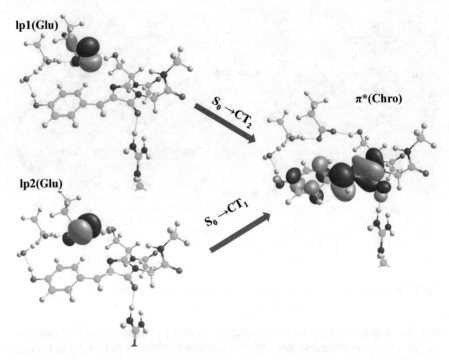

Fig. 8 Orbitals involved in the transitions reaching the charge-transfer states upon excitation in GFP

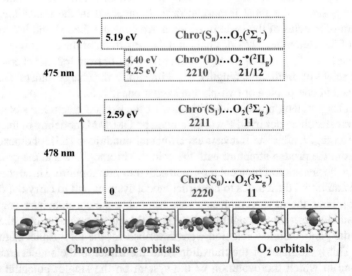

Fig. 9 Orbitals and their occupancies in multiconfigurational wavefunctions upon modeling photo-bleaching in GFP. The bottom panels illustrate the orbitals computed for the model system. The nature of electronic states is clarified by the notes in the rectangular frames. Occupation numbers of the chromophore or oxygen orbitals are specified in each case. The charge-transfer states are distinguished by using the red color

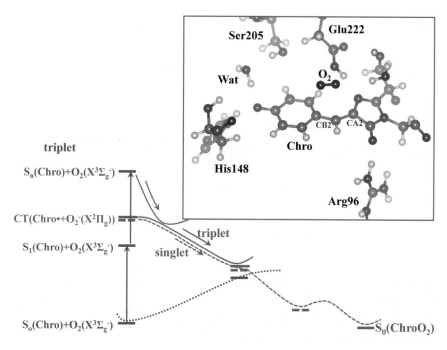

Fig. 10 The energy diagram explaining initial steps in the GFP photobleaching reaction for the model system shown in the inset

Figure 10 shows the computed energy profiles for the model system drawn in the inset.

As only the charge-transfer (CT) states are populated upon photoexcitation, the system evolves on the triplet state potential energy surface until the intersection point with the lowest energy singlet state surface is reached. Further evolution of the system in the singlet state finally leads to the decomposition of the chromophore due to cleavage of the bridging CA2-CB2 bond [56], as illustrated in Fig. 11.

7 Concluding Remarks

As promised in Introduction, we described in this chapter applications of the QM/MM theory to only one, but very important photo-active system, namely, the green fluorescent protein (GFP). We show that various aspects of GFP properties are described by constructing model molecular systems subdivided to QM- and MM-parts and using the flexible effective fragment version [4, 5] of the QM/MM method. Different options of quantum chemical approaches can be applied in the QM part of this technique. Besides conventional simulations of structures on the ground state potential energy surfaces, which are carried out with DFT methods, and calculations of vertical

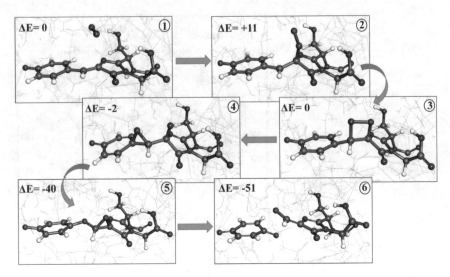

Fig. 11 Intermediates in the photoinduced reaction of the chromophore with the oxygen molecule in GFP after the electron transfer from Chro to O_2 and switch to the singlet state potential energy surface. The final point 6 corresponds to the reaction products, i.e. the benzoquinone molecule and the enolate ion bound to the peptide backbone of the protein. The energies relative to the level of the triplet state reactants are given in kcal/mol

transition energies with relatively simple TD-DFT and CI methods, more sophisticated CASSCF-based approaches are employed. The latter, however, require a careful analysis of the orbitals and compositions of multiconfigurational wavefunctions involved in different electronic states.

It is interesting to note that QM/MM modeling properties of only GFP macromolecule is not an overcrowded field. Besides the references mentioned in Introduction [3, 20, 23], we should point out to the papers describing the application of the ONIOM version of QM/MM for calculations of separate steps in the GFP chromophore maturation [57–59] and modeling GFP decarboxylation reaction [60]. It is hard to compare the performance of different QM/MM versions, e.g., of the flexible effective fragment potential method and ONIOM. We can comment that the attempts to model the separate steps in GFP maturation [57–59] led to the profiles with the energy barriers much higher than those expected from the kinetic measurements [32] and those obtained in the flexible effective fragment QM/MM [33]. This failure may be assigned to either unsuccessful choice of reaction coordinates or to drawbacks of the applied technique.

We think that from the methodological side, GFP can serve as a suitable testing system for evaluation of different QM/MM approaches, novel versions of which certainly will be developed in the future. This protein molecule is fairly simple, it contains no metal ions, the beta-barrel structure is firm enough, many of GFP properties are firmly established in the multiple experimental studies. Sooner or later, the known properties of GFP will be comprehensively characterized computationally by perfect quantum-based methods, and novel prospects will be predicted.

Acknowledgements The authors thank Prof. Anna I. Krylov, Dr. Igor V. Polyakov and Dr. Maria G. Khrenova for a fruitful collaboration on the fluorescent proteins projects. This works described in this paper were partly supported by the Russian Science Foundation (project #17-13-01051).

References

1. Rodriguez EA, Campbell RE, Lin JY et al (2017) The growing and glowing toolbox of fluorescent and photoactive proteins. Trends Biochem Sci 42:111–129. https://doi.org/10.1016/j.tibs.2016.09.010
2. Weber W, Helms V, McCammon JA, Langhoff PW (1996) Shedding light on the dark and weakly fluorescence states of green fluorescent protein. Proc Natl Acad Sci USA 96:6177–6182. https://doi.org/10.1073/pnas.96.11.6177
3. Filippi C, Buda F, Guidoni L, Sinicropi A (2012) Bathochromic shift in green fluorescent protein: a puzzle for QM/MM approaches. J Chem Theory Comput 8:112–124. https://doi.org/10.1021/ct200704k
4. Grigorenko BL, Nemukhin AV, Topol IA, Burt SK (2002) Modeling of biomolecular systems with the quantum mechanical and molecular mechanical method based on the effective fragment potential technique: proposal of flexible fragments. J Phys Chem A 106:10663–10672. https://doi.org/10.1021/jp026464w
5. Nemukhin AV, Grigorenko BL, Topol IA, Burt SK (2003) Flexible effective fragment QM/MM method: validation through the challenging tests. J Comput Chem 24:140–1420. https://doi.org/10.1002/jcc.10309
6. Zimmer M (2002) Green fluorescent protein (GFP): applications, structure, and related photophysical behavior. Chem Rev 102:759–782. https://doi.org/10.1021/cr010142r
7. Craggs TD (2009) Green fluorescent protein: structure, folding and chromophore maturation Chem Soc Rev 38:2865–2875. https://doi.org/10.1039/b903641p
8. van Thor JJ (2009) Photoreactions and dynamics of the green fluorescent protein. Chem Soc Rev 38:2935–2950. https://doi.org/10.1039/b820275n
9. Meech SR (2009) Excited state reactions in fluorescent proteins. Chem Soc Rev 38:2922–2934. https://doi.org/10.1039/b820168b
10. Remington SJ (2011) Green fluorescent protein: a perspective. Protein Sci 20:1509–1519. https://doi.org/10.1002/pro.684
11. Lippincott-Schwartz J, Patterson GH (2009) Photoactivatable fluorescent proteins for diffraction-limited and super-resolution imaging. Trends Cell Biol 19:555–565. https://doi.org/10.1016/j.tcb.2009.09.003
12. Bourgeous D, Adam V (2012) Reversible photoswitching in fluorescent proteins: a mechanistic view. IUBMB Life 64:482–491. https://doi.org/10.1002/iub.1023
13. Chattoraj M, King BA, Bublitz GU, Boxer SG (1996) Ultra-fast excited state dynamics in green fluorescent protein: multiple states and proton transfer. Proc Natl Acad Sci USA 93:8362–8367. https://doi.org/10.1073/pnas.93.16.8362
14. Martin ME, Negri F, Olivucci M (2004) Origin, nature, and fate of the fluorescent state of the green fluorescent protein chromophore at the CASPT2//CASSCF resolution. J Am Chem Soc 126:5452–5464. https://doi.org/10.1021/ja037278m
15. Send R, Suomivuori CM, Kaila VRI, Sundholm D (2015) Coupled-cluster studies of extensive green fluorescent protein models using the reduced virtual space approach. J Phys Chem B 119(7):2933–2945. https://doi.org/10.1021/jp5120898

16. Takaba K, Tai Y, Eki H et al (2019) Subatomic resolution X-ray structures of green fluorescent protein. IUCrJ 6:387–400. https://doi.org/10.1107/S205225251900246X

17. Chang J, Romei MG, Boxer SG (2019) Structural evidence of photoisomerization pathways in fluorescent proteins. J Am Chem Soc 141:15504–15508. https://doi.org/10.1021/jacs.9b08356

18. Grigorenko BL, Nemukhin AV, Polyakov IV, Morozov DI, Krylov AI (2013) First-principles characterization of the energy landscape and optical spectra of green fluorescent protein along the A→I→B proton transfer route. J Am Chem Soc 135:11541–11549. https://doi.org/10.1021/ja402472y

19. Karelina M, Kulik HJ (2017) Systematic quantum mechanical region determination in M/MM simulation. J Chem Theory Comput 13:563–576. https://doi.org/10.1021/acs.jctc.6b01049

20. Sinicropi A, Andruniow T, Ferré N, Basosi R, Olivucci M (2005) Properties of the emitting state of the green fluorescent protein resolved at the CASPT2//CASSCF/CHARMM level. J Am Chem Soc 127:11534–11535. https://doi.org/10.1021/ja045269n

21. Laino T, Nifosì R, Tozzini V (2004) Relationship between structure and optical properties in green fluorescent proteins: a quantum mechanical study of the chromophore environment. Chem Phys 298(1–3):17–28. https://doi.org/10.1016/j.chemphys.2003.10.040

22. Hasegawa J-Y, Fujimoto K, Swerts B, Miyahara T, Nakatsuji H (2007) Excited states of GFP chromophore and active site studied by the SAC-CI method: effect of protein-environment and mutations. J Comput Chem 28(15):2443–2452. https://doi.org/10.1002/jcc.20667

23. Daday C, Curutchet C, Sinicropi A, Mennucci B, Filippi C (2015) Chromophore–protein coupling beyond nonpolarizable models: understanding absorption in green fluorescent protein. J Chem Theory Comput 11(10):4825–4839. https://doi.org/10.1021/acs.jctc.5b00650

24. Bravaya KB, Khrenova MG, Grigorenko BL, Nemukhin AV, Krylov AI (2011) Effect of protein environment on electronically excited and ionized states of the green fluorescent protein chromophore. J Phys Chem B 115(25):8296–8303. https://doi.org/10.1021/jp2020269

25. Schwabe T, Beerepoot MTP, Olsen JMH, Kongsted J (2015) Analysis of computational models for an accurate study of electronic excitations in GFP. Phys Chem Chem Phys 17(4):2582–2588. https://doi.org/10.1039/C4CP04524F

26. Zech A, Ricardi N, Prager S, Dreuw A, Wesolowski TA (2018) Benchmark of excitation energy shifts from frozen-density embedding theory: introduction of a density-overlap-based applicability threshold. J Chem Theory Comput 14:4028–4040. https://doi.org/10.1021/acs.jctc.8b00201

27. Gordon MS, Freitag MA, Bandyopadhyay P et al (2001) The effective fragment potential method: a QM-based MM approach to modeling environmental effects in chemistry. J Phys Chem A 105:293–397. https://doi.org/10.1021/jp002747h

28. Gordon MS, Fedorov DG, Pruitt SR, Slipchenko LV (2012) Fragmentation methods: a root to accurate calculations on large systems. Chem Rev 112:632–672. https://doi.org/10.1021/cr200093j

29. Gurunathan PK, Acharya A, Ghosh D et al (2016) Extension of the effective fragment potential method to macromolecules. J Phys Chem B 120:6562–6574. https://doi.org/10.1021/acs.jpcb.6b04166

30. Schmidt MW, Baldridge KK, Boatz JA et al (1993) General atomic and molecular electronic structure system. J Comput Chem 14:1347–1363. https://doi.org/10.1002/jcc.540141112

31. Heim R, Prasher DC, Tsien RY (1994) Wavelength mutations and posttranslational autoxidation of green fluorescent protein. Proc Natl Acad Sci USA 91(26):12501–12504. https://doi.org/10.1073/pnas.91.26.12501

32. Reid BG, Flynn GC (1997) Chromophore formation in green fluorescent protein. Biochemistry 36(22):6786–6791. https://doi.org/10.1021/bi970281w

33. Grigorenko BL, Krylov AI, Nemukhin AV (2017) Molecular modeling clarifies the mechanism of chromophore maturation in the green fluorescent protein. J Am Chem Soc 139(30):10239–10249. https://doi.org/10.1021/jacs.7b00676

34. Barondeau DP, Putnam CD, Kassmann CJ, Tainer JA, Getzoff ED (2003) Mechanism and energetics of green fluorescent protein chromophore synthesis revealed by trapped intermediate structures. Proc Natl Acad Sci USA 100:12111–12116. https://doi.org/10.1073/pnas.2133463100

35. Barondeau DP, Kassmann CJ, Tainer JA, Getzoff ED (2005) Understanding GFP chromophore biosynthesis: controlling backbone cyclization and modifying post-translational chemistry. Biochemistry 44:1960–1970. https://doi.org/10.1021/bi0479205

36. Wood TI, Barondeau DP, Hitomi C et al (2005) Defining the role of arginine 96 in green fluorescent protein fluorophore biosynthesis. Biochemistry 44(49):16211–16220. https://doi.org/10.1021/bi051388j

37. Barondeau DP, Kassmann CJ, Tainer JA, Getzoff ED (2006) Understanding GFP posttranslational chemistry: structures of designed variants that achieve backbone fragmentation, hydrolysis, and decarboxylation. J Am Chem Soc 128(14):4685–4693. https://doi.org/10.1021/ja0566351

38. Barondeau DP, Kassmann CJ, Tainer JA, Getzoff ED (2007) The case of the missing ring: radical cleavage of a carbon−carbon bond and implications for GFP chromophore biosynthesis. J Am Chem Soc 129(11):3118–3126. https://doi.org/10.1021/ja063983u

39. Zhang L, Patel HN, Lappe JW, Wachter RM (2006) Reaction progress of chromophore biogenesis in green fluorescent protein. J Am Chem Soc 128(14):4766–4772. https://doi.org/10.1021/ja0580439

40. Wachter RM (2007) Chromogenic cross-link formation in green fluorescent protein. Acc Chem Res 40(2):120–127. https://doi.org/10.1021/ar040086r

41. Pouwels LJ, Zhang L, Chan NH, Dorrestein PC, Wachter RM (2008) Kinetic isotope effect studies on the de novo rate of chromophore formation in fast- and slow-maturing GFP variants. Biochemistry 47(38):10111–10122. https://doi.org/10.1021/bi8007164

42. Pletneva NV, Pletnev VZ, Lukyanov KA et al (2010) Structural evidence for a dehydrated intermediate in green fluorescent protein chromophore biosynthesis. J Biol Chem 285(21):15978–15984. https://doi.org/10.1074/jbc.M109.092320

43. Bartkiewicz M, Kazazić S, Krasowska JA et al (2018) Non-fluorescent mutant of green fluorescent protein sheds light on the mechanism of chromophore formation. FEBS Lett 592(9):1516–1523. https://doi.org/10.1002/1873-3468.13051

44. Nemukhin AV, Grigorenko BL, Khrenova MG, Krylov AI (2019) Computational challenges in modeling of representative bioimaging proteins: GFP-like proteins, flavoproteins, and phytochromes. J Phys Chem B 123:6133–6149. https://doi.org/10.1021/acs.jpcb.9b00591

45. Brejc K, Sixma TK, Kitts PA et al (1997) Structural basis for dual excitation and photoisomerization of the aequorea victoria green fluorescent protein. Proc Natl Acad Sci USA 94:2306–2311. https://doi.org/10.1073/pnas.94.6.2306

46. Granovsky AA (2011) Extended multi-configuration quasi-degenerate perturbation theory: the new approach to multi-state multi-reference perturbation theory. J Chem Phys 13(21):214113–214127. https://doi.org/10.1063/1.3596699

47. Grimme S (2003) Improved second-order møller–plesset perturbation theory by separate scaling of parallel- and antiparallel-spin pair correlation energies. J Chem Phys 118(20):9095–9102. https://doi.org/10.1063/1.1569242

48. Rhee YM, Head-Gordon M (2007) Scaled second-order perturbation corrections to configuration interaction singles: efficient and reliable excitation energy methods. J Phys Chem A 111(24):5314–5326. https://doi.org/10.1021/jp068409j

49. Nadal-Ferret M, Gelabert R, Moreno M, Lluch JM (2015) Transient low-barrier hydrogen bond in the photoactive state of green fluorescent protein. Phys Chem Chem Phys 17(46):30876–30888. https://doi.org/10.1039/c5cp01067e

50. Shinobu A, Agmon N (2017) Proton wire dynamics in the green fluorescent protein. J Chem Theory Comput 13(1):353–369. https://doi.org/10.1021/acs.jctc.6b00939

51. van Thor JJ, Gensch T, Hellingwerf KJ, Johnson LN (2002) Phototransformation of green fluorescent protein with UV and visible light leads to decarboxylation of glutamate 222. Nature Struct Biol 9:37–41. https://doi.org/10.1038/nsb739

52. Bell AF, Stoner-Ma D, Wachter RM, Tonge PJ (2003) Light-driven decarboxylation of wild-type green fluorescent protein. J Am Chem Soc 125:6919–6926. https://doi.org/10.1021/ja034588w

53. Grigorenko BL, Nemukhin AV, Morozov DI, Polyakov IV, Bravaya KB, Krylov AI (2012) Toward molecular-level characterization of photo-induced decarboxylation of the green fluorescent protein: accessibility of the charge-transfer states. J Chem Theory Comput 8:1912–1920. https://doi.org/10.1021/ct300043e

54. Ivanic J (2003) Direct configuration interaction and multiconfigurational self-consistent-field method for multiple active spaces with variable occupations I method. J Chem Phys 119:9364–9376. https://doi.org/10.1063/1.1615954

55. Shaner NC, Steinbach PA, Tsien RY (2005) A guide to choosing fluorescent proteins. Nat Methods 2:905–909. https://doi.org/10.1038/nmeth819

56. Grigorenko BL, Nemukhin AV, Polyakov IV, Khrenova MG, Krylov AI (2015) A light-induced reaction with oxygen leads to chromophore decomposition and irreversible photobleaching in GFP-type proteins. J Phys Chem B 119(17):5444–5452. https://doi.org/10.1021/acs.jpcb.5b02271

57. Ma Y, Zhang H, Sun Q, Smith SC (2016) New insights on the mechanism of cyclization in chromophore maturation of wild-type green fluorescence protein: a computational study. J Phys Chem B 120(24):5386–5394. https://doi.org/10.1021/acs.jpcb.6b04406

58. Ma Y, Yu J-G, Sun Q, Li Z, Smith SC (2015) The mechanism of dehydration in chromophore maturation of wild-type green fluorescent protein: a theoretical study. Chem Phys Lett 631–632:42–46. https://doi.org/10.1016/j.cplett.2015.04.061

59. Ma Y, Sun Q, Smith SC (2017) The mechanism of oxydation in chromophore maturation of wild-type green fluorescent protein: a theoretical study. Phys Chem Chem Phys 19(20):12942–12952. https://doi.org/10.1039/c6cp07983k

60. Ding L, Chung LW, Morokuma K (2013) Reaction mechanism of photoinduced decarboxylation of the photoactivatable green fluorescent protein: an ONIOM(QM:MM) study. J Phys Chem B 117:1075–1084. https://doi.org/10.1021/jp3112952

DNA Photodamage and Repair: Computational Photobiology in Action

Antonio Francés-Monerris, Natacha Gillet, Elise Dumont, and Antonio Monari

Abstract DNA is constantly exposed to external and metabolic stress agents, including the solar radiation and in particular the UV portion of the electromagnetic spectrum. Such source of stress can induce photochemical modification of the structure of DNA and of its basic components, i.e. the nucleobases. DNA lesions may ultimately lead to genomic instability, mutations, and even to carcinogenesis. Hence, cells dispose of complex biochemical repair pathways in charge of remove the DNA lesions and avoid their accumulation. In this Chapter, we present the complexity of the DNA lesion chemical and structural space, also complicated by the intricate coupling with the biological relevant signaling pathways. Through some relevant examples, we will show how proper multiscale simulation protocols can provide a unified picture of the complex phenomena and hence answer biological relevant questions, paving the way to a veritable computational photobiology approach.

Keywords Photobiology · Multiscale Modeling · DNA lesions · Classical molecular dynamics · DNA repair · Excited States methods

1 Introduction: DNA Lesions and Their Biological Effects

The fundamental biological role of nucleic acids, and DNA in particular, as the repository of the genetic information, the regulator of its transduction and expression,

A. Francés-Monerris · A. Monari (✉)
Université de Lorraine and CNRS, LPCT UMR 7019, 5400 Nancy, France
e-mail: antonio.monari@univ-lorraine.fr

A. Francés-Monerris
Departamento de Química Física, Universitat de València, 46100 Burjassot, Spain

N. Gillet · E. Dumont
Univ Lyon, ENS de Lyon, CNRS UMR 5182, Université Claude Bernard Lyon 1, Laboratoire de Chimie, 69342 Lyon, France

E. Dumont
Institut Universitaire de France, 1 rue Descartes, 75005 Paris, France

© Springer Nature Switzerland AG 2021
T. Andruniów and M. Olivucci (eds.), *QM/MM Studies of Light-responsive Biological Systems*, Challenges and Advances in Computational Chemistry and Physics 31,
https://doi.org/10.1007/978-3-030-57721-6_7

and the vector of its replication is well established. However, such a role requires that the molecular building blocks that compose the biopolymer should possess a number of crucial properties. In particular, DNA constituents should experience a fairly high stability to avoid the accumulation of chemical modifications [1, 2], i.e. lesions, which ultimately could compromise the biological function leading to mutagenesis or, more generally speaking, to genomic instability. However, this affirmation should be moderated since, and from a totally reductionist point of view, evolution itself may ultimately be tracked back to genomic mutation.

However, DNA is constantly exposed to an impressive number of stress sources (Fig. 1) coming both from exogenous and endogenous agents [3–6]. As a paradigmatic example, one may cite reactive oxygen species (ROS), related to oxidative stress, and produced either by healthy or pathologic metabolic pathways, or by external agents such as drugs or ionizing radiation [3, 6–13]. DNA is also constantly exposed to the action of electromagnetic radiation, such as visible or UV light, that may also trigger photophysical or photochemical reactive channels, and hence lead to the accumulation of photoinduced lesions, which will constitute the main focus of the present Chapter.

In addition, the synergic effects between two or more stress factors can also strongly increase the level and complexity of the harmful effects produced to the DNA. Such a synergy will not only enhance the possible occurrence of a given type of lesion, but may even open totally new chemical pathways, thus leading to the emergence of different damages. An illustrative example is the case of DNA photosensitization [14, 15], in which a drug absorbing infra-red, visible or near UV

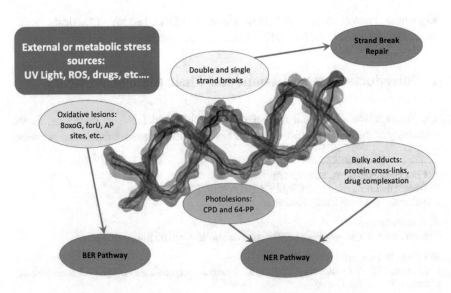

Fig. 1 Schematic view of DNA lesions type and the corresponding repair pathway

wavelengths, can interact with DNA and subsequently trigger intricate photochemical channels, normally involving either energy- or electron-transfer phenomena, ultimately leading to the appearance of DNA photolesions.

To cope with this hostile environment, evolution has adopted two strategies: on the one hand, the components of nucleic acids have been selected to maximize their photo and chemical stability, on the other hand, highly efficient DNA repair mechanisms have been developed to restore nucleic acids' functional structure, thereby reducing dangerous accumulations of DNA damage [16] (Fig. 1). When the number of lesions overcomes a given threshold, irreversibly compromising the viability of the cell, complex cellular signaling pathways are invoked to trigger apoptosis.

Despite the efficiency of all those strategies, the accumulation of DNA lesions has been directly related to genomic instability and also to tumorigenesis. In this respect, photolesions are again particularly significant since their accumulation in the skin cells, mostly due to unprotected sun exposure, has been directly correlated to the development of malignant tumors such as melanoma [17–21].

However, the induction of a high amount of DNA lesions to induce malignant cells death is also at the base of cancer chemotherapy [22–26] protocols. Indeed, the mechanism of action of one of the first, and still widely used, chemotherapeutic agents, cis-platinum [27–29], is based on the formation of covalent bonds with the DNA backbone [30]. Cis-platinum, like most classical chemotherapeutic agents, has a very poor selectivity, and hence is plagued by heavy secondary effects seriously limiting the quality of life of the patients.

Alternative therapeutic strategies, based on the use of photosensitizers [31–35], are developed and constitute the domain of photodynamic therapy [36–40] (PDT) or light-assisted chemotherapy (LAC) [41–44]. Obviously, the enhanced selectivity will be due to the local application of light that is intended to be restricted only to the area of the cancer lesion. Although attractive, PDT and LAC still suffer from severe drawbacks sometimes limiting their application. Furthermore, absorption in the red or infrared portion of the electromagnetic spectrum, the therapeutic window, is necessary to maximize the light penetration into the biological tissues and treat non-superficial lesions [45].

The fascinating complexity of DNA damages is already apparent from these preliminary considerations. Furthermore, the global picture is also complicated by the fact that the chemical space spanned by the different lesions is extremely large, comprising chemical modifications happening at the DNA backbone or at the level of the nucleic acid bases [4, 8, 46, 47].

Oxidative DNA lesions may result either in single- or double-strand breaks [48–51] or in the chemical modification of the DNA bases [46, 52, 53]. Due to the favorable oxidation potential, lesions involving purines, and guanine in particular, are the most common oxidatively generated damages. This includes the famous 8-oxo-guanine (8OxoG) resulting by the action of singlet oxygen ($^1O_2$). Oxidative pathways also lead to other and more complex outcomes, such as the production of the so called apurinic/apyrimidinic (AP) or abasic sites in which the nucleobase is totally cleaved from the sugar moiety [54–56]. Interestingly, AP sites also represent an intermediate in DNA repair processes [57]. Oxidatively generated lesions are also commonly produced via DNA photosensitization when the photophysical pathways triggered by the sensitizers lead to the production of $^1O_2$.

As opposed to oxidative and photosensitized lesions, intra-strand base dimerization is the main outcome of DNA photolesions, especially in case of direct UV absorption. Because of the more favorable topology and energetic landscape of the involved excited states, pyrimidines, and thymine in particular, are the most vulnerable nucleobases. As far as thymine photodimers are concerned, two main photoproducts should be taken into account: the so-called cyclobutane pyrimidine dimers (CPDs) and the pyrimidine (6-4) pyrimidone photoproduct (64-PP) [58–62]. While CPDs are by far the most abundant form of photodamage; 64-PP is however characterized by a very high mutagenicity making it extremely dangerous.

The interaction with external drugs leads also to the production of other less common, yet important, lesions [44, 63–65]. Those include the formation of covalently bound, and in some cases bulky, adducts via a chemical coupling to either the backbone or the nucleobases. In some cases, inter-strand cross-links may also be produced, via the formation of a covalent bridge between the 5′- and 3′-strand. Interestingly, inter-strand cross-link may also be produced photochemically by the action of photodissociable drugs [64].

Finally, highly toxic lesions, gaining more and more recognition, are due to DNA–protein cross-links, in which an oxidized nucleobase (guanine) forms a covalent bond with aminoacids, such as lysine or asparagine, containing an amine in the lateral chain [46, 66]. The same chemical process may also lead to the cross-link between DNA and primary amines, such as putrescine, spermine, or spermidine that are widely present in the nuclei [67, 68]. Furthermore, since DNA at rest is coiled around histones bearing a high density of lysine and asparagine residues, the relevance of such lesions cannot be underestimated.

The toxicity of the different DNA lesions varies considerably, and, at first approximation, depends on the interplay between the frequency of occurrence, the mutagenic potential, and the repair efficiency. Thus, DNA repair enzymes, that should be flexible enough to deal with the very complicated chemical space spanned by the DNA lesions, are one of the key players assuring genome stability in all living organism. In addition, the specific DNA repair machineries varies considerably among different organisms, and especially between eukaryotes and prokaryotes. As an example, in bacteria the repair of DNA photolesions proceeds through photolyases, remarkably exploiting light absorption to trigger the process. Those enzymes are however absent in mammals and humans, even if their evolutionary analysis points to their relation with chryptochromes, i.e. flavoproteins responsible for the regulation of circadian rhythms [69, 70].

When DNA lesions involve only one strand, the undamaged complementary strand may efficiently serve as a template to guide the repair process and minimize the errors. Two main repair pathways are known: base excision repair (BER) [71–75] is active in the case of damages localized on only one nucleobase, while nucleotide excision repair (NER) [76–79] is instead used for bulky lesions or intra-strand dimers, such as photolesions. The main difference between the two pathways is due to the fact that in BER only the lesioned nucleotide is excised, in a two-step process leading to the formation of an AP intermediate, while in NER a larger portion of the oligomer, usually some tens of nucleotides, is excised and substituted. Notably, the BER pathways due to its relative simplicity is performed by a much more reduced number of proteins compared to the NER pathways, for which a complex cascade of event starting with the lesion recognition and continuing with the recruitment and activation of the proper repair enzymes is necessary.

In the case of double-strand breaks or inter-strand cross-links, neither NER nor BER are applicable since the complementary strand cannot act as a template. Instead, three different repair mechanisms *non-homologous end-joining, microhomology-mediated end joining,* and *homologous recombination* are invoked [80, 81]. The double-strand breaks repair mechanisms are considerably more complex than the BER pathway, and a complete discussion of its functioning is clearly out of the scope of the present contribution. However, it is interesting to point out that the three mechanisms differ in terms of the rate of repair and are differentially activated as a function of the magnitude of the external stress factors and of the accumulation of the double-strand breaks. Note that even when globally the repair pathways coincide between different organisms, the specific actors, i.e. the recognition and repair proteins, may vary considerably in terms of both their sequence and tertiary structure, hence leading to important differences in the repair rates [82].

One of the key factors dictating the repair efficiency is the recognition of the damaged strand. As a consequence, the structural modification induced by the different lesions on the canonical DNA conformations is crucial to rationalize these processes [83, 84]. The organization of nuclear DNA, especially in the case of eukaryotes, is quite complex due to the super-coiling of the nucleic acid in nucleosomes and chromatin [85, 86]. This organization leads to important consequences for repair, indeed regions of high or low compaction may be characterized by different repair rates, while the chromatin organization may change in the course of the cellular cycles, also due to epigenetic signaling. Indeed, epigenetic globally refers to all the reversible modification of the DNA structure or organization that are governing the expression, or the silencing, of specific genes. Moreover, epigenetic regulations are based on complex cross-talks between the chemical modification of the DNA bases, mainly the methylation of cytosine, the acetylation of the histones, and finally the modification of the DNA compaction level in chromatin [87–89].

In addition, DNA lesions may be concentrated in a relatively spatial restricted area giving rise to the so-called DNA cluster lesions [82, 90, 91], i.e. the combination of two or more lesions appearing in between one or two helical turns. As compared to isolated damages, cluster lesions may behave in a correlate way, and hence induce structural deformation of DNA that can significantly impede the repair rate [83, 84].

This brief and non-exhaustive introduction, clearly demonstrates that the world of DNA lesions and photolesions is fascinating yet extremely complex. To achieve a proper rationalization of all the different intertwined relevant phenomena, a proper multi-scale approach is needed, that should be able to take into account all the different effects at molecular and systemic level. In particular, molecular modeling and simulation are fundamental to provide an atomistic and electronic resolution of the different phenomena coming into play, and hence discriminate between the different processes, identify the causalities between them, untangle the different cross-talks, and predict or rationalize the final outcomes.

The computational study of isolated or hydrogen bonded nucleobases, used as model systems of the extended DNA strands, have been and are still actively pursued. As a non-exhaustive example, we may cite the analysis of the electronic factors providing the specificity of the DNA hydrogen-bonding network that has been performed using energy decomposition analysis tools. The studies conducted both on Watson and Crick paired nucleobases [92, 93] and non-canonical structures such as guanine-quadruplexes (G4) [93–96] have clearly highlighted the role of cooperativity in hydrogen bonding. In addition, the study of isolated systems performed at a high quantum chemical level has also allowed for a constant development and improvement of force fields for molecular dynamics (MD) simulations, whose quality is nowadays well established and allows for a proper description of the most relevant structural parameters of DNA in different environments [97–100]. Most of the force fields development has been performed by reproducing the structural and dynamic properties of solvated, and relatively short, DNA oligomers. Some considerable efforts will be necessary in the future to improve the description of DNA in complex surroundings and most specifically DNA in chromatin environment. However, modern force fields have already allowed a good representation of the complex interactions taking place between nucleic acids and proteins, either for compaction, replication, or repair [101–103].

The study of isolated model systems has not been restricted to the ground state phenomena, but on the contrary, it has also dealt with the determination of the excited states' potential energy surfaces of the key DNA constituents to reveal their specific photophysics and photochemistry. These approaches have allowed, on the one hand, to clarify the reasons for the remarkable DNA photo stability [104–109], and on the other hand, to unravel, most often in combination with spectroscopic experiments, the molecular basis behind DNA photolesions' appearance and evolution [110–114].

However, the crude approximations underlying this approach have shown their limitations, once again both from an experimental and a theoretical point of view [52]. For obvious reasons, all the events related to the complex and heterogeneous environment of real biological structures are out of reach of simple model systems. Indeed, substantial differences in the distribution of reaction products between solvated nucleotides and oligomeric DNA have been observed for both oxidative and photoinduced DNA lesions. The development of efficient multiscale methods, in particular at hybrid quantum mechanical/molecular mechanical (QM/MM) level, has allowed to rationalize such occurrence with an electronic scale resolution.

However, the development of QM/MM algorithms to model DNA reactivity and photophysics is far from being trivial, and adds specific challenges to the usually complex hybrid procedures. Some of the specific difficulties in applying QM/MM methods to nucleic acids may indeed derive from the presence of π-stacked units, the nucleobases, that give rise to a collective behavior, hence necessitating either the enlargement of the QM partition [115] or the use of specific semi-empirical and phenomenological Hamiltonians, such as the Frenkel approach [116, 117]. Moreover,

DNA oligonucleotides are also floppy molecules, characterized by a very large flexibility of the backbone, that leads to the problem of a proper statistical sampling of the conformational space, while its structure depends on a complex interplay between weak non-covalent interactions, such as π-stacking and hydrogen bonds, that should be treated on the same footing. DNA oligonucleotides are also highly charged species due to the phosphate backbones, hence they strongly interact with counterions in solution. The DNA-counterion mutual influence questions, in some cases, the use of non-polarizable force fields, and, as it will be illustrated in the following, the possibility to open novel reactive channels as a result of specific interactions with the counterions, hence, these phenomena should be properly modelled.

Despite these difficulties, the use of QM/MM algorithms has also allowed achieving some clear insights into DNA repair mechanisms, in particular obtaining highly precise and statistically converged free energy profile for enzymatic reactions, also including photoactivated mechanisms in the case of bacterial systems [118–122]. However, less systematic studies are devoted to global structural deformations induced by the different classes of DNA lesions and their consequence on the repair efficiency, as well as to the coupling between cluster lesions and their effects on the recognition by repair machineries. To answer such questions, an efficient and effective sampling of the conformational space of DNA strands, both solvated and in interaction with repair enzymes, should be performed [83, 84]. In other words, it is necessary to enlarge the focus from purely electronic effects to structural and dynamic perturbations.

In the following sections of this chapter we provide some example dealing with the study of DNA photolesions production, either by direct light absorption or through photosensitization, and with the related DNA repair mechanisms. We show that proper multiscale protocols are nowadays available, and that by carefully choosing the right level of theory the description of the entirety of the complex mechanisms into play is available. We demonstrate that the journey starting from the description of light/matter interaction, proceeding through the study of photochemical evolutions, and achieving with the elucidation of the biological outcome, has become accessible.

Hence, and in addition to the relevant interest of DNA lesions per se, this chapter also illustrates the maturity of molecular modeling and simulation, that is nowadays completely assuming its role as a real computational microscope. Such a maturity has, in our opinion, paved the way to the emergence and the development of a novel, and complementary, scientific approach, that should be properly recognized, i.e. computational molecular photobiology. Indeed, computational photobiology has been gradually constructed by the enormous efforts of a large number of researchers from different fields and by the impressive methodological development of the last decades coupled with the computational power increase displayed by modern computers. It has also stemmed from the change of paradigm that has switched the focus from the simple molecular modeling and simulation to the multiscale description of the interplay between complex phenomena and complex environments. This chapter provides illustrative examples of computational photobiology into action, clearly highlighting the novel and fascinating possibilities it can offer to the scientific community.

2   DNA Photolesions by Direct UV Light Absorption

2.1   Cyclobutane Pyrimidine Dimers and Pyrimidine (6-4) Pyrimidone Photolesions

The formation of CPDs and 64-PPs adducts (see Fig. 2) accounts for 75% of the DNA photolesions produced by direct sunlight absorption. Since those lesions have been unambiguously related to the development of skin cancer, also due to their high mutagenicity [123, 124], the comprehension of these photoprocesses is of utmost importance. Formally, CPDs are produced through a [2+2] cycloaddition between the C5=C6 double bond of two adjacent pyrimidine nucleobases [125], whereas 64-PPs are the result of a Paternò–Büchi reaction between the carbonyl group of one monomer and the C5=C6 double bond of the adjacent pyrimidine nucleobase (see Fig. 2) [15]. Other photodimerizations involving purine/purine [126–130] and purine/pyrimidine [131–133] nucleobases have also been profusely reported in the literature, although its occurrence is in general less probable than the CPD one. Due to the inherent self-protecting phototostability of DNA [1, 2], the photochemical quantum yields of CPDs are very low, in the order of less than ten lesions per 10^4 base pairs [134], depending on irradiation wavelengths [135] and the flanking bases [136–138]. In particular, CPDs are the dominant photolesion at UVA wavelengths [139]

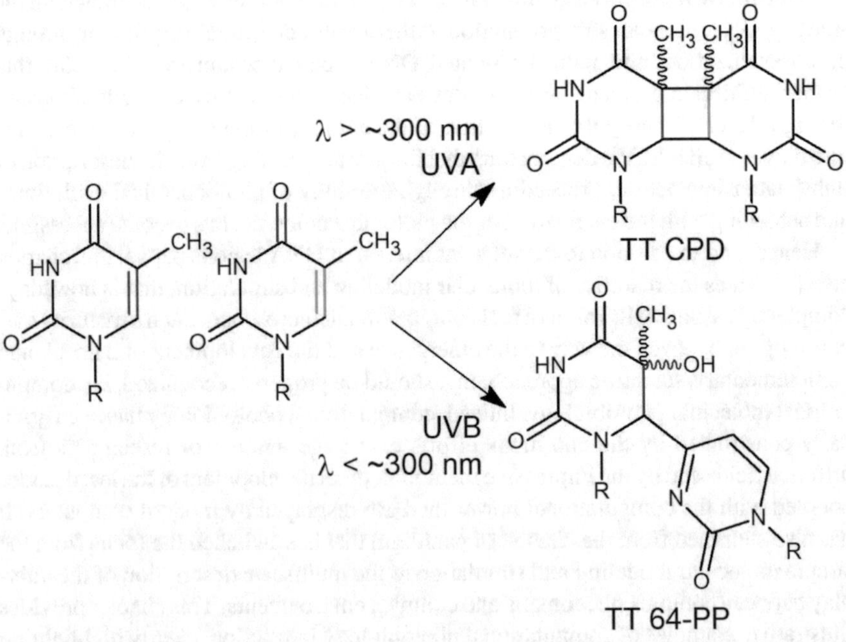

Fig. 2 Schematic view of CPD and 64-PP lesions. As an example, only the thymine-thymine photoreactions are shown

while 64-PP only occur in the UVB range [135]. The timescale of the two processes are also strikingly different: whereas CPDs occur at the picosecond regime [140, 141], 64-PPs are produced at the millisecond scale [142]. The ionization potential of the adjacent nucleobase to the TT site influences the T<>T yield [138]: higher ionization potentials increase the quantum yield of the photoproduct most likely due to the reduced possibility to form intra-strand charge-transfer states, thus favoring the formation of $^1\pi$, $\pi*$ excitons which mainly drive CPD photoprocesses [112]. In contrast to CPDs, 64-PPs are mediated by inter-strand charge-transfer states [112].

Alternative explanations regarding the nature of the excited states that participate in the ultrafast T<>T formation have been provided by non-adiabatic molecular dynamics in the gas phase performed by Rauer et al. [113]. The authors found that the photoreaction takes place via doubly excited states π^2, $\pi*^2$, in which two electrons are redistributed from a π orbital to a virtual $\pi*$ orbital, populated by internal conversion from the bright singlet excited state. No triplet states were populated in the set of trajectories run in the study.

Even though it can be considered as a relative novel branch of computational chemistry and biology, computational photobiology has already played a key role in the understanding of the complex events described above. DNA photochemistry is, in general, very sensitive to the biological environment [115, 143–145]; therefore, the use of QM/MM methodology is crucial to capture its influence and achieve the description of photoinduced phenomena in cellular environments.

CPD and 64-PP lesions require certain spatial proximity of both involved pyrimidines, a structural arrangement which is far from the canonical DNA disposition, where the separation between π-stacked nucleobases is larger. Those closer packed arrangements are often labelled as pro-reactive structures since they determine the photoreaction product in a great extent. By using the QM/MM approach, Conti et al. [146] have studied the combination of electronic and structural factors that favor the formation of either T<>T CPD and 64-PP. Whereas the excited-state energy profile for the former is barrierless leading to a conical intersection (CI) with the ground state, as also noticed by Merchán, Serrano-Andrés and coworkers [147, 148], the 64-PP reaction pathway displays an energy barrier lower than 0.3 eV, compatible with the lower yields and longer timescales recorded for this channel [146]. An alternative explanation for these experimental observations has been provided by Giussani et al. by means of QM/MM calculations using the CASPT2//CASSCF protocol to determine the photochemical landscape [149]. The authors found that the reactive CI (between a π, $\pi*$ thymine-thymine charge-transfer state and the ground state) lies at a higher energy (~0.5 eV) with respect to a non-reactive CI.

Conti et al. [146] also studied the specific role of the geometrical factors in driving the occurrence of photolesions, emphasizing the fact that the photoreaction is the result of a cooperation of a number of specific spatial parameters (like stacking and shift distances and sugar puckering angles, among others) coupled with certain electronic factors. For example, partial localization of the excited state on the 3′ thymine moiety tends to increase the coupling between the electron density of both C5=C6 bonds, thus favoring the production of CPDs. On the other hand, the population of a charge-transfer state where the electron density is redistributed from the $5'$ to the $3'$ thymine favors the 64-PP production [146].

As already stated, the quantum yield of CPDs and 64-PPs production is influenced by the flanking bases surrounding the reactive pyrimidine, mainly due to the formation of intra-strand charge-transfer states by charge injection from the flanking base. Hence the ionization potentials [150] of the latter nucleobases play a major role [138, 151, 152]. These phenomena were rationalized in 2019 by Lee and Matsika [136], who assessed the specific influence of guanine, the DNA nucleobase with the lowest ionization potential, as a flanking base for the CPD formation. The authors concluded that the G->T charge-transfer state is the most stable one, whereas the $5'$ position of the flanking base also stabilizes the charge-transfer states that drive the photoprocess. Interestingly, the same stabilizing effect, is also induced by the presence of the epigenetic relevant 5-methylcytosine that hence favors CPD formation.

The formation of DNA lesions and photolesions in less conventional DNA structures, such as G4s [153, 154], have also been studied with computational and experimental tools. In guanine quadruplexes, the larger flexibility of the external loops coupled to the rigid guanine core allows the production of the uncommon *anti* CPD, a lesion highly hindered in regular DNA duplexes, but detected in guanine quadruplexes [155]. Regarding the production of 64-PP with thiolated bases, a unusual DNA probe and a potential photosensitizator, a recent study by Zou et al. provides the molecular basis for the formation of the corresponding adduct between 4-thyopyrimidine and thymine, through an intermediate thietane structure analogous to the oxetane species in conventional thymine-thymine 64-PP [156].

2.2 Hydrogen/Proton Transfer Processes in Watson–Crick Base Pairs

In stark contrast to the intrastrand CPD and 64-PP lesions, which are either extremely badly repaired or highly mutagenic, the interplay between opposite strands in DNA duplexes open photochemical processes that are mainly leading to photostability [157–161]. These mechanisms are triggered by inter-strand charge-transfer states where the purine nucleobase acts as electron donor and the complementary pyrimidine nucleobase as the electron acceptor (see Fig. 3a). The charge separation induces then a proton transfer from the former to the latter, yielding a neutral intermediate that decays to the ground state enabling the back electron transfer from the pyrimidine to the purine nucleobase, also accompanied by the back transfer of the same or a different proton, inducing photostability and tautomerization, respectively [162–167]. Such processes have been styled as excited-state hydrogen transfer (ESHT) [165] and globally belong to the realm of proton-coupled electron transfer (PCET) processes [168–170].

Whereas this mechanism could be operative in both G-C and A-T base pairs, only the former has been unambiguously identified in model oligonucleotides in solution

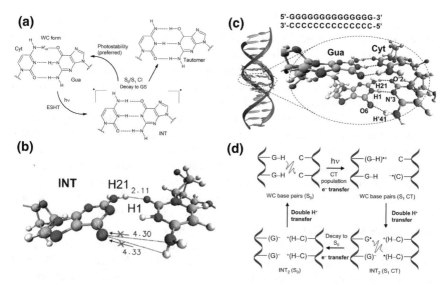

Fig. 3 **a** The molecular structures of Watson and Crick paired G and C nucleobases highlighting the proton transfer leading to tautomerization or photostability. **b** representative snapshot from the QM/MM MD trajectories evidencing the rotation of the amine group impeding tautomerization. **c** Enlarged QM partition including 4 base pairs and allowing the observation of the four proton transfer (FPT) schematically represented in (**d**). Adapted with permission from Ref. [115]—Published by the Royal Society of Chemistry under a Creative Common License

[171]. This fact could be explained in terms of the energy barriers that have to be overcome to access the inter-strand charge-transfer states, relatively high for the A-T base pair [166], whereas the whole photoprocess is barrierless or nearly barrierless for the G-C couple [162, 164, 165]. Moreover, it has been found that the proton transfer events detected in G-C base pairs in solution mostly lead to photostability, with the tautomerization yield estimated at ~10% [161]. Indeed, for oligonucleotides in solution, no tautomer formation is observed [159, 172]. The effect of the biological environment suppressing the tautomerization channels was already captured by the work of Groenhof et al. in 2007 [164], in which the trajectories from QM/MM dynamics simulations performed on a DNA double strand exciting a single G-C base pair, invariably resulted in the recovery of the canonical DNA structure.

Francés-Monerris et al. provided a complementary rationalization of the lack of tautomerization by performing QM/MM excited-state simulations on a dG·dC oligonucleotide in solution [115] (see also Fig. 3). The authors found that the energy released during the proton transfer step significantly weakens the Watson and Crick H-bonding pattern between guanine and cytosine. This dynamical effect induces out-of-plane distortions of the C amino group hence preventing the tautomerization and thereby preserving the canonical DNA structure. In the same work, and by significantly enlarging the QM partition up to four stacked base-pairs, the authors also documented a totally novel type of proton transfer event in DNA named four-proton transfer (FPT, Fig. 3c, d). The former process is directly driven by the complex heterogeneous environment, and is in particular due to the interaction of the charge-transfer state with a nearby Na$^+$ ion [115]. As a result, two protons, one on each G-C adjacent base pair, are concertedly transferred to the respective cytosine moieties. Analogously to the traditional ESHT mechanism, the intermediate formed during the FPT decays to the ground state, triggering the back transfer of the protons toward the respective guanine moieties. This example showcases the relevance of a proper treatment of the biological environment, since only the combined description of the multicromophoric DNA strand and of the heterogeneous surroundings, in combination with a dynamic approach, can reveal novel phenomena such as the FPT mechanism.

In addition to interstrand charge transfer states, intrastrand charge transfer states, i.e. excitation producing charge separation between π-stacked guanine and cytosine on the same strand, have also been highlighted as capable of induce proton transfer events in alternated (dG)·(dC) duplexes, since guanine remains positively charged and cytosine negatively charged as a consequence of the charge-transfer process [158, 159, 173]. The charge separation drives inter-strand proton transfers via base pairs either in cationic or anionic radical states, depending on which guanine-cytosine couple is involved in the proton transfer. Interestingly, inter-strand charge-transfer states have lower excitation energies in alternating nucleotide sequences [174]. Those states present a much flatter potential energy surface along the proton transfer reaction coordinate, as opposed to interstrand charge transfer states. Hence, they can be invoked to justify the longer relaxation time of alternating DNA sequences as compared to non-alternating ones [175], for which interstrand charge transfer states should be dominant due to the respective energetic order.

2.3   Photo-Induced Electron Transfer in DNA

As on outcome of the excitation the energy gained by light absorption can also eject one electron from a photoexcited DNA. The obtained "hole", or radical cation, migrates rapidly along the DNA strands thanks to the π-stack arrangement of nucleobases that maximizes the electronic coupling. In addition to direct light absorption, photoinduced charge-formation and charge-migration are also produced as an outcome of photosensitization by using photo-oxidizable molecules, having the advantage to displace the absorption to the visible spectrum [176, 177]. Charge-transfer in DNA is highly sequence dependent [178] as the four nucleobases do not have the same properties: guanine presents the lowest ionization potential and hence a G-G sequence constitutes a charge trap; conversely, thymine do not stabilize the positive charges but can be involved in tunneling processes [179]. Indeed, long range charge transfer in DNA appears very specific, with a combination of different decay functions, depending on the donor–acceptor distance. This particularity has been recently pointed out by Beratan, and explained considering a switch between different electron transfer mechanisms (superexchange tunneling, flickering resonant transport or multistep hopping) [180].

The simulation of charge migration along DNA strands thus represents a considerable scientific challenge for molecular modeling and simulation. First, the three key charge transfer parameters (energy gap between donor and acceptor, electronic coupling, and reorganization energy) must be determined with high accuracy. Several high-level quantum calculations have focused on this issue, also providing a consistent benchmark for further calculations [181, 182]. However, they are limited in terms of the strand size, conformational sampling and explicit description of the solvent or the ions, preventing the proper study of the long-range diffusion of the hole along DNA. Bittner has also analyzed charge migration and delocalization in DNA strands by using quantum dynamic, based on a simplified excitonic Hamiltonian, notably highlighting that while adenosine exciton remains stable over hundredths of fs, exciton delocalized over thymine rapidly decompose into mobile hole/electron pairs [183]. To overcome the main computational limitations, the combination of classical simulations and QM/MM schemes appears essential and allows to take into account the whole DNA fragment and its environment. To decrease the computational cost, the fragment orbital approaches (FO) [184], in which only the highest occupied molecular orbital (HOMO) of the nucleobases are considered, is particularly appealing. Another strategy involves semi-empirical computational approach such as FO-DFTB/MM and allows to simulate the charge propagation in an unbiased way [185]; this approach coupled to a QM/MM protocol allowed to build a bottom-up theoretical model able to push the simulation time limit to 1 μs [186]. Beyond the sequence dependence, the conformation also modifies the redox properties of DNA. For example, the guanine-rich G4 structure is a very efficient trap for radical cations which have a longer lifetime with respect to B-DNA [187]. In addition to the fundamental role played in the domain of DNA lesions, DNA charge-transfer properties, coupled with its structural tunability, are extremely appealing for the development of smart materials for molecular electronics. However, such an approach will require a much better understanding of the long-range charge- and energy-transfer properties of DNA arrangements to allow for a finely tuning of the conduction properties.

3 Modeling DNA Photosensitization

3.1 Classic Photosensitization: The Paradigmatic Case of Benzophenone

The number of potential DNA photosensitizers is extremely large and it includes drugs, pollutants, and their metabolites. In addition, the great majority of phototherapeutic drugs may be considered as photosensitizers ultimately producing $^1O_2$ [188], upon light absorption and intersystem crossing via the "Type II" photosensitization [189]. Transition metal complexes, especially based on heavy ions such as Ru, Re, and Ir, are also known to induce photosensitization and in some instances have also been proposed for clinical trials [42, 43]. Globally speaking, the mechanisms inducing photosensitization are different, and comprise energy- and electron-transfer phenomena as well as other more complex photoreactive pathways including H-abstraction from DNA constituents.

Among the many possible photosensitizers, benzophenone (BP, Fig. 4), despite its limited therapeutic applicability, due to its absorption in the UVA range, occupies a special role. Because of the efficient population of its triplet manifold, BP may trigger different competitive photosensitization pathways, the most important one being triplet energy transfer to a nearby thymine, as it has been nicely reviewed by Miranda's group [15]. In addition, the inherent photophysics of isolated BP is far from being innocent and trivial, especially due to its three-state quasi-degeneracy involving the first excited singlet and two triplet states [190]. Hence, in this context BP also represents a most suitable example to illustrate the combination of different simulation protocols and level of theories necessary in computational photobiology.

The native photophysics of BP has been studied by high-level QM methods based on complete active space self-consistent field (CAS-SCF) and its second order perturbation theory correction (CAS-PT2), notably by Sergentu et al. [190]. The potential energy landscape of BP is dominated by an extended region of quasi-degeneracy between the S_1 state (mainly of $n\pi^*$ character) and the first two triplet state, i.e. T_1 ($n\pi^*$) and T_2 (π^*). Those findings are in agreement with the experimental almost

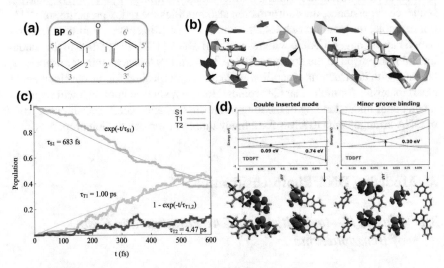

Fig. 4 a Molecular structure of benzophenone and **b** the double insertion (left) and minor groove binding (right) interaction modes with double stranded DNA as obtained from classical MD. **c** Time evolution of the isolated benzophenone excited state manifold population as obtained from CAS-SCF non-adiabatic dynamics. **d** Potential energy surfaces over the general coordinate describing the triplet–triplet energy transfer for double insertion (left) and minor groove binding (right) obtained from QM/MM simulations. Adapted with permission from Ref. [191] Copyright 2013 American Chemical Society, Ref. [192] Copyright 2015 American Chemical Society, and Ref. [193] Copyright 2016 American Chemical Society

unitary intersystem crossing [194–196], but do not provide a clear mechanistic picture of the photophysical processes.

To provide a definitive answer CAS-SCF non-adiabatic molecular dynamics, in the state-hopping framework, and including spin–orbit coupling (SOC), have been performed for gas-phase BP [193] (Fig. 4). This allows to follow the time evolution of the population of the different electronic states over a statistically relevant number of semi-classical trajectories, hence mimicking the behavior of a nuclear wavepacket. The non-adiabatic dynamics confirms a fast and efficient intersystem crossing, indeed at 600 fs a majority of the trajectories is already in the triplet manifold. Interestingly, while unsurprisingly the population of T_1 is predominant, a persistent population of T_2, reaching up to 10% and behaving like a real spectroscopic state rather than an intermediate can be observed [193]. The persistence of the T_2 population is due to an equilibrium between T_1 and T_2, characterized by constant forward and backward hops between the two states [193].

Coming back to the study of DNA photosensitization, the first problem to face is the absence of any experimental structure for the DNA/BP aggregate. In this context, molecular modeling and simulation has been fundamental in providing the required answer. Indeed, by using classical MD simulations two stable interaction modes between BP and a model poly-d(AT) decamer have been identified as reported in Fig. 4 [191]. While one of the modes was based on a well described minor-groove interaction, the second one, styled double-insertion, was observed for the first time. The latter is characterized by the ejection of a full base pair from the DNA double helix and its substitution by the sensitizer. Both modes are stable and persistent in the timescale, hundredths of ns, spanned by the classic molecular dynamics. Subsequently, the binding free energy have been determined [197], by using free energy perturbation (FEP) methodology [198]. The minor groove binding appeared as favored since the corresponding aggregate is stabilized by about 2 kcal/mol compared to the solvated system.

Since the coupling between BP and the DNA nucleobases differs significantly between the two interaction modes, the triplet–triplet energy transfer, i.e. the photo-sensitization, was studied for both conformations using a hybrid QM/MM approach to model the effects of the nucleic acid environment. The photophysical pathway was described taking into account a general coordinate based on the linear interpola-tion between the equilibrium geometries of the triplet states centered on BP and on the nearest thymine, respectively. The potential energy surfaces of the most relevant excited states were calculated at CAS-PT2 and TD-DFT (Fig. 4) level pointing out an almost barrierless energy transfer from T_1 in case of the double inserted mode, while the minor groove binding presents a considerable barrier of about 0.9 eV [192]. Even though the latter value should be considered as an upper bound due to the use of a linear interpolation coordinate, such an energetic penalty already indicates that the photophysical transfer from the minor groove binding mode, which happens to be the thermodynamically favored one, is much less competitive as compared to the one from the double inserted mode. Hence, the question of the efficient triplet transfer remains unanswered. However, as suggested by the non-adiabatic dynamics, the participation of the benzophenone T_2 state should also be taken into account, and indeed a barrierless pathway connecting the triplet state centered on BP to the one on thymine exists for this state both in the minor groove and the double inserted mode [192]. Therefore, the significant population of T_2 state should be considered as the key photophysical factor allowing an efficient DNA photosensitization by triplet–triplet energy transfer. We stress out that this comprehensive picture has been obtained only by the use of combined multiscale modeling and simulation protocols, including the determination of binding free energy, non-adiabatic dynamics, and high-level QM/MM methodologies.

The DNA environment also plays a peculiar role in modulating a minor BP sensitization pathway, namely hydrogen abstraction. Indeed, the triplet state of BP induces the homolytic breaking of C-H bonds, a property that is also exploited in photocatalysis and polymer science [199]. Once again, the two interactions modes differ considerably: double inserted BP may extract a hydrogen from a thymine methyl group, while groove bound BP attacks the backbone sugar to ultimately produce strand breaks. However, when considering only a minimal model system composed of BP in its triplet state and the reactive DNA counterpart, i.e. the nucleobase or the sugar respectively, activation energies of less than 10 kcal/mol are obtained and both reactions are exergonic [200]. Hence, the question of why such a channel is largely uncommon in the BP sensitization landscape inevitably arises. The answer, once again, comes from a proper multiscale treatment including the analysis of the structure and dynamics of the BP/DNA aggregates. Indeed, the MD trajectories for both double insertion and groove binding reveal that the distribution of distances between BP carbonyl and the reactive hydrogen peaks at around 9 Å. Thus, only a marginal population of distances are compatible with the formation of a prereactive complex. In this sense, one can conclude that DNA actually implements a sort of self-protection, maintaining BP at sufficient large distances from its reactive moieties to significantly diminish the occurrence of a reaction that, from an electronic point of view, should be fast and favorable [200].

3.2 DNA Trojan Horses

In the previous subsection we have pointed out how the DNA macromolecular environment can play an efficient protective role. The opposite scenario may however emerge when DNA lesions act as internal photosensitizers. The presence of such "Trojan horses", significantly expanding the possibility of DNA damages formation, has been experimentally proven by Miranda's group for the very well-known 64-PP photocrosslink. Indeed, the former lesion contains one pyrimidone (Pyo) unit (Fig. 5) that absorbs light in the UVA range, undergoes a relatively efficient intersystem crossing, and triggers triplet–triplet energy transfer towards nearby thymine [202] ultimately resulting in CPD production via a triplet state path [114]. The experimental results, obtained using a modified nucleobase composed of Pyo subunit (dPyo) instead of the full 64-PP lesion, have also been rationalized via molecular modeling

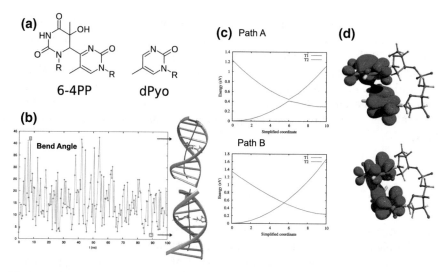

Fig. 5 a Molecular formula of the 64-PP and the corresponding chromophoric unit dPyo used to build the artificial nucleobase. **b** Time series of the total bending of the dPyo containing DNA oligomer together with two representative snapshots corresponding to the highest and lowest bending, respectively. **c** Potential energy surface along the triplet–triplet transfer coordinate for two representative starting conformations, note in Path B the presence of a crossing between the two surfaces. **d** Natural transition orbitals describing the unpaired electrons and calculated at the crossing point in Path B showing the delocalization over both unit and the overlap between the functions. Adapted with permission from Ref. [201]. Copyright 2015 WILEY–VCH Verlag GmbH & Co. KGaA, Weinheim

and simulation [201] (Fig. 5). In particular, by using classical MD, it has been highlighted that dPyo artificial nucleobase induces only very minor perturbation in the structure of the DNA double helix, and in particular is not altering the coupling due to the π-stacking arrangement. The triplet–triplet energy transfer has also been studied using the same QM/MM protocol as employed for BP sensitization, showing that, even though the global driving force appears smaller, the energy transfer is still possible and favorable [201] as reported in Fig. 5. Furthermore, in some snapshots the coupling between the two chromophores, i.e. Pyo and thymine, is found to be extremely high, and gives raise to discontinuities along the potential energy surface that are reminiscent of conical intersections (Fig. 5c, d). Such critical points should funnel the triplet transfer, hence contributing to the increase of the efficiency of the full photophysical process [201].

More recently, the same "Trojan horse" effect has been evidenced for formyluracil (ForU), a lesion resulting from the oxidization of thymine [203]. Once again, molecular modeling and simulation have clarified the full photophysical pathways, providing a detailed mechanistic study of the intersystem crossing events that mediate the triplet population, in particular pointing out its feasibility also due to the absence of significant energy barriers blocking this channel [204]. A feasible mechanism for the triplet–triplet energy transfer between ForU and thymine has also been characterized by means of CAS-PT2 calculations, highlighting the relatively small energy barrier <0.23 eV determined for this process. Finally, the effects of the environment, i.e. the DNA double-strand, have been taken into account using a combination of classical MD and QM/MM calculations, showing that the perturbation of the DNA structural parameters by ForU is modest or even absent, while the conditions for an efficient intersystem crossing and energy transfer hold in the biological environment [204].

3.3 DNA Artificial Nucleobases

The concept of "Trojan Horse" lesion may be related to the more general concept of DNA photosensitization by artificial nucleobases that may be exploited for therapeutic purposes in the framework of PDT and LAC. Indeed, in this respect Peng's group has reported the synthesis and characterization of a modified nucleobase containing nitroimidazole, i.e. a photolabile leaving group [44] (Fig. 6) that can induce the formation of DNA interstrand cross-link upon irradiation. The interest of such an approach is due to the fact that it directly tackles one of the major limitations

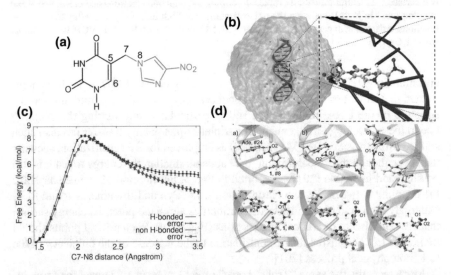

Fig. 6 a Molecular formula of the modified nucleobase (thymine) including a nitroimidazole leaving group and **b** the representation of the simulation box used for the classical MD and the QM/MM simulations including the solvated oligonucleotide bearing the modified nucleobase. **c** free energy profile obtained at QM/MM DFT level for the dissociation of the nitroimidazole in the radical anion system with a without the formation of a H-bond between the leaving group and a nearby adenine. **d** representative snapshots of the reactants, transition state, and product regions for the two conformations, i.e. H-bonded (upper panel) and non H-bonded (lower panel). Adapted with permission from Ref. [205]. Copyright 2019 American Chemical Society

of PDT, i.e. its dependence to oxygen activation, which severely reduces its applicability in hypoxic solid tumors. The precise mechanism of action has been recently assessed by Francés-Monerris et al. [205] using once again a multiscale protocol combining high level QM calculations with QM/MM and MD simulations (Fig. 6).

In particular, the PESs of the relevant low-lying excited states show that the photodissociation of the nitroimidazole leaving group from the excited neutral manifold requires overcoming a too high energy barrier hence this possibility can be safely discarded. On the contrary, the photodissociation proceeding from the radical anion appears favorable having a barrier of only 0.6 eV, therefore the first event following light absorption should result in charge migration along the DNA. Classical MD simulations have provided the structural parameters of a DNA oligomer bearing the modified nucleobase, showing the global maintaining of the canonical DNA structure and the very limited deformations. In addition, they have also provided starting configurations used for a subsequent study of the photodissociation process via a QM/MM potential of mean force (PMF) procedure that has provided the free energy profile taking into account both environmental and dynamical factors. A small barrier of about 8 kcal/mol has been highlighted, however the backwards reaction, i.e. the recombination of the leaving nitroimidazole with the artificial nucleobase, is also possible and could constitute the key factor limiting the global photochemical efficiency. Interestingly, it has been shown that two conformations are present differing by the formation of a hydrogen bond between nitroimidazole and a nearby adenine. Even if the hydrogen bond seems to have almost no effect on the dissociation activation energy it can limit the diffusion of the leaving group and hence favor recombination, as is also evident from the corresponding PMF, once again pointing to the subtle equilibrium between different environmental factors in driving the global efficiency of photoinduced processes in nucleic acids and DNA.

4 DNA Photolesions Repair

Different repair strategies to remove photolesions exist and are essential to assure viable conditions for both animal and vegetal organisms. Indeed, knocking out the photolesions repair enzymes results in death or high impairing diseases such as *Xeroderma pigmentosa* [206, 207]. In the following, we present different DNA repair pathways, also focusing on the complex interplay between the structural modifications induced by the lesions and the repair efficiency and on the role played by computational photobiology in contributing to their rationalization.

4.1   DNA Repair by Photoliases

DNA photolyases are flavoenzymes using sunlight to specifically repair 64-PP or CPD DNA damages in many organisms including bacteria, plants, insects, amphibians, etc. [208, 209]. Their efficiency relies on a photo-induced electron transfer from a photoexcited reduced flavin adenine dinucleotide ($FADH^-$) buried in the active site and the oxidized damage, preliminary flipped-out from the DNA helical structure [210, 211]. Many experiments have been performed to elucidate the repair mechanism and the specificity of these proteins. The structures of several photolyase/DNA complexes have also been resolved by X-Ray crystallography [210, 212, 213], and have helped to pinpoint the residues involved in the enzymatic process. Femtosecond transient spectroscopy studies have also elucidated the catalytic cycles including the obtention of the rate constants for both CPD and 64-PP repair, as schematized in Fig. 7 [214]. The repair of CPD only involves the reduced FAD and the DNA lesion: the electron transfer from the former to the latter induces the spontaneous and fast (few ps) C–C bonds breaking leading to the recovery of the two adjacent thymine nucleobases.

While the conversion of CPDs takes place with a quantum yield exceeding 80% [215], the 6-4 photolyase is way less efficient and has a quantum yield of about 10%. In the latter case, and in addition to the photoinduced electron transfer, a proton exchange, involving a histidine in the active site occurs and competes with the futile back electron transfer to FAD. Mutagenesis studies have also illustrated the higher complexity of the 64-PP repair machinery compared to the CPD one: if three mutations are sufficient to convert a 6-4 photolyase into an efficient CPD repair enzyme, the reverse conversion has not been obtained yet, even summing up 11 mutations in the active site [216].

Classical MD simulations were conducted to understand how photolyases identify a damaged site and flip it into the active site. In CPD photolyase, a biased classical MD approach was used to study the flipping-out mechanism with or without the protein [217]. The free energy calculation was based on umbrella sampling including

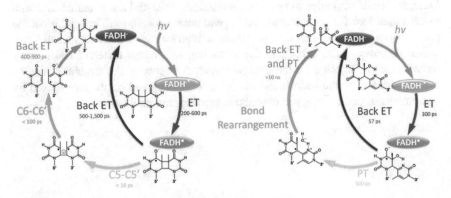

Fig. 7   CPD photolyase and 6-4 photolyase catalytic cycles

Hamiltonian replica exchange protocol. Both the DNA deformation and the interaction with the protein decrease the free energy penalty required to obtain an extrahelical conformation of the two damaged thymine nucleobases. The presence of an arginine in the DNA binding region also possibly helps to stabilize the damage in the extrahelical conformation by interacting with the backbone phosphate [218]. Computational approaches also helped to characterize the non-covalent interactions taking place in the active site of 6-4 photolyases. Indeed, positively-charged amino acids in the DNA pocket play a crucial role in the repair mechanism: a lysine (K246) characterized in the active site of the *Drosophila melanogaster* 6-4 photolyase interacts with the damage and can favor its reduction, whereas arginine (R421), keeps the lesion locked in the active site [219]. These interactions have been further characterized by DFT and hybrid DFT QM/MM calculations in the homologous active site of *Xenopus laevis* 6-4 photolyase: once again the positive group of the arginine interacts with the negatively charged phosphate groups of the damage and form a cation-π interaction with the 3' nucleobase flanking the 64-PP [220]. However, some 6-4 photolyases, such as PhrB, do not present the homologous lysine or arginine and their activity is dependent on the concentration of divalent cations (Ca^{2+} and Mg^{2+}). The structure of the PhrB/DNA complex has not been resolved yet, however classical MD simulations starting form homologous models have been performed using the crystallographic structure of PhrB alone and the DNA damaged fragment from the *Drosophilia melanogaster* photolyase repairing complex. These simulations reveal that two Mg^{2+} enter in the active site and are locked by the interaction with two aspartates, whose mutations to asparagine decrease the DNA-repair activity. The interaction of the cations with the DNA damage has also been evidenced and is fundamental in assuring the positioning of the lesion in the active site, playing a similar role to the positively charged lysine and arginine residues in *Drosophilia melanogaster*'s photolyase [221] (see Fig. 8).

When the photodamage is stabilized in the active site, the catalytic repair can start. Despite the precise rate constants obtained thanks to femtosecond time-resolved spectroscopy [214]; the chemical mechanisms remain partially unclear. It is widely accepted that the first step consists in an electron transfer from the photoactivated FADH⁻ to the CPD or 64-PP damages [222]. However, the nature, and consequently the rate, of the charge transfer appears as structure-dependent. Indeed, in the whole photolyase family (including also cryptochrome proteins), the FAD cofactor adopts an unusual U-shape where the isoalloxazine ring and the adenine moiety are fairly stacked, hence allowing energy and charge transfer between the two π-systems [223]. It is also well established that light absorption produces excitation that are localized on the isoalloxazine moiety [224]. The surrounding environment and the conformation assumed by the DNA damage in the enzyme pocket induce different electrostatic field tuning the charge transfer pathway and mechanism [225]. The different rates observed experimentally between the three classes of CPD-Photolyase indicate the coexistence of several electron-transfer mechanisms (see also Fig. 9): a direct tunneling through adenine is dominant in Class I CPD photolyase whereas a two-step hopping pathway, with a transient negatively-charged adenine, is favored in class II. Class III presents an intermediate behavior [214]. Using models from

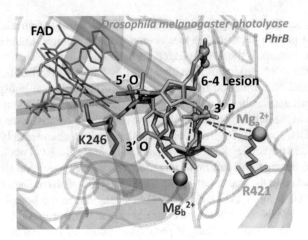

Fig. 8 Comparison between the active site of Drosophilia melanogaster 6-4 photolyase (yellow) and PhrB (cyan). K246 and R421 belong to the Dros. mel. Photolyase crystallographic structure (3CVU) and classical simulation have shown that K246 side chain moves toward 3'P. The Mg2+ cations belong to the PhrB structure after classical MD simulation. Reproduced with permission from Ref. [221]. Copyright 2019 Federation of European Biochemical Societies

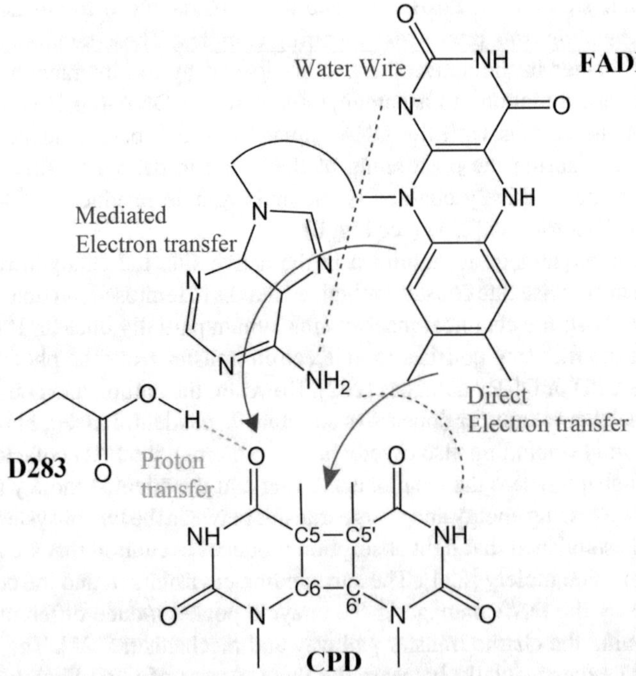

Fig. 9 Different mechanisms for the CPD repair in CPD-Photolyase: direct electron transfers between FAD and damage methyl group (red), tunneling transfer through adenine (purple), transfer involving a water wire and proton transfer (blue), or electron transfer from FAD with proton exchange with D283 (green)

class I CPD Photolyase, multiscale studies combining classical MD with the calculation of excitation energies and electronic coupling support a superexchange mechanism which does not involve adenine but the methyl group of the isoalloxazine ring [226, 227]. Moreover, wavepacket quantum dynamics (QD) describing the electron-transfer process, which has been determined solving the time-dependent Schrödinger equation with a finite-difference approach at the DFT/GGA/B3LYP level of theory [228], suggests that adenine polarizes the wave function to drive the transfer toward the CPD lesion damage. The hopping mechanism has also been considered by means of hybrid QM/MM calculations with a high level, multireference description of the excited states including adiabatic diagram correction ADC(2) and CASSCF [229]. The comparison of a set of four CPD photolyases from mesophiles and (hyper)-thermophiles species, based on classical MDs and electron tunneling analysis, shows that the adenine moiety significantly contributes to the electronic coupling of the electron transfer to the damage in the most rigid proteins [230]. At higher temperature, however, water molecules can compete with adenine for the tunneling pathway.

In addition, a third mechanism has been computationally described, using CASSCF and CASPT2/MM levels of theory in the form of a PCET involving water molecules bridging the adenine and the 5′ thymine of the photolesion [231], in which the C-C bond cleavage is favored by the enhanced hydrogen bond network. The crucial role of a strong hydrogen bond to the 5′- damaged thymine has also been explored by means of QM/MM MD [218]. In this case, the QM partition encompassed the damage and one active site glutamate, and the simulations describe a proton transfer from the glutamate to the lesion, followed by the asynchronous bond cleavage, starting with the C5-C5′ breaking.

The repair of 64-PPs requires covalent bond rearrangement which involves the C-O cleavage and formation. The activation of the C-O bond is triggered by proton transfer from a histidine placed in the active site resulting in the reduction of the 64-PP, although the exact mechanism remains unclear. Different hypotheses have been proposed based on one- or two-photon absorption, positively charged or neutral active site histidine, concerted or sequential PCET, oxetane or water intermediates (see Fig. 10). QM/MM calculations have been proved a very efficient tool to discriminate the most likely mechanism(s) by the determination and comparison of different energy barriers. In general, only one photoexcitation is considered in the mechanism, but in 2010 Sadeghian et al. [232] reported a relaxed potential energy surface at DFT/MM level for a mechanism involving an oxetane intermediate and two electron transfers from the FAD which allow to overcome the barrier present in the ground state. Nevertheless, this hypothesis involving two consecutive excitations is not consistent with the experiments and previous computational studies [118, 233]. Indeed, due to the natural photon density, the lifetime of an intermediate between two-photon absorption should be around 1 ms; while the spectroscopic signature of such intermediate has not yet been experimentally detected.

Another crucial issue for the elucidation of the repair mechanism is the protonation of the active site histidine, which can be protonated on Nδ, Nε or on both nitrogen atoms. Historically, the positively-charged histidine has been preferred, as it should be a better proton donor [234]. However, the determination of electronic

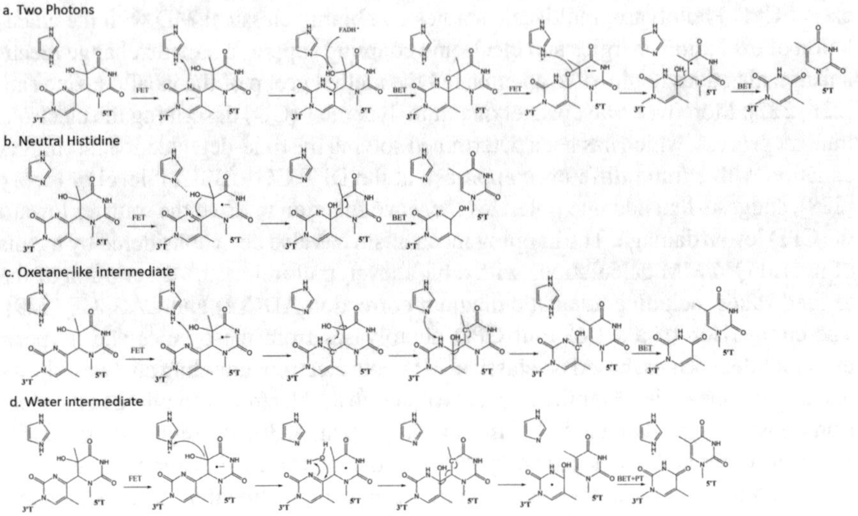

a. Two Photons

b. Neutral Histidine

c. Oxetane-like intermediate

d. Water intermediate

Fig. 10 Different mechanisms for the 64 pp repair in 6-4 photolyase starting with an electron transfer from excited FADH⁻ (FET), involving proton transfer (PT) or bond activation thanks to hydrogen bonds between the active site histidine and the damage, and a backward electron transfer (BET)

paramagnetic resonance (EPR) parameters using the DFT(B3LYP)/MM method and their comparison with the experimental values, provides no clear preference to any protonation state, even though both *pKa* and structural MM studies motivated the authors to suggest a neutral histidine, while another active site histidine is supposed to be doubly protonated [235]. High level XMCQDT2-CASSCF calculations of the excitation energies of the active site complex (FADH⁻, histidine and DNA damage) also provide evidences that support a neutral histidine state, as the protonated one would cause a spontaneous PCET which would prevent the repair mechanism [236]. The inclusion of the environment (protein, DNA, water and counterions) in a similar CASSCF-CASPT2/MM study slightly counterbalances this conclusion [237]. These calculations showed that the presence of a positively-charged histidine decreases the energy of the reduced damage state by 7–20 kcal/mol, although the dead-end PCET is still possible. On the contrary, classical MD simulations [221, 238] show that the double protonated histidine is more consistent with the experimental data concerning both structure and reactivity. Indeed, the RMSD of the histidine and the active site as well as the His-damage or His-FAD distances (among others) in *Drosophilia melanogaster* photolyase MD are more consistent with the crystallographic data in the presence of positively charged histidine. In the simulation of PhrB photoliases, the behavior of Mg^{2+} cations described earlier is in agreement with mutagenesis studies only if a double protonated histidine is considered [221, 239]. Moreover, the calculation of the repair of the Dewar photoproduct in a model of the 6-4 photolyase active site described at DFT/continuum level gives a lower energy profile in the presence a protonated histidine than a neutral one [240].

Faraji and Dreuw performed an exhaustive analysis of the repair mechanism using DFT/MM optimizations, which is summarized in ref [241]. They concluded that a one-electron transfer from $FADH^-$, followed by a proton transfer from the positively charged histidine and the simultaneous intramolecular OH transfer through an oxetane-like intermediate, is the most feasible mechanism. The QM/MM approach of Dokainish et al. [242], based on the optimization of a large QM partition (982 atoms at DFT –UM062X level after a first optimization of a smaller QM zone -166 atoms), validates a quite different mechanism involving the formation of a water molecule during the OH transfer instead of an oxetane-like intermediate and a rate-limiting barrier of 13.4 kcal/mol. When the thymine-thymine adduct is replaced by a thymine-cytosine photoproduct, a similar mechanism is obtained with an activation barrier of 20.4 kcal/mol and involving the intermediate formation of a free NH_3 group, while the energy of the azetidine intermediate is prohibitive (about 60 kcal/mol) [243].

Thus, despite these substantial efforts in computational studies, which assessed the one-photon mechanism with a doubly protonated histidine, the characterization of the 64-PP repair mechanism still remains under debate. The different results raise the issue of the definition of the QM zone versus the QM level of theory, as a higher level of theory would usually imply a smaller QM partition. Another issue concerns the dynamical behavior of the photolyase active site. The proton transfer and the bond reorganization are expected to occur at ~500 ps and more than 10 ns, respectively, which would let sufficient time to the confined environment to relax with regards to the charge state and the bond arrangement, pointing to the need of an appropriate sampling of the active site structural reorganization.

4.2 Interplay Between Structural Deformations and Repair: 64-PP Versus CPD

In addition to the elucidation of the chemical repair mechanisms, using either static techniques or *ab initio* MD sampling, the complex panorama induced by the structural flexibility of damaged DNA oligonucleotides should be properly dealt with since it can strongly affect the repair rates. Because of the involved large-scale rearrangements, the use of extended sampling via classical MD is compulsory and the case of DNA photolesions, i.e. CPD and 64-PP, is a most paradigmatic example.

Indeed, the repair rate and toxicity of the two photo-lesions are extremely different, and while 64-PP is much more easily fixed by the NER apparatus, it is also extremely mutagenic [47]. By using extended classical dynamics exceeding the µs time-scale, it has been shown that DNA duplexes containing thymine-thymine CPD (T<>T) lesions are quite rigid. Indeed, both their global bending and the solvent accessible surface area (SASA), i.e. an indicator of the compaction of the structure, happen to be virtually indistinguishable from the ones typical of undamaged B-DNA strands [59].

On the contrary, 64-PP experiences a veritable polymorphism, and is constantly oscillating between bent and straight forms, as well as between hollow and compact structures (Fig. 11). This polymorphism, evidenced by the appearance of different maxima in the distributions of the relevant structural indicators (Fig. 11), is also accompanied by a major reorganization of the arrangements and π-stacking couplings involving the nucleobases surrounding the lesion, also leading to a strong alteration of the original Watson and Crick pairing [59]. The results of the MD simulations have allowed to reconcile contrasting experimental results, that where evidencing bent or straight strands, respectively. More importantly, they provide a rationale for the different repair rate and toxicity. Indeed, while T<>T may easily escape the recognition by the repair machinery due to the limited structural modifications, its rigidity

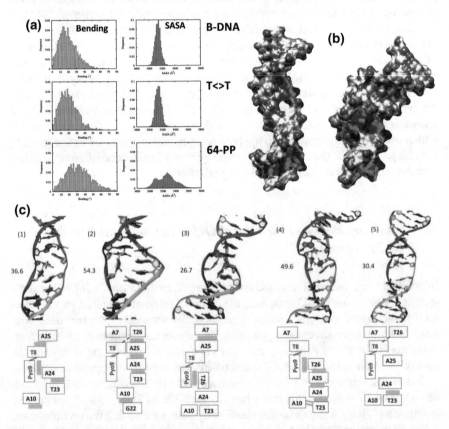

Fig. 11 a The distribution of the global bending (red) and SASA (blue) for undamaged B-DNA and the corresponding strands containing T<>T and 64-PP lesions respectively. **b** Surface representation of two representative snapshots for 64-PP showing the coexistence of straight and bent, as well as compact and hollow structures. **c** Cartoon representation of representative conformations observed during the MD simulations together with the schematic representation of the interaction patterns experienced (green boxes represent π-stacking, blue line hydrogens bonds). Adapted with permission from Ref. [59]. Copyright 2017 Oxford University Press

induces replication fork blocking, hence strongly limiting its mutagenicity. The much more structurally distorted 64-PP, on the contrary, is readily recognized by the NER machinery and the repair process is efficiently triggered. However, its flexibility and polymorphisms and in particular the disruption of the canonical nucleobases pairing, may be related to a higher rate of replication errors and hence mutations.

The rationale behind the more efficient repair of 64-PP, as compared to CPD, has been clearly provided; however, other fine effects should be properly rationalized. In particular, it has been demonstrated that in human skin cells, i.e. fibroblast and keratinocytes, CPD lesions are not repaired with the same efficiency [244]. In particular, the presence of a cytosine at the 5' position of the CPD strongly increase the repair rate, hence C<>T and C<>C are removed more efficiently than T<>C and T<>T [244]. While no evident structural differences between the isolated oligonucleotides have been evidenced so far to justify this behavior, their interaction with the NER recognition factor DDB2 should be the base of subtle, yet significant, differences. Apart from the interest per se, the study of the interplay between structural modification and DNA lesion repair rates is also crucial because it proves how the use of the proper computational methodologies and protocols allows to rationalize not only chemical or biochemical processes but also data issued directly from cellular assays.

5   Conclusion and Perspectives

The fundamental biological role of DNA and the impact of DNA lesions in cell viability or in the development of highly debilitating pathology have clearly stimulated a wide interest by the scientific community and by computational chemistry and biophysics. This is also justified by the complexity and the subtle interplay with both the inhomogeneous environment and the diverse signaling pathways happening at cellular level. Hence, the rationalization of DNA lesions, and of the induction of photolesions in particular, represents a clear scientific challenge that requires proper and finely tailored multiscale approaches, but ultimately allows to bring answer to fundamental biological questions.

From this standpoint, DNA has been a paradigmatic system for computational biochemistry, with years of intensive efforts by the molecular dynamics community to propose charge-fixed or polarizable force fields that capture the subtle dynamics of the B-DNA backbone at the μs scale. This activity has also taken advantage of high-level quantum mechanics methods to quantify the cooperative non-covalent interactions, hydrogen bonding and π-stacking, that allows DNA to adopt, most often, a B-helix structure. In the same respect, the transition between different double helical form (A and Z) and towards non-canonical arrangements, such as G4s, has also been taken into account.

The formation of photolesions and the subsequent recognition and repair by dedicated enzymes has profited from intensive research efforts by several pioneer groups that have used advanced computational methods to determine potential or free energy surfaces related to mechanical (such as damage extrusion) or electronic events (electron- or charge-transfer). These protocols have shown an increasing degree of complexity tackling both the isolated model systems and the macromolecular environments. Hybrid QM/MM simulations have particularly proven their capability to complement experimental data, notably by clarifying via the exploration of the complex energy landscapes the occurrence of complex chemical reactivity such as the rationalization of various transport phenomena based on charge, electron, or hydrogen (proton) transfer. However, important methodological developments are needed to tackle some of the key difficulties of hybrid methods regarding to the specific properties of DNA. In this respect, while the significant enlargement of the QM partition, made possible by the advancement of electronic structure methods, should allow to properly take into account the π-stacked multichromophores, the coupling with the vibrational degrees of freedom, and in particular the floppiness of the backbone, necessitate a careful consideration of the statistical sampling. In the same aspect the development of (polarizable) force field able to better describe the behavior of highly charged species, will be extremely beneficial for the QM/MM community.

A real complementarity and the interaction between simulation and experiments has been built in the last two decades, and while leading to significant progress in the field, it has also allowed the clear emergence of molecular computational biology approaches. Notably, the accuracy in the description of transfer phenomena, has also allowed the treatment of DNA-based materials with potential applications in the field of organic and molecular optoelectronics.

Despite the impressive success of the simulations of DNA lesions and the answer that have been brought to the community, one main limitation of computational approaches is somehow due to the consideration of model systems presenting regular sequences, such as poly(dG), poly(dG·dC), poly(dA) and poly(dA·dT), hence limiting the understanding of subtle structural and electronic sequence-dependent effects. Undoubtedly, such an approach is due to the relatively high computational cost necessary for the study of different sequences whose complexity grows combinatorially, and by the related difficulty in the analysis of the complex data obtained. As in many scientific fields, the interplay between the development of more efficient computational protocols and the machinery of artificial intelligence will certainly provide consistent breakthroughs in the nearby future, both in term of increased computational power and analytical rationalization of complex and multivaried series of data.

Another aspect that will be crucial in the following years and that should be strongly improved is the analysis of the role of complex non-canonical environment on the production and outcome of DNA lesions. This will include the study of the role of DNA lesions happening in the nucleosomal environment, i.e. with the oligomer wrapped around histone proteins. Apart from the size of the systems, important complexity arises by the interplay with intrinsically disordered structures, the histone tails, and by the opening of novel, and the tuning of existent photochemical pathways, such as tautomerization [115] or protein-DNA cross-link formation [245], by the strongly inhomogeneous and highly charged nucleosomal environment.

Finally, a fascinating, yet at the moment rather unexplored field, will be the study of the role of DNA lesions, and photolesions, in the control of gene expression. Recent experimental [154] and computational [153] results, indeed point to an important role of DNA lesions in modulating the epigenetic regulation.

Despite the limitations that should be carefully taken into account, it is evident that the computational study of DNA lesions also constitutes a paradigmatic example in which by using proper multiscale approaches, and by dealing with both the electronic and structural effects on an equal footing, the answer to relevant biological questions becomes possible. This in turn brings an unprecedented and complementary point of view able to achieve an atomic and molecular scale resolution of crucial phenomena, and in the long term will also lead to the definition of protective strategies or to the rational development of novel DNA-interacting photodrugs. On the other hand it is also clear that to provide a full picture of DNA photobiology, one should go beyond a pure QM/MM description and instead the former should be coupled with extended MD simulations, and as a perspective mesoscopic models, that while losing the electronic resolution will provide insights in the longer scale biological relevant outcomes ultimately produced by the interaction of nucleic bases with light. In other words, we strongly believe that the study of DNA has strongly contributed to the affirmation of molecular computational photobiology as a fundamental approach to complement the more classical experimental determination in chemistry and molecular biology. It has also contributed to settle down some of the principles to which a reliable computational photobiology should adhere, namely the strong interplay with experimental sciences, and a fully multiscale approach aiming at providing a unified description of different temporal and spatial scales.

In this respect, we believe that, also by profiting from the impressive methodological development, the time for computational photobiology to come on and shine is just in front of us.

References

1. Beckstead AA, Zhang Y, de Vries MS, Kohler B (2016) Life in the light: nucleic acid photoproperties as a legacy of chemical evolution. Phys Chem Chem Phys 18:24228–24238
2. Serrano-Andrés L, Merchán M (2009) Are the five natural DNA/RNA base monomers a good choice from natural selection?. A photochemical perspective. J Photochem Photobiol C Photochem Rev 10:21–32
3. Klaunig JE, Kamendulis LM, Hocevar BA (2010) Oxidative Stress and oxidative damage in carcinogenesis. Toxicol Pathol 38:96–109
4. Sinha RP, Häder D-P (2002) UV-induced DNA damage and repair: a review. Photochem Photobiol Sci 1:225–236
5. Kryston TB, Georgiev AB, Pissis P, Georgakilas AG (2011) Role of oxidative stress and DNA damage in human carcinogenesis. Mutat Res Fundam Mol Mech Mutagen 711:193–201
6. Salmon TB (2004) Biological consequences of oxidative stress-induced DNA damage in Saccharomyces cerevisiae. Nucleic Acids Res 32:3712–3723
7. Hamanaka RB, Chandel NS (2010) Mitochondrial reactive oxygen species regulate cellular signaling and dictate biological outcomes. Trends Biochem Sci 35:505–513

8. Cadet J, Ravanat J-L, Martinez GR et al (2006) Singlet oxygen oxidation of isolated and cellular DNA: product formation and mechanistic insights. Photochem Photobiol 82:1219–1225

9. Miyamoto S, Di Mascio P (2014) Lipid hydroperoxides as a source of singlet molecular oxygen. Subcell Biochem 77:3–20

10. Finkel T (2011) Signal transduction by reactive oxygen species. J Cell Biol 194:7–15

11. Cadet J, Wagner JR (2013) DNA base damage by reactive oxygen species, oxidizing agents, and UV radiation. Cold Spring Harb Perspect Biol 5:a012559-(1–18).

12. Factors E (2016) Skin stress response pathways. Springer

13. Valko M, Leibfritz D, Moncol J et al (2006) Free radicals and antioxidants in normal physiological functions and human disease. Int J Biochem Cell Biol 39:1–41

14. Epe B (2012) DNA damage spectra induced by photosensitization. Photochem Photobiol Sci 11:98–106

15. Cuquerella MC, Lhiaubet-Vallet V, Cadet J, Miranda MA (2012) Benzophenone photosensitized DNA damage. Acc Chem Res 45:1558–1570

16. Evans TC, Nichols NM (2008) DNA repair enzymes. Annu Rev Biochem 57:1–12

17. Huang XX, Bernerd F, Halliday GM (2009) Ultraviolet A within sunlight induces mutations in the epidermal basal layer of engineered human skin. Am J Pathol 174:1534–1543

18. Ziegler A, Leffell DJ, Kunala S et al (1993) Mutation hotspots due to sunlight in the p53 gene of nonmelanoma skin cancers. Proc Natl Acad Sci U S A 90:4216–4220

19. Drouin R, Therrien JP (1997) UVB-induced cyclobutane pyrimidine dimer frequency correlates with skin cancer mutational hotspots in p53. Photochem Photobiol 66:719–726

20. Brash DE, Rudolph JA, Simon JA et al (1991) A role for sunlight in skin cancer: UV-induced p53 mutations in squamous cell carcinoma. Proc Natl Acad Sci U S A 88:10124–10128

21. Taylor JS (1994) Unraveling the Molecular pathway from sunlight to skin cancer. Acc Chem Res 27:76–82

22. Curtin NJ (2012) DNA repair dysregulation from cancer driver to therapeutic target. Nat Rev Cancer 12:801–817

23. Sage E, Harrison L (2011) Clustered DNA lesion repair in eukaryotes: relevance to mutagenesis and cell survival. Mutat Res Fundam Mol Mech Mutagen 711:123–133

24. Goldstein M, Kastan MB (2015) The DNA damage response: implications for tumor responses to radiation and chemotherapy. Annu Rev Med 66:129–143

25. Breugom AJ, Swets M, Bosset JF et al (2015) Adjuvant chemotherapy after preoperative (chemo)radiotherapy and surgery for patients with rectal cancer: a systematic review and meta-analysis of individual patient data. Lancet Oncol 16:200–207

26. Crafton SM, Salani R (2016) Beyond chemotherapy: an overview and review of targeted therapy in cervical cancer. Clin Ther 38:449–458

27. Schabel FMJ, Trader MW, Laster WRJ et al (1979) cis-Dichlorodiammineplatinum(II): combination chemotherapy and cross-resistance studies with tumors of mice. Cancer Treat Rep 63:1459–1473

28. Lehane DE, Bryan RN, Horowitz B et al (1983) Intraarterial Cis-Platinum Chemotherapy for Patients with Primary and Metastatic Brain Tumors. Cancer Drug Deliv 1:69–77

29. Peckham MJ, Horwich A, Hendry WF (1985) Advanced seminoma: Treatment with cis-platinum-based combination chemotherapy or carboplatin (JM8). Br J Cancer 52:7–13

30. Cerón-Carrasco JP, Jacquemin D, Cauët E (2012) Cisplatin cytoxicity: a theoretical study of induced mutations. Phys Chem Chem Phys 14:12457–12464

31. Howell ST, Cardwell LA, Feldman SR (2018) A review and update of phototherapy treatment options for psoriasis. Curr Dermatol Rep 7:43–51

32. Schneider LA, Hinrichs R, Scharffetter-Kochanek K (2008) Phototherapy and photochemotherapy. Clin Dermatol 26:464–476

33. Yoon SM, Park YG, Park ES, Choi CY (2017) Application of nipple-areolar complex impending necrosis with light-emitting diode treatment for immediate breast reconstruction after nipple-sparing mastectomy. Med Lasers 6:5–9

34. Ortiz-Salvador JM, Pérez-Ferriols A (2017) Phototherapy in Atopic dermatitis. Springer, Cham, pp 279–286

35. Colin Theng TS, Tan ES-T (2018) Phototherapy in Pigmentary disorders. Springer, Cham, pp 235–252

36. Henderson BW, Dougherty TJ (1992) How does photodynamic therapy work? Photochem Photobiol 55:145–157

37. Zenkevich E, Sagun E, Knyukshto V et al (1996) Photophysical and photochemical properties of potential porphyrin and chlorin photosensitizers for PDT. J Photochem Photobiol B Biol 33:171–180

38. Koshi E, Mohan A, Rajesh S, Philip K (2011) Antimicrobial photodynamic therapy: an overview. J Indian Soc Periodontol 15:323

39. Dai T, Huang YY, Hamblin MR (2009) Photodynamic therapy for localized infections-State of the art. Photodiagnosis Photodyn Ther 6:170–188

40. Allison RR, Moghissi K (2013) Oncologic photodynamic therapy: clinical strategies that modulate mechanisms of action. Photodiagnosis Photodyn Ther 10:331–341

41. Reessing F, Szymanski W (2017) Beyond photodynamic therapy: Light-activated cancer chemotherapy. Curr Med Chem 24:4905–4950

42. Askes SHC, Reddy GU, Wyrwa R et al (2017) Red light-triggered CO release from Mn2(CO)10 using triplet sensitization in polymer nonwoven fabrics. J Am Chem Soc 139:15292–15295

43. Bonnet S, Limburg B, Meeldijk JD et al (2011) Ruthenium-decorated lipid vesicles: light-induced release of [Ru(terpy)(bpy)(OH2)]2+ and thermal back coordination. J Am Chem Soc 133:252–261

44. Kuang Y, Sun H, Blain JC, Peng X (2012) Hypoxia-selective DNA interstrand cross-link formation by two modified nucleosides. Chem A Eur J 18:12609–12613

45. Higgins SLH, Brewer KJ (2012) Designing red-light-activated multifunctional agents for the photodynamic therapy. Angew Chemie Int Ed 51:11420–11422

46. Cadet J, Delatour T, Douki T et al (1999) Hydroxyl radicals and DNA base damage. Mutat Res Mol Mech Mutagen 424:9–21

47. Cadet J, Sage E, Douki T (2005) Ultraviolet radiation-mediated damage to cellular DNA. Mut Res Fund Mol Mec Mut 571:3–17

48. Scott SP, Pandita TK (2006) The cellular control of DNA double-strand breaks. J Cell Biochem 99:1463–1475

49. Mehta A, Haber JE (2014) Sources of DNA double-strand breaks and models of recombinational DNA repair. Cold Spring Harb Perspect Biol 6:a016428

50. McKinnon PJ, Caldecott KW (2007) DNA Strand break repair and human genetic disease. Annu Rev Genomics Hum Genet 8:37–55

51. Cannan WJ, Pederson DS (2016) Mechanisms and consequences of double-strand DNA Break formation in chromatin. J Cell Physiol 231:3–14

52. Dumont E, Monari A (2015) Understanding DNA under oxidative stress and sensitization: the role of molecular modeling. Front Chem 3:43

53. Dumont E, Grüber R, Bignon E et al (2016) Probing the reactivity of singlet oxygen with purines. Nucleic Acids Res 44:56–62

54. Hazel RD, Tian K, de Los SC (2008) NMR solution structures of bistranded abasic site lesions in DNA. Biochemistry 47:11909–11919

55. Gelfand CA, Plum GE, Grollman AP et al (1998) Thermodynamic consequences of an abasic lesion in duplex DNA are strongly dependent on base sequence. Biochemistry 37:7321–7327

56. Ayadi L, Coulombeau C, Lavery R (2000) The impact of abasic sites on DNA flexibility. J Biomol Struct Dyn 17:645–653

57. Mol CD, Izumi T, Mitra S, Talner JA (2000) DNA-bound structures and mutants reveal abasic DNA binding by APE1 DNA repair and coordination. Nature 403:451–456

58. Lo HL, Nakajima S, Ma L et al (2005) Differential biologic effects of CPD and 6–4PP UV-induced DNA damage on the induction of apoptosis and cell-cycle arrest. BMC Cancer 5:135

59. Dehez F, Gattuso H, Bignon E et al (2017) Conformational polymorphism or structural invariance in DNA photoinduced lesions: implications for repair rates. Nucleic Acids Res 45:3654–3662
60. Park CJ, Lee JH, Choi BS (2007) Functional insights gained from structural analyses of DNA duplexes that contain UV-damaged photoproducts. Photochem Photobiol 83:187–195
61. Sage E, Drouin R, Rouabhia M (2005) Chemical sequencing profiles of photosensitized DNA damage. In: From DNA photolesions to mutations, skin cancer and cell death. The Royal Society of Chemistry, pp 15–31
62. Ikehata H, Ono T (2011) The mechanisms of UV mutagenesis. J Radiat Res 52:115–125
63. Clauson C, Scharer OD, Niedernhofer L (2013) Advances in understanding the complex mechanisms of DNA interstrand cross-link repair. Cold Spring Harb Perspect Biol 5:a012732–a012732
64. Derheimer FA, Hicks JK, Paulsen MT et al (2009) Psoralen-induced DNA interstrand cross-links block transcription and induce p53 in an ataxia-telangiectasia and rad3-related-dependent manner. Mol Pharmacol 75:599–607
65. Bignon E, Dršata T, Morell C et al (2017) Interstrand cross-linking implies contrasting structural consequences for DNA: insights from molecular dynamics. Nucleic Acids Res 45:2188–2195
66. Tretyakova NY, Groehler A, Ji S (2015) DNA-protein cross-links: formation, structural identities, and biological outcomes. Acc Chem Res 48:1631–1644
67. Bignon E, Chan C-H, Morell C, et al (2017) Molecular dynamics insights into polyamine–DNA binding modes: implications for cross-link selectivity. Chem A Eur J 23
68. Silerme S, Bobyk L, Taverna-Porro M et al (2014) DNA-polyamine cross-links generated upon one electron oxidation of DNA. Chem Res Toxicol 27:1011–1018
69. Sancar A (2000) Cryptochrome: The second photoactive pigment in the eye and its role in circadian photoreception. Annu Rev Biochem 69:31–67
70. Mei Q, Dvornyk V (2015) Evolutionary history of the photolyase/cryptochrome superfamily in eukaryotes. PLoS ONE 10:e0135940
71. Krokan HE, Bjørås M (2013) Base excision repair. Cold Spring Harb Perspect Biol 5:1–22
72. Robertson AB, Klungland A, Rognes T, Leiros I (2009) Base excision repair: the long and short of it. Cell Mol Life Sci 66:981–993
73. Wallace SS (2014) Base excision repair: a critical player in many games. DNA Repair (Amst) 19:14–26
74. Mantha AK, Sarkar B, Tell G (2014) A short review on the implications of base excision repair pathway for neurons: relevance to neurodegenerative diseases. Mitochondrion 16:38–49
75. David SS, O'Shea VL, Kundu S (2007) Base-excision repair of oxidative DNA damage. Nature 447:941–950
76. Sancar A, Tang MS (1993) Nucleotide excision repair. Photochem Photobiol 57:905–921
77. Hu J, Adebali O, Adar S, Sancar A (2017) Dynamic maps of UV damage formation and repair for the human genome. Proc Natl Acad Sci U S A 114:6758–6763
78. Schärer OD (2013) Nucleotide excision repair in Eukaryotes. Cold Spring Harb Perspect Biol 5:a012609
79. Marteijn JA, Lans H, Vermeulen W, Hoeijmakers JHJ (2014) Understanding nucleotide excision repair and its roles in cancer and ageing. Nat Rev Mol Cell Biol 15:465–481
80. Mladenov E, Magin S, Soni A, Iliakis G (2013) DNA double-strand break repair as determinant of cellular radiosensitivity to killing and target in radiation therapy. Front Oncol 3:113
81. Mladenov E, Magin S, Soni A, Iliakis G (2016) DNA double-strand-break repair in higher eukaryotes and its role in genomic instability and cancer: cell cycle and proliferation-dependent regulation. Semin Cancer Biol 37–38:51–64
82. Georgakilas AG, Bennett PV, Wilson DM, Sutherland BM (2004) Processing of bistranded abasic DNA clusters in γ-irradiated human hematopoietic cells. Nucleic Acids Res 32:5609–5620
83. Gattuso H, Durand E, Bignon E et al (2016) Repair rate of clustered abasic DNA lesions by human endonuclease: molecular bases of sequence specificity. J Phys Chem Lett 7:3760–3765

84. Bignon E, Gattuso H, Morell C et al (2016) Correlation of bistranded clustered abasic DNA lesion processing with structural and dynamic DNA helix distortion. Nucleic Acids Res 44:8588–8599

85. Han C, Srivastava AK, Cui T et al (2015) Differential DNA lesion formation and repair in heterochromatin and euchromatin. Carcinogenesis 37:129–138

86. Price BD, D'Andrea AD (2013) Chromatin remodeling at DNA double-strand breaks. Cell 152:1344–1354

87. Geiman TM, Robertson KD (2002) Chromatin remodeling, histone modifications, and DNA methylation—how does it all fit together? J Cell Biochem 87:117–125

88. Deans C, Maggert KA (2015) What do you mean, "Epigenetic"? Genetics 199:887–896

89. Hognon C, Besancenot V, Gruez A et al (2019) Cooperative effects of cytosine methylation on DNA structure and dynamics. J Phys Chem B 123:7365–7371

90. Georgakilas AG, O'Neill P, Stewart RD (2013) Induction and repair of clustered DNA lesions: what do we know so far? Radiat Res 180:100–109

91. Georgakilas AG, Bennett PV, Sutherland BM (2002) High efficiency detection of bi-stranded abasic clusters in gamma-irradiated DNA by putrescine. Nucleic Acids Res 30:2800–2808

92. Fonseca Guerra C, Bickelhaupt FM, Baerends EJ (2002) Orbital interactions in hydrogen bonds important for cohesion in molecular crystals and mismatched pairs of DNA bases. Cryst Growth Des 2:239–245

93. Fonseca Guerra C, Bickelhaupt FM, Snijders JG, Baerends EJ (1999) The nature of the hydrogen bond in DNA base pairs: the role of charge transfer and resonance assistance. Chem A Eur J 5:3581–3594

94. Fonseca Guerra C, Bickelhaupt FM, Snijders JG, Baerends EJ (2000) Hydrogen bonding in DNA base pairs: reconciliation of theory and experiment. J Am Chem Soc 122:4117–4128

95. Zaccaria F, Paragi G, Fonseca Guerra C (2016) The role of alkali metal cations in the stabilization of guanine quadruplexes: why K+ is the best. Phys Chem Chem Phys 18:20895–20904

96. Szolomájer J, Paragi G, Batta G et al (2011) 3-Substituted xanthines as promising candidates for quadruplex formation: computational, synthetic and analytical studies. New J Chem 35:476–482

97. Dans PD, Ivani I, Hospital A et al (2017) How accurate are accurate force-fields for B-DNA? Nucleic Acids Res 45:4217–4230

98. Perez A, Marchán I, Svozil D et al (2007) Refinement of the AMBER force field for nucleic acids: improving the description of alphaγConformers. Biophys J 92:3817–3829

99. Galindo-Murillo R, Robertson JC, Zgarbová M et al (2016) Assessing the current state of amber force field modifications for DNA. J Chem Theory Comput 12:4114–4127

100. Zgarbová M, Šponer J, Otyepka M et al (2015) Refinement of the sugar-phosphate backbone torsion beta for AMBER force fields improves the description of Z- and B-DNA. J Chem Theory Comput 11:5723–5736

101. Nygaard M, Terkelsen T, Vidas Olsen A et al (2016) The mutational landscape of the oncogenic MZF1 SCAN domain in cancer. Front Mol Biosci 3:78

102. Lambrughi M, De Gioia L, Gervasio FL et al (2016) DNA-binding protects p53 from interactions with cofactors involved in transcription-independent functions. Nucleic Acids Res 44:9096–9109

103. Etheve L, Martin J, Lavery R (2016) Protein-DNA interfaces: A molecular dynamics analysis of time-dependent recognition processes for three transcription factors. Nucleic Acids Res 44:9990–10002

104. Nenov A, Giussani A, Fingerhut BP et al (2015) Spectral lineshapes in nonlinear electronic spectroscopy. Phys Chem Chem Phys 17:30925–30936

105. Gustavsson T, Improta R, Markovitsi D (2010) DNA/RNA: Building blocks of life under UV irradiation. J Phys Chem Lett 1:2025–2030

106. Giussani A, Segarra-Martí J, Roca-Sanjuán D, Merchán M (2015) Excitation of nucleobases from a computational perspective I: Reaction paths. Top Curr Chem 355:57–98

107. Merchan M, Gonzalez-Luque R, Climent T et al (2006) Unified model for the ultrafast decay of pyrimidine nucleobases. J Phys Chem B 110:26471–26476

108. Barbatti M, Aquino AJA, Szymczak JJ et al (2010) Relaxation mechanisms of UV-photoexcited DNA and RNA nucleobases. Proc Natl Acc Sci 107:21453–21458

109. Serrano-Andres L, Merchan M, Borin AC (2006) Adenine and 2-aminopurine: paradigms of modern theoretical photochemistry. Proc Natl Acad Sci U S A 103:8691–8696

110. Giussani A, Conti I, Nenov A, Garavelli M (2017) Photoinduced formation mechanism of the thymine-thymine (6–4) adduct in DNA; a QM(CASPT2//CASSCF):MM(AMBER) study. Faraday discussions

111. Francés-Monerris A, Segarra-Martí J, Merchán M, Roca-Sanjuán D (2016) Theoretical study on the excited-state π-stacking versus intermolecular hydrogen-transfer processes in the guanine–cytosine/cytosine trimer. Theor Chem Acc 135:1–15

112. Banyasz A, Douki T, Improta R et al (2012) Electronic excited states responsible for dimer formation upon UV absorption directly by thymine strands: joint experimental and theoretical study. J Am Chem Soc 134:14834–14845

113. Rauer C, Nogueira JJ, Marquetand P, González L (2016) Cyclobutane thymine photodimerization mechanism revealed by nonadiabatic molecular dynamics. J Am Chem Soc 138:15911–15916

114. Roca-Sanjuan D, Olaso-Gonzalez G, Gonzalez-Ramirez I et al (2008) Molecular basis of DNA photodimerization: intrinsic production of cyclobutane cytosine dimers. J Am Chem Soc 130:10768–10779

115. Francés-Monerris A, Gattuso H, Roca-Sanjuán D et al (2018) Dynamics of the excited-state hydrogen transfer in a (dG)·(dC) homopolymer: intrinsic photostability of DNA. Chem Sci 9:7902–7911

116. Padula D, Jurinovich S, Di Bari L, Mennucci B (2016) Simulation of electronic circular dichroism of nucleic acids: from the structure to the spectrum. Chem A Eur J 22:17011–17019

117. Gattuso H, Assfeld X, Monari A (2015) Modeling DNA electronic circular dichroism by QM/MM methods and Frenkel Hamiltonian. Theor Chem Acc 134:36

118. Faraji S, Dreuw A (2014) Physicochemical mechanism of light-driven DNA repair by (6–4) photolyases. Annu Rev Phys Chem 65:275–292

119. Lenz SAP, Wetmore SD (2017) QM/MM study of the reaction catalyzed by alkyladenine DNA glycosylase: examination of the substrate specificity of a DNA repair enzyme. J Phys Chem B 121:11096–11108

120. Sebera J, Hattori Y, Sato D et al (2017) The mechanism of the glycosylase reaction with hOGG1 base-excision repair enzyme: Concerted effect of Lys249 and Asp268 during excision of 8-oxoguanine. Nucleic Acids Res 45:5231–5242

121. Aranda J, Roca M, López-Canut V, Tuñón I (2010) Theoretical study of the catalytic mechanism of DNA-(N4-Cytosine)- methyltransferase from the bacterium proteus vulgaris. J Phys Chem B 114:8467–8473

122. Aranda J, Attana F, Tuñón I (2017) Molecular Mechanism of Inhibition of DNA Methylation by Zebularine. ACS Catal 7:1728–1732

123. Cadet J, Douki T (2018) Formation of UV-induced DNA damage contributing to skin cancer development. Photochem Photobiol Sci 17:1816–1841

124. Ikehata H, Mori T, Kamei Y et al (2019) Wavelength- and tissue-dependent variations in the mutagenicity of cyclobutane pyrimidine dimers in mouse skin. Photochem Photobiol. https://doi.org/10.1111/php.13159

125. Alzueta OR, Cuquerella MC, Miranda MA (2019) Triplet energy transfer versus excited state cyclization as the controlling step in photosensitized bipyrimidine dimerization. J Org Chem 84:13329–13335

126. Clingen PH, Jeremy R, Davies H (1997) Quantum yields of adenine photodimerization in poly (deoxyadenylic acid) and DNA. J Photochem Photobiol B Biol 38:81–87

127. Banyasz A, Martinez-Fernandez L, Ketola TM et al (2016) Excited state pathways leading to formation of adenine dinners. J Phys Chem Lett 7:2020–2023

128. Spata VA, Matsika S (2013) Bonded excimer formation in π-stacked 9-methyladenine dimers. J Phys Chem A 117:8718–8728

129. Olaso-González G, Merchán M, Serrano-Andrés L (2009) The role of adenine excimers in the photophysics of oligonucleotides. J Am Chem Soc 131:4368–4377

130. Conti I, Nenov A, Hoefinger S et al (2015) Excited state evolution of DNA stacked adenines resolved at the CASPT2//CASSCF/Amber level: from the bright to the excimer state and back. Phys Chem Chem Phys 17:7291–7302

131. Zhao X, Nadji S, Kao JL-F, Taylor J-S (1996) The Structure of d(TpA)*, the Major Photoproduct of Thymidylyl-(3'–5')-Deoxyadenosine. Nucleic Acids Res 24:1554–1560

132. Davies RJH, Malone JF, Gan Y et al (2007) High-resolution crystal structure of the intramolecular d(TpA) thymine–adenine photoadduct and its mechanistic implications. Nucleic Acids Res 35:1048–1053

133. Bose SN, Davies RJ, Sethi SK, McCloskey JA (1983) Formation of an adenine-thymine photo adduct in the deoxydinucleoside monophosphate d(TpA) and in DNA. Science (80–) 220:723 LP–725

134. Douki T, Cadet J (2001) Individual determination of the yield of the main UV-induced dimeric pyrimidine photoproducts in DNA suggests a high mutagenicity of CC photolesions. Biochemistry 40:2495–2501

135. Madsen MM, Jones NC, Nielsen SB, Hoffmann SV (2016) On the wavelength dependence of UV induced thymine photolesions: a synchrotron radiation circular dichroism study. Phys Chem Chem Phys 18:30436–30443

136. Lee W, Matsika S (2019) Role of charge transfer states into the formation of cyclobutane pyrimidine dimers in DNA. Faraday Discuss 216:507–519

137. Pan Z, McCullagh M, Schatz GC, Lewis FD (2011) Conformational control of thymine photodimerization in purine-containing trinucleotides. J Phys Chem Lett 2:1432–1438

138. Pan Z, Hariharan M, Arkin JD et al (2011) Electron donor-acceptor interactions with flanking purines influence the efficiency of thymine photodimerization. J Am Chem Soc 133:20793–20798

139. Cadet J, Mouret S, Ravanat JL, Douki T (2012) Photoinduced damage to cellular DNA: direct and photosensitized reactions. In: Photochemistry and photobiology. Blackwell Publishing Ltd, pp 1048–1065

140. Schreier WJ, Schrader TE, Koller FO et al (2007) Thymine dimerization in DNA is an ultrafast photoreaction. Science (80–) 315:625–629

141. Schreier WJ, Kubon J, Regner N et al (2009) Thymine dimerization in DNA model systems: cyclobutane photolesion is predominantly formed via the singlet channel. J Am Chem Soc 131:5038–5039

142. Marguet S, Markovitsi D (2005) Time-resolved study of thymine dimer formation. J Am Chem Soc 127:5780–5781

143. Marquetand P, Nogueira JJ, Mai S et al (2017) Challenges in simulating light-induced processes in DNA. Molecules 22:49

144. Reiter S, Keefer D, De Vivie-Riedle R (2018) RNA environment is responsible for decreased photostability of uracil. J Am Chem Soc 140:8714–8720

145. Markovitsi D, Talbot F, Gustavsson T et al (2006) Complexity of excited-state dynamics in DNA. Nature 441:E7

146. Conti I, Martínez-Fernández L, Esposito L et al (2017) Multiple electronic and structural factors control cyclobutane pyrimidine dimer and 6–4 thymine-thymine photodimerization in a DNA duplex. Chem s Eur J 23:15177–15188

147. Serrano-Pérez JJ, González-Ramírez I, Coto PB et al (2008) Theoretical insight into the intrinsic ultrafast formation of cyclobutane pyrimidine dimers in UV-irradiated DNA: thymine versus cytosine. J Phys Chem B 112:14096–14098

148. Gonzalez-Ramirez I, Roca-Sanjuan D, Climent T et al (2011) On the photoproduction of DNA/RNA cyclobutane pyrimidine dimers. Theor Chem Acc 128:705–711

149. Giussani A, Conti I, Nenov A, Garavelli M (2018) Photoinduced formation mechanism of the thymine-thymine (6–4) adduct in DNA; A QM(CASPT2//CASSCF):MM(AMBER) study. Faraday Discuss 207:375–387

150. Roca-Sanjuan D, Rubio M, Merchan M, Serrano-Andres L (2006) Ab initio determination of the ionization potentials of DNA and RNA nucleobases. J Chem Phys 125

151. Cannistraro VJ, Taylor J-S (2009) Acceleration of 5-methylcytosine deamination in cyclobutane dimers by G and its Implications for UV-induced C-to-T mutation hotspots. J Mol Biol 392:1145–1157

152. Lee W, Matsika S (2018) Photochemical formation of cyclobutane pyrimidine dimers in DNA through electron transfer from a flanking base. ChemPhysChem 19:1568–1571

153. Hognon C, Gebus A, Barone G, Monari A (2019) Human DNA telomeres in presence of oxidative lesions: The crucial role of electrostatic interactions on the stability of guanine quadruplexes. Antioxidants 8:337

154. Fleming AM, Zhu J, Howpay Manage SA, Burrows CJ (2019) Human NEIL3 gene expression regulated by epigenetic-like oxidative DNA modification. J Am Chem Soc 141:11036–11049

155. Lee W, Matsika S (2017) Conformational and electronic effects on the formation of anti cyclobutane pyrimidine dimers in G-quadruplex structures. Phys Chem Chem Phys 19:3325–3336

156. Zou X, Sun Z, Zhao H, Zhang CY (2019) Mechanistic insight into photocrosslinking reaction between triplet state 4-thiopyrimidine and thymine. Phys Chem Chem Phys 21:21305–21316

157. Improta R, Santoro F, Blancafort L (2016) Quantum Mechanical Studies on the Photophysics and the Photochemistry of Nucleic Acids and Nucleobases. Chem Rev 116:3540–3593

158. Zhang Y, Li X-B, Fleming AM et al (2016) UV-induced proton-coupled electron transfer in cyclic DNA miniduplexes. J Am Chem Soc 138:7395–7401

159. Zhang Y, De La Harpe K, Beckstead AA et al (2015) UV-induced proton transfer between DNA strands. J Am Chem Soc 137:7059–7062

160. Bucher DB, Schlueter A, Carell T, Zinth W (2014a) Watson-Crick base pairing controls excited-state decay in natural DNA. Angew Chemie Int Ed 53:11366–11369

161. Rottger K, Marroux HJB, Grubb MP et al (2015) Ultraviolet absorption induces hydrogen-atom transfer in G center dot C Watson-Crick DNA base pairs in solution. Angew Chemie Int Ed 54:14719–14722

162. Sobolewski AL, Domcke W (2004) Ab initio studies on the photophysics of the guanine-cytosine base pair. Phys Chem Chem Phys 6:2763–2771

163. Sobolewski AL, Domcke W, Hattig C (2005) Tautomeric selectivity of the excited-state lifetime of guanine/cytosine base pairs: the role of electron-driven proton-transfer processes. Proc Natl Acad Sci 102:17903–17906

164. Groenhof G, Schäfer LV, Boggio-Pasqua M et al (2007) Ultrafast deactivation of an excited cytosine-guanine base pair in DNA. J Am Chem Soc 129:6812–6819

165. Sauri V, Gobbo JP, Serrano-Pérez JJ et al (2013) Proton/hydrogen transfer mechanisms in the guanine-cytosine base pair: Photostability and tautomerism. J Chem Theory Comput 9:481–496

166. Gobbo JP, Saurí V, Roca-Sanjuán D et al (2012) On the deactivation mechanisms of adenine-thymine base pair. J Phys Chem B 116:4089–4097

167. Frances-Monerris A, Segarra-Marti J, Merchan M, Roca-Sanjuan D (2016) Theoretical study on the excited-state pi-stacking versus intermolecular hydrogen-transfer processes in the guanine-cytosine/cytosine trimer. Theor Chem Acc 135:31

168. Hammes-Schiffer S, Stuchebrukhov AA (2010) Theory of coupled electron and proton transfer reactions. Chem Rev 110:6939–6960

169. Hammes-Schiffer S (2015) Proton-coupled electron transfer: moving together and charging forward. J Am Chem Soc 137:8860–8871

170. Hammes-Schiffer S (2012) Proton-coupled electron transfer: classification scheme and guide to theoretical methods. Energy Environ Sci 5:7696–7703

171. Röttger K, Marroux HJB, Chemin AFM et al (2017) Is UV-induced electron-driven proton transfer active in a chemically modified A·T DNA base pair? J Phys Chem B 121:4448–4455

172. Bucher DB, Schlueter A, Carell T, Zinth W (2014b) Watson-Crick base pairing controls excited—state decay in natural DNA. Angew Chemie Int Ed 53:11366–11369

173. Martinez-Fernandez L, Improta R (2018) Photoactivated proton coupled electron transfer in DNA: insights from quantum mechanical calculations. Faraday discussion accepted. https://doi.org/10.1039/C7FD00195A

174. Ko C, Hammes-Schiffer S (2013) Charge-transfer excited states and proton transfer in model guanine-cytosine DNA duplexes in water. J Phys Chem Lett 4:2540–2545

175. De La Harpe K, Crespo-Hernández CE, Kohler B (2009) Deuterium isotope effect on excited-state dynamics in an alternating GC oligonucleotide. J Am Chem Soc 131:17557–17559

176. Fujitsuka M, Majima T (2017) Charge transfer dynamics in DNA revealed by time-resolved spectroscopy. Chem Sci 8:1752–1762

177. Lewis FD, Kalgutkar RS, Wu Y et al (2000) Driving force dependence of electron transfer dynamics in synthetic DNA hairpins. J Am Chem Soc 122:12346–12351

178. Núñez ME, Hall DB, Barton JK (1999) Long-range oxidative damage to DNA: Effects of distance and sequence. Chem Biol 6:85–97

179. Kanvah S, Joseph J, Schuster GB et al (2010) Oxidation of DNA: Damage to nucleobases. Acc Chem Res 43:280–287

180. Beratan DN (2019) Why are DNA and protein electron transfer so different? Annu Rev Phys Chem 70:71–97

181. Siriwong K, Voityuk AA (2012) Electron transfer in DNA. Wiley Interdiscip Rev Comput Mol Sci 2:780–794

182. Peluso A, Caruso T, Landi A, Capobianco A (2019) The dynamics of hole transfer in DNA. Molecules 24:4044

183. Bittner ER (2007) Frenkel exciton model of ultrafast excited state dynamics in AT DNA double helices. J Photochem Photobiol A Chem 190:328–334

184. Lee MH, Brancolini G, Gutiérrez R et al (2012) Probing charge transport in oxidatively damaged DNA sequences under the influence of structural fluctuations. J Phys Chem B 116:10977–10985

185. Kubar T, Elstner M (2013) A hybrid approach to simulation of electron transfer in complex molecular systems. J R Soc Interf 10

186. Wolter M, Elstner M, Kleinekathöfer U, Kubař T (2017) Microsecond simulation of electron transfer in DNA: bottom-up parametrization of an efficient electron transfer model based on atomistic details. J Phys Chem B 121:529–549

187. Banyasz A, Martínez-Fernández L, Balty C et al (2017) Absorption of low-energy UV radiation by human telomere G-quadruplexes generates long-lived guanine radical cations. J Am Chem Soc 139:10561–10568

188. Allison RR, Downie GH, Cuenca R et al (2004) Photosensitizers in clinical PDT. Photodiagnosis Photodyn Ther 1:27–42

189. Yano S, Hirohara S, Obata M et al (2011) Current states and future views in photodynamic therapy. J Photochem Photobiol C Photochem Rev 12:46–67

190. Sergentu D-C, Maurice R, Havenith RWA et al (2014) Computational determination of the dominant triplet population mechanism in photoexcited benzophenone. Phys Chem Chem Phys 16:25393–25403

191. Dumont E, Monari A (2013) Benzophenone and DNA: evidence for a double insertion mode and its spectral signature. J Phys Chem Lett 4:4119–4124

192. Dumont E, Wibowo M, Roca-Sanjuán D et al (2015) Resolving the benzophenone DNA-photosensitization mechanism at QM/MM level. J Phys Chem Lett 6:576–580

193. Marazzi M, Mai S, Roca-Sanjuán D et al (2016) Benzophenone ultrafast triplet population: revisiting the kinetic model by surface-hopping dynamics. J Phys Chem Lett 7:622–626

194. Soep B, Mestdagh J-M, Briant M et al (2016) Direct observation of slow intersystem crossing in an aromatic ketone, fluorenone. Phys Chem Chem Phys 18:22914–22920

195. Tamai N, Asahi T, Masuhara H (1992) Intersystem crossing of benzophenone by femtosecond transient grating spectroscopy. Chem Phys Lett 198:413–418

196. Aloïse S, Ruckebusch C, Blanchet L et al (2008) The benzophenone S1(n, π*) →T1(n, π *) states intersystem crossing reinvestigated by ultrafast absorption spectroscopy and multivariate curve resolution. J Phys Chem A 112:224–231

197. Gattuso H, Dumont E, Chipot C et al (2016) Thermodynamics of DNA: sensitizer recognition. Characterizing binding motifs with all-atom simulations. Phys Chem Chem Phys 18:33180–33186

198. Chipot C, Pohorille A (2007) Free energy calculations: theory and applications in chemistry and biology. Springer

199. Pastor-Pérez L, Barriau E, Frey H et al (2008) Photocatalysis within hyperbranched polyethers with a benzophenone core. J Org Chem 73:4680–4683

200. Marazzi M, Wibowo M, Gattuso H et al (2016) Hydrogen abstraction by photoexcited benzophenone: consequences for DNA photosensitization. Phys Chem Chem Phys 18:7829–7836

201. Bignon E, Gattuso H, Morell C et al (2015) DNA photosensitization by an "insider": photophysics and triplet energy transfer of 5-Methyl-2-pyrimidone deoxyribonucleoside. Chem A Eur J 21:11509–11516

202. Vendrell-Criado V, Rodríguez-Muñiz GM, Cuquerella MC et al (2013) Photosensitization of DNA by 5-methyl-2-pyrimidone deoxyribonucleoside: (6–4) photoproduct as a possible trojan horse. Angew Chemie Int Ed 52:6476–6479

203. Aparici-Espert I, Garcia-Lainez G, Andreu I et al (2018) Oxidatively generated lesions as internal photosensitizers for pyrimidine dimerization in DNA. ACS Chem Biol 13:542–547

204. Francés-Monerris A, Hognon C, Miranda MA et al (2018) Triplet photosensitization mechanism of thymine by an oxidized nucleobase: from a dimeric model to DNA environment. Phys Chem Chem Phys 20:25666–25675

205. Francés-Monerris A, Tuñón I, Monari A (2019) Hypoxia-selective dissociation mechanism of a nitroimidazole nucleoside in a DNA environment. J Phys Chem Lett 10:6750–6754

206. Goyal JL, Rao VA, Srinivasan R, Agrawal K (1994) Oculocutaneous manifestations in xeroderma pigmentosa. Br J Ophthalmol 78:295 LP–297

207. Butt FMA, Moshi JR, Owibingire S, Chindia ML (2010) Xeroderma pigmentosum: a review and case series. J Cranio-Maxillofacial Surg 38:534–537

208. Sancar A (2003) Structure and function of DNA photolyase and cryptochrome blue-light photoreceptors. Chem Rev 103:2203–2237

209. Kavakli IH, Ozturk N, Gul S (2019) DNA repair by photolyases. In: Advances in protein chemistry and structural biology. Academic Press Inc., pp 1–19

210. Mees A, Klar T, Gnau P et al (2004) Crystal structure of a photolyase bound to a CPD-like DNA lesion after in situ repair. Science (80–) 306:1789–1793

211. Essen LO (2006) Photolyases and cryptochromes: common mechanisms of DNA repair and light-driven signaling? Curr Opin Struct Biol 16:51–59

212. Maul MJ, Barends TRM, Glas AF et al (2008) Crystal structure and mechanism of a DNA (6–4) photolyase. Angew Chemie Int Ed 47:10076–10080

213. Kiontke S, Geisselbrecht Y, Pokorny R et al (2011) Crystal structures of an archaeal class II DNA photolyase and its complex with UV-damaged duplex DNA. EMBO J 30:4437–4449

214. Zhang M, Wang L, Zhong D (2017) Photolyase: dynamics and mechanisms of repair of sun-induced DNA damage. Photochem Photobiol 93:78–92

215. Liu Z, Wang L, Zhong D (2015) Dynamics and mechanisms of DNA repair by photolyase. Phys Chem Chem Phys 17:11933–11949

216. Yamada D, Dokainish HM, Iwata T et al (2016) Functional conversion of CPD and (6–4) photolyases by mutation. Biochemistry 55:4173–4183

217. Knips A, Zacharias M (2017) Both DNA global deformation and repair enzyme contacts mediate flipping of thymine dimer damage. Sci Rep 7:41324

218. Masson F, Laino T, Rothlisberger U, Hutter J (2009) A QM/MM investigation of thymine dimer radical anion splitting catalyzed by DNA photolyase. ChemPhysChem 10:400–410

219. Jepsen KA, Solov'yov IA, (2017) On binding specificity of (6–4) photolyase to a T(6–4)T DNA photoproduct. Eur Phys J D 71:155

220. Terai Y, Sato R, Yumiba T et al (2018) Coulomb and CH-φ interactions in (6–4) photolyase-DNA complex dominate DNA binding and repair abilities. Nucleic Acids Res 46:6761–6772

221. Ma H, Holub D, Gillet N et al (2019) Two aspartate residues close to the lesion binding site of Agrobacterium (6-4) photolyase are required for Mg 2+ stimulation of DNA repair. FEBS J 286:1765–1779

222. Zhong D (2015) Electron Transfer Mechanisms of DNA Repair by Photolyase. Annu Rev Phys Chem 66:691–715

223. Park HW, Kim ST, Sancar A, Deisenhofer J (1995) Crystal structure of DNA photolyase from Escherichia coli. Science (80-) 268:1866–1872

224. Harbach PHP, Schneider M, Faraji S, Dreuw A (2013) Intermolecular coulombic decay in biology: the initial electron detachment from FADH- in DNA photolyases. J Phys Chem Lett 4:943–949

225. Liu Z, Guo X, Tan C et al (2012) Electron tunneling pathways and role of adenine in repair of cyclobutane pyrimidine dimer by DNA photolyase. J Am Chem Soc 134:8104–8114

226. Antony J, Medvedev DM, Stuchebrukhov AA (2000) Theoretical study of electron transfer between the photolyase catalytic cofactor FADH- and DNA thymine dimer. J Am Chem Soc 122:1057–1065

227. Prytkova TR, Beratan DN, Skourtis SS (2007) Photoselected electron transfer pathways in DNA photolyase. Proc Natl Acad Sci U S A 104:802–807

228. Acocella A, Jones GA, Zerbetto F (2010) What is adenine doing in photolyase? J Phys Chem B 114:4101–4106

229. Lee W, Kodali G, Stanley RJ, Matsika S (2016) Coexistence of Different Electron-Transfer Mechanisms in the DNA Repair Process by Photolyase. Chem A Eur J 22:11371–11381

230. Rousseau BJG, Shafei S, Migliore A et al (2018) Determinants of photolyase's DNA repair mechanism in mesophiles and extremophiles. J Am Chem Soc 140:2853–2861

231. Wang H, Chen X, Fang W (2014) Excited-state proton coupled electron transfer between photolyase and the damaged DNA through water wire: aphoto-repair mechanism. Phys Chem Chem Phys 16:25432–25441

232. Sadeghian K, Bocola M, Merz T, Schütz M (2010) Theoretical study on the repair mechanism of the (6–4) photolesion by the (6-4) photolyase. J Am Chem Soc 132:16285–16295

233. Faraji S, Zhong D, Dreuw A (2016) Characterization of the intermediate in and identification of the repair mechanism of (6–4) Photolesions by photolyases. Angew Chemie Int Ed 55:5175–5178

234. Li J, Liu Z, Tan C et al (2010) Dynamics and mechanism of repair of ultraviolet-induced (6–4) photoproduct by photolyase. Nature 466:887–890

235. Condic-Jurkic K, Smith AS, Zipse H, Smith DM (2012) The protonation states of the active-site histidines in (6–4) potolyase. J Chem Theory Comput 8:1078–1091

236. Domratcheva T (2011) Neutral histidine and photoinduced electron transfer in DNA photolyases. J Am Chem Soc 133:18172–18182

237. Moughal Shahi AR, Domratcheva T (2013) Challenges in computing electron-transfer energies of DNA repair using hybrid QM/MM models. J Chem Theory Comput 9:4644–4652

238. Dokainish HM, Kitao A (2016) Computational assignment of the histidine protonation state in (6–4) photolyase enzyme and its effect on the protonation step. ACS Catal 6:5500–5507

239. Ma H, Zhang F, Ignatz E et al (2017) Divalent cations increase DNA repair activities of bacterial (6–4) photolyases. In: Photochemistry and photobiology. Blackwell Publishing Inc., pp 323–330

240. Ai YJ, Liao RZ, Chen SL et al (2011) Repair of DNA dewar photoproduct to (6–4) photoproduct in (6–4) photolyase. J Phys Chem B 115:10976–10982

241. Faraji S, Dreuw A (2017) Insights into light-driven DNA repair by photolyases: challenges and opportunities for electronic structure theory. Photochem Photobiol 93:37–50

242. Dokainish HM, Yamada D, Iwata T et al (2017) Electron fate and mutational robustness in the mechanism of (6–4)photolyase-mediated DNA repair. ACS Catal 7:4835–4845

243. Dokainish HM, Kitao A (2018) Similarities and differences between thymine(6–4)thymine/cytosine DNA lesion repairs by photolyases. J Phys Chem B 122:8537–8547

244. Mouret S, Charveron M, Favier A et al (2008) Differential repair of UVB-induced cyclobutane pyrimidine dimers in cultured human skin cells and whole human skin. DNA Repair (Amst) 7:704–712

245. Chan C-H, Monari A, Ravanat J-L, Dumont E (2019) Probing interaction of a trilysine peptide with DNA underlying formation of guanine–lysine cross-links: insights from molecular dynamics. Phys Chem Chem Phys 21:23418–23424

Index

© Springer Nature Switzerland AG 2021
T. Andruniów and M. Olivucci (eds.), *QM/MM Studies of Light-responsive Biological Systems*, Challenges and Advances in Computational Chemistry and Physics 31,
https://doi.org/10.1007/978-3-030-57721-6

Printed in the United States
by Baker & Taylor Publisher Services